D1261503

An Introduction to
Industrial Chemistry

This book is dedicated to
the memory of my Father,
John Arthur Alan Heaton

An Introduction to Industrial Chemistry

Third edition

Edited by

Alan Heaton

Reader in Industrial Chemistry
School of Pharmacy and Chemistry
Liverpool John Moores University

BLACKIE ACADEMIC & PROFESSIONAL
An Imprint of Chapman & Hall

London · Glasgow · New York · Tokyo · Melbourne · Madras

Published by
Blackie Academic & Professional, an imprint of Chapman & Hall
Wester Cleddens Road, Bishopbriggs, Glasgow G64 2NZ

Chapman & Hall, 2–6 Boundary Row, London SE1 8HN, UK

Blackie Academic & Professional, Wester Cleddens Road, Bishopbriggs, Glasgow G64 2NZ, UK

Chapman & Hall GmbH, Pappelallee 3, 69469 Weinheim, Germany

Chapman & Hall USA, 115 Fifth Avenue, Fourth Floor, New York NY 10003, USA

Chapman & Hall Japan, ITP-Japan, Kyowa Building, 3F, 2-2-1 Hirakawacho, Chiyoda-ku, Tokyo 102, Japan

DA Book (Aust.) Pty Ltd, 648 Whitehorse Road, Mitcham 3132, Victoria, Australia

Chapman & Hall India, R. Seshadri, 32 Second Main Road, CIT East, Madras 600 035, India

First edition 1984
Second edition 1991
This edition 1996

© 1996 Chapman & Hall

Typeset in 10/12 pt Times by AFS Image Setters Ltd, Glasgow
Printed in England by Clays Ltd, St Ives plc, Bungay, Suffolk

ISBN 0 7514 0272 9

A catalogue record for this book is available from the British Library

Library of Congress Catalog Card Number: 95–80422

∞ Printed on permanent acid-free text paper, manufactured in accordance with ANSI/NISO Z39.48-1992 and ANSI/NISO Z39.48-1984 (Permanence of Paper)

Preface
to the Third Edition

Following the success of the first two editions of this book in which the core subject matter has been retained, we have taken the opportunity to add substantial new material, including an additional chapter on that most important activity of the chemical industry, research and development. Topical items such as quality, safety and environmental issues also receive enhanced coverage.

The team of authors for this edition comprises both those revising and updating their chapters and some new ones. The latter's different approach to the subject matter is reflected in the new titles: Organisational Structures – A Story of Evolution (chapter 5) and Environmental Impact of the Chemical Industry (chapter 9). The chapter on Energy retains its original title but different approach of the new authors is evident.

We have updated statistics and tables wherever possible and expanded the index. We hope readers find the brief 'pen pictures' of authors to be interesting.

It is worth stressing again that this book is designed to be used with its companion volume – *The Chemical Industry*, 2nd Edition, ed. Alan Heaton (referred to as Volume 2) – for a complete introduction to the chemical industry.

Thanks are due to all contributors and to my wife Joy for typing my contributions.

Alan Heaton

Contents

10   Chlor-alkali products   289
Steve Kelham

11   Catalysts and catalysis   309
John Pennington

12 Petrochemicals

350

John Pennington

Contributors

Dr. Derek Bew   *Formerly of ICI Petrochemicals & Plastics Division, Wilton*
Derek Bew obtained his M.Sc. in 1951 and Ph.D. in Organic Chemistry in 1954. He then joined ICI's Billingham Division and spent periods in Research and Market Development, Process Development and Plant Management and then in Project Management in the Technical Department. There followed an extended period in the Research and Technology Department working on Process Economics and Evaluation. Derek retired from ICI in 1990.

Dr. Will Bland   *Department of Chemistry and Applied Chemistry, Kingston University, Penrhyn Road, Kingston-upon-Thames, Surrey KT21 1EE*
Will Bland worked for three years as a research chemist in industry, obtained his B.Sc. and Ph.D. from the University of Leicester and is currently Acting Associate Head of the School of Applied Chemistry at Kingston University as well as being Course Director of the B.Sc. (Honours) Degree in Environmental Science. His main teaching and research interests are in Industrial Chemistry, Resources and the Environment. He is a Fellow of the Royal Society of Chemistry and a member of the Committee of Tertiary Education Group.

Dr. Alan Heaton   *School of Pharmacy and Chemistry, Liverpool John Moores University, Byrom Street, Liverpool L4 3AF*
Alan Heaton worked in the coal-tar chemicals industry as a young man whilst studying for the Grad. R.I.C. qualification. He obtained his Ph.D. in Organofluorine Chemistry at Durham University in 1967. Following a brief spell as Lecturer in Organic Chemistry at the University of Salford, he joined Liverpool Polytechnic in 1969, where he was given responsibility for developing courses in the area of Industrial Chemistry. He is now Reader in Industrial Chemistry at Liverpool John Moores University and carries out research in the areas of organofluorine chemistry and pesticides.
   Since 1971 he has been a Tutor, Tutor Counsellor and Consultant to the Open University. He is a nationally-elected member of the Council of the Royal Society of Chemistry and serves on the Awards Committee of the Society of Chemical Industry as well as being on a number of committees of each of these societies at regional and local levels.

Dr. Andrew S. Hursthouse   *Department of Chemistry and Chemical Engineering, University of Paisley, High Street, Paisley PA1 2BE*
Andrew Hursthouse is a geochemist and a senior lecturer in the Department of Chemistry and Chemical Engineering, University of Paisley. His research interests cover: environmental analysis, the mobility of inorganic and organic species in the environment, the biogeochemistry of metallic and organic compounds and the environmental impact of waste discharges. He has acted as a consultant for industrial and public bodies on a range of environmental problems and manages the Centre for Particle Characterisation and Analysis and R&D, problem solving facility for industry within the Faculty of Science and Technology.

Mr. Steve Kelham   *Process Development Group, Research and Technology Department, ICI Chemicals and Polymers, PO Box 8, The Heath, Runcorn WA7 4QD*
Steve Kelham has had over 20 years of experience of the Chlor-Alkali Industry. He graduated from Cambridge University in 1968 and joined ICI as a Chemical Engineer. He has been primarily involved in process design, plant troubleshooting, commissioning and business development activities linked to chlor-alkali production from mercury, diaphragm and membrane cells at plants around the world. Currently he is managing process development activities within the R&T area in ICI at Runcorn, Cheshire.

Mr. Ted Laird   *Department of Chemistry and Applied Chemistry, Kingston University, Penrhyn Road, Kingston-upon-Thames, Surrey KT21 1EE*
Ted Laird took his B.Sc. chemistry degree plus his Ph.D. in physical-organic chemistry at Southampton University. After 2 years in the RAF he joined ICI to work in a physical chemistry laboratory. He worked on plant problems at their Wilton Works and then moved on to BNFL to carry out similar work.
   He became a chartered chemical engineer and has taught mainly industrial chemistry at Kingston Polytechnic and University for the last 27 years before recently retiring.

Mrs. Jo McCloskey   *Business School, Liverpool John Moores University, 98 Mount Pleasant, Liverpool L3 5UZ*
Jo McCloskey has studied at universities in Ireland and Scotland. She worked in the public sector before embarking on an academic career. She taught in Ireland and Africa before coming to England. She came to Liverpool Business School after having taught at Leicester Business School. Currently, she is Principal Lecturer and Head of Business Policy and Marketing.

Her research interests and recent publications have been in the areas of environmental management and marketing. She is an experienced marketing and management consultant, having completed various projects in Europe, U.S.A. and Africa.

Mr. John Pennington   *4 Bessacarr Avenue, Willerby, Hull HU10 6JA*
After graduation and a spell in research at Cambridge University, John Pennington joined the Research and Development laboratories attached to a manufacturing site for bulk organic chemicals near Hull, now owned by BP Chemicals. The work primarily involved factory support and new process development, in both laboratory and pilot plant, but occasionally more speculative research. John progressed in a technical capacity, latterly as a company-wide 'internal consultant' until (early) retirement (at the end of 1988).

Dr. Richard Szczepanski   *Infochem Computer Services Ltd., South Bank Technopark, 90 London Road, London SE1 6LN*
Richard Szczepanski is a director of Infochem Computer Services Ltd., a consultancy specializing in physical property data and software for chemical engineering and petroleum engineering applications. His main area of work is in modelling phase and chemical equilibria. Dr. Szczepanski was formerly a Project Leader at the BP Research Centre and a lecturer in chemical engineering at Imperial College, London.

Conversion factors

1 tonne (metric ton) = 1000 kilograms = 2205 pounds
                                                      = 0·984 tons
1 ton   = 1016 kilograms = 2240 pounds
                                                      = 1·016 tonnes

Mass

1 litre   = 0·220 gallons (U.K. or Imperial) = 1 cubic metre
1 gallon = 4·546 litres
1 gallon = 1·200 U.S. gallons = 0·00455 cubic metres
1 barrel = 42 U.S. gallons = 35 gallons = 0·159 cubic metres

(Densities of crude oil vary, but 7·5 barrels per tonne is an accepted average figure.)

1 cubic metre = 35·31 cubic feet
1 cubic foot   = 0·02832 cubic metres

Volume

1 atmosphere = 1·013 bar = 14·696 pounds per square inch
                    = $1·013 \times 10^5$ newtons per square metre
                    = $1·013 \times 10^5$ pascal

Pressure

Degrees Centigrade = 0·556 (degrees Fahrenheit − 32)
Degrees Fahrenheit = 1·80 (degrees Centigrade) + 32
Degrees Kelvin   = degrees Centigrade + 273

Temperature

1 therm   = 100 000 British thermal units
1 British thermal unit = 0·252 kilocalories = 1·055 kilojoules
1 kilocalorie   = 4·184 kilojoules
1 kilowatt hour = 3600 kilojoules = 859·8 kilocalories
                  = 3412 British thermal units.

Energy

1 horsepower = 0·746 kilowatts
1 kilowatt   = 1·34 horsepower

Power

Nomenclature of organic compounds

Common or trivial name	Systematic (or IUPAC) name	Structure	
Paraffin	Alkane	—	**(a) Classes of compounds**
Cycloparaffins or Naphthenes	Cycloalkanes	—	
Olefins	Alkenes	—	
Acetylenes	Alkynes	—	
Methacrylates	2-Methylpropenoates	$CH_2{=}C{-}CO_2R$ $\vert$ CH_3	
Ethylene	Ethene	$CH_2{=}CH_2$	**(b) Individual compounds**
Propylene	Propene	$CH_3CH{=}CH_2$	
Styrene	Phenylethene	$\langle\bigcirc\rangle{-}CH{=}CH_2$	
Acetylene	Ethyne	$H{-}C{\equiv}C{-}H$	
Isoprene	2-Methylbuta-1, 3-diene	$CH_2{=}C{-}CH{=}CH_2$ $\vert$ CH_3	
Ethylene oxide	Oxirane	$CH_2{-}CH_2$ with O bridge	
Propylene oxide	1-Methyloxirane	$CH_3{-}CH{-}CH_2$ with O bridge	
Methyl iodide	Iodomethane	CH_3I	
Methyl chloride	Chloromethane	CH_3Cl	
Methylene dichloride	Dichloromethane	CH_2Cl_2	
Chloroform	Trichloromethane	$CHCl_3$	
Carbon tetrachloride	Tetrachloromethane	CCl_4	
Vinyl chloride	Chloroethene	$CH_2{=}CH{-}Cl$	
Ethylene dichloride	1, 2-Dichloroethane	$ClCH_2CH_2Cl$	
Allyl chloride	3-Chloropropene	$CH_2{=}CH{-}CH_2{-}Cl$	
Chloroprene	2-Chlorobuta-1, 3-diene	$CH_2{=}C{-}CH{=}CH_2$ $\vert$ Cl	
Epichlorohydrin	1-Chloromethyloxirane	$ClCH_2CH{-}CH_2$ with O bridge	
Ethylene glycol	Ethane-1, 2-diol	$HOCH_2CH_2OH$	

Propargyl alcohol	Prop-2-yn-1-ol	$H-C{\equiv}C-CH_2OH$
Allyl alcohol	Prop-2-en-1-ol	$CH_2{=}CH-CH_2OH$
iso-Propanol	2-Propanol	CH_3CHCH_3 $\quad\ \ \vert$ $\quad\ \ OH$
Glycerol	Propane-1,2,3-triol	$HOCH_2-CH-CH_2OH$ $\qquad\quad\ \vert$ $\qquad\quad OH$
sec-Butanol	2-Butanol	$CH_3CHCH_2CH_3$ $\quad\ \ \vert$ $\quad\ \ OH$
Pentaerythritol	2,2-Di (hydroxymethyl) propane-1,3-diol	$\qquad\quad CH_2OH$ $\qquad\quad \vert$ $HOCH_2-C-CH_2OH$ $\qquad\quad \vert$ $\qquad\quad CH_2OH$
Lauryl alcohol	Dodecanol	$CH_3(CH_2)_{10}CH_2OH$
Acetone	Propanone	CH_3COCH_3
Methylisobutyl ketone	4-Methylpentan-2-one	$CH_3COCH_2CHCH_3$ $\qquad\qquad\quad \vert$ $\qquad\qquad\quad CH_3$
Formaldehyde	Methanal	$HCHO$
Acetaldehyde	Ethanal	CH_3CHO
Chloral	2,2,2-Trichloroethanal	Cl_3CCHO
Propionaldehyde	Propanal	CH_3CH_2CHO
Acrolein	Propenal	$CH_2{=}CHCHO$
Butyraldehyde	Butanal	$CH_3CH_2CH_2CHO$
Formic acid	Methanoic acid	HCO_2H
Methyl formate	Methyl methanoate	HCO_2CH_3
Acetic acid	Ethanoic acid	CH_3CO_2H
Acetic anhydride	Ethanoic anhydride	$(CH_3CO)_2O$
Peracetic acid	Perethanoic acid	CH_3CO_3H
Vinyl acetate	Ethenyl ethanoate	$CH_2{=}CHO_2CCH_3$
Acrylic acid	Propenoic acid	$CH_2{=}CH-CO_2H$
Dimethyl oxalate	Dimethyl ethanedioate	CO_2CH_3 $\vert$ CO_2CH_3
Propionic acid	Propanoic acid	$CH_3CH_2CO_2H$
Methyl methacrylate	Methyl 2-methylpropenoate	$CH_2{=}C-CO_2CH_3$ $\qquad\ \ \vert$ $\qquad\ \ CH_3$
Maleic acid	cis-Butenedioic acid	
Maleic anhydride	cis-Butenedioic anhydride	

Citric acid	2-Hydroxypropane-1,2,3-tricarboxylic acid	$$\begin{array}{c} CH_2CO_2H \\ | \\ HO-C-CO_2H \\ | \\ CH_2CO_2H \end{array}$$
Methyl laurate	Methyl dodecanoate	$CH_3(CH_2)_{10}CO_2CH_3$
Stearic acid	Octadecanoic acid	$CH_3(CH_2)_{16}CO_2H$
Acrylonitrile	Propenonitrile	$CH_2=CH-CN$
Adiponitrile	Hexane-1,6-dinitrile	$NC-(CH_2)_6-CN$
Urea	Carbamide	H_2NCONH_2
Ketene	Ethenone	$CH_2=C=O$

Toluene	Methylbenzene	
Aniline	Phenylamine	
Cumene	iso-Propylbenzene	
Benzyl alcohol	Phenylmethanol	
o-Xylene	1,2-Dimethylbenzene	
m-Xylene	1,3-Dimethylbenzene	
p-Xylene	1,4-Dimethylbenzene	
Phthalic acid	Benzene-1,2-dicarboxylic acid	
Isophthalic acid	Benzene-1,3-dicarboxylic acid	
Terephthalic acid	Benzene-1,4-dicarboxylic acid	

o-Toluic acid   2-Methylbenzoic acid

p-Toluic acid   4-Methylbenzoic acid

p-Tolualdehyde   4-Methylbenzaldehyde

Benzidine   4,4'-Biphenyldiamine

Furfural   2-Formylfuran

(c) Additional compounds   HFA 134a   1,1,1,2-Tetrafluoroethane CF_3CH_2F
                                LTBE   Ethyl t-butyl ether $CH_3CH_2OC(CH_3)_3$
                                MTBE   Methyl t-butyl ether $CH_3O\,C(CH_3)_3$
                                TAME   t-Amyl methyl ether $H_3C—C-OCH_3$
                                                                         $H_3C\quad CH_2CH_3$

Editorial Introduction

Chemistry is a challenging and interesting subject for academic study. Its principles and ideas are used to produce the chemicals from which all manner of materials and eventually consumer products are manufactured. The diversity of examples is enormous, ranging from cement to iron and steel, and on to modern plastics which are so widely used in the packaging of consumer goods and in the manufacture of household items. Indeed life as we know it today could not exist without the chemical industry. Its contribution to the saving of lives and relief of suffering is immeasurable; synthetic drugs such as those which lower blood pressure (e.g. β-blockers), attack bacterial and viral infections (e.g. antibiotics such as the penicillins and cephalosporins) and replace vital natural chemicals which the body is not producing due to some malfunction (e.g. insulin, some vitamins), are particularly noteworthy in this respect. Effect chemicals also clearly make an impact on our everyday lives. Two examples are the use of polytetrafluoroethylene (polytetrafluoroethene Teflon™ or Fluon™) to provide a non-stick surface coating for cooking utensils, and silicones which are used to ease the discharge of bread from baking tins. It should also be noted that the chemical industry's activities have an influence on all other industries, either in terms of providing raw materials or chemicals for quality control analyses and to improve operation, and to treat boiler water, cooling water and effluents. The general public is increasingly interested in the operations of the chemical industry, in its concern both about the safety of chemicals and the operation of chemical plant.

Industrial chemistry is a topic of growing interest and importance for all chemistry students. Indeed a survey[1] of all U.K. departments which offer a degree course in chemistry showed that almost two-thirds included some industrial chemistry in their courses and several offered a full degree in this subject.

Industrial chemistry is characterized by the very broad nature of the subject, spanning as it does several different disciplines. Apart from chemistry it includes topics such as organization and management of a company, technical economics, chemical engineering and environmental pollution control, and it would not be complete without an in-depth study of several particular sectors of the chemical industry. The latter would be selected as a representative cross-section of the entire industry.

Clearly a comprehensive treatment of all these topics would hardly be possible even in a full degree in industrial chemistry, let alone as an option or

The importance of industrial chemistry

part of a chemistry degree course. Nor would this be appropriate. An understanding of the basic aspects of, and an appreciation of, the language of some of the above topics, and their linking with physico-chemical principles in the manufacture of chemicals is required.

The growing interest in the study of industrial chemistry in undergraduate courses clearly requires the availability of suitable accompanying student textbooks. In some areas of the subject excellent monographs are available (these are detailed in the 'further reading' section at the end of each chapter). However, for a broad introductory treatment to the whole of industrial chemistry they are too detailed and therefore inappropriate. Books presenting an overall introduction to industrial chemistry are few and far between. The few valiant attempts indicate the difficulties, since they either (a) attempt to be too comprehensive and are therefore somewhat superficial in the treatment of certain topics[2], (b) tend to be rather a catalogue of factual material[3], or (c) adopt an entirely different approach—that of process development—leading to the coverage of rather different topics[4]. A more recent two-volume publication[5] has much merit but its very high cost puts it beyond the reach of students.

The aim of writing this textbook and its companion volume, *The Chemical Industry*, is to provide a readable introduction to the very broad subject of industrial chemistry (or chemical technology) in two books of reasonable length. Although the texts are aimed primarily at chemists, much of the material will be of value to first- and second-year students studying for degrees in chemical engineering. Finally those graduates about to enter industry after taking a 'pure' chemistry degree course should find them a useful introduction to industry, and therefore provide a bridge between their academic studies and their first employment. Our approach has been to use a team of specialist authors comprising both practising industrialists and teachers. We hope to convey the challenge, excitement, and also the difficulties which are involved in chemicals manufacture. Emphasis will be given to factors vital in the production process. Examples are the economics, engineering and pollution-control aspects. However, it should be made clear that our aim is just to introduce these subjects and not provide an extensive course in them. This should equip the reader—assuming he or she is a chemist—not only with a broader appreciation of their subject but also with the understanding to enable them to converse with chemical engineers and technical economists. This is vital in the very large projects undertaken by the major chemical companies where co-operation within a team comprising scientific and commercial personnel of several different disciplines is essential. Those readers requiring a more detailed study are directed to the bibliography at the end of the appropriate chapter. Physico-chemical principles will be integrated with the above aspects, where appropriate. The interplay, and often the compromise, between these various factors will be discussed and emphasized. Even within the chemistry itself there are often substantial differences between college and industrial chemistry. The general area of oxidation reactions is a good example. In the academic situation sophisticated, expensive, or even toxic reagents are often used, e.g. osmium tetroxide, whereas industrially the same reaction will almost certainly be carried out catalytically and using the cheapest reagent of all—air. Another important point of difference is that in college decisions are usually made in a situation where the required background information is virtually complete. For example, assigning a particular mechanism to a reaction is usually carried out on the basis of a substantial amount of experimental supporting evidence. Although one can never prove a chosen mechanism correct, the more supporting evidence that is collected the more confidence one has in it. The manager in industry, in contrast, is invariably working in a situation of only limited information. Thus he may have to decide which of several new projects to back, on the basis of financial evaluations projected up to ten years ahead. The uncertainty in the figures is considerable since assumptions have to be made on the interest rates, inflation, taxation, etc. Nevertheless he has to use his managerial ability to come to a decision in a situation of limited availability of reliable information. Particular sectors of the chemical industry are largely dealt with in more detail in the companion volume, *The Chemical Industry, 2nd Edition,* by Alan Heaton.

Statistics

Production statistics and prices are extensively used in certain chapters in this book in order to illustrate points such as scale of operation, or comparison of companies and national chemical industries. National and international statistics, due to their extensive nature, take some considerable time to collect and collate, and may refer to a period which ended a few years before the date of their publication. We therefore recommend that the reader overcomes these difficulties by regarding all figures not as absolute or currently correct but more as being indicative of orders of magnitude, trends or relative positions. In other words it is the conclusions which we can draw from the figures which are important rather than the individual statistics themselves. If current figures are of interest, however, they can often be gleaned from journals such as *Chemical and Engineering News, Chemical Marketing Reporter* or *European Chemical News.* More specific sources are detailed at the end of each chapter.

There are a number of points which should be borne in mind when considering statistics—particularly those relating to economic comparisons. Firstly, in terms of national and international statistics items included under the term, say 'chemicals' may vary from country to country. In other words there is not a single standard classification system. The information may also be incomplete since it may cover only the larger companies, and in some countries there may be a legal obligation to provide it whereas in others it may be purely voluntary. Similarly, sales of chemicals compared either internationally or by individual multinational corporations (a term which covers all large chemical companies) can be significantly affected by the currency exchange rate chosen in order to obtain all figures in U.S. dollars or U.K. pounds. Because of its fluctuating nature, depending on the time chosen for the interconversion, it could favour some companies or countries and adversely affect others. This reinforces the suggestion made above to treat figures on a relative rather than an absolute basis.

Although figures relating to companies and national chemical industries are available for most of the world, those relating to the communist bloc of countries (the former USSR, Eastern Europe and China) are either difficult to obtain or are of limited reliability. For this reason these countries have not been considered in the text.

Costing details relating to specific processes are also difficult to obtain. Whilst the reluctance of companies to make these known, for commercial reasons, can to a large extent be appreciated, this does leave an important gap when putting together case studies for use in teaching. It must be acknowledged that U.S. companies are rather more forthcoming in this respect than their European counterparts. Fortunately journals like *European Chemical News* and particularly *Chemical and Process Engineering* do publish such information from time to time.

Units and nomenclature

There is a general trend in science towards a more systematic approach to both units and nomenclature, i.e. naming specific chemical compounds. For units the S.I. (Système International) system has been widely introduced, and science students in Europe are brought up on this. However, in industry a range of non-S.I. units are used. For example, weights may be expressed as short tons (2000 lbs), metric tons or tonnes (1000 kg or 2205 lbs) or long tons or tons (2240 lbs) or even (particularly in the U.S.A.) millions of pounds. It is therefore necessary to be bi- or even multilingual and to assist in this conversion factors are given at the beginning of this book.

There are arguments for and against whichever units are used but we have chosen to standardize on tonnes for weight (tonnes), degrees centigrade for temperature (°C), and atmospheres for pressure (atm). Both pounds sterling (£) and U.S. dollars ($) are used for monetary values because of the volatility of their exchange rates over the last two decades. Billions are U.S. billions, i.e. one thousand millions.

In naming chemical compounds the systematic IUPAC system is increasingly used in educational establishments. However in many areas of chemistry, e.g. natural products, trivial names are still far more important, as indeed they are in the chemical industry. Again it is desirable to be bilingual. To assist in this trivial names are used in this book, but the IUPAC name is usually given in brackets afterwards. A reference table for the two systems of naming compounds is also provided at the front of the book. Since only trivial names are used in the index in this book, this conversion table should be used to obtain the trivial name from its systematic counterpart.

A selection of some of the important sources of information on the chemical industry and its major processes is given below. These should be used in conjunction with the more specific references given at the end of each chapter. **General bibliography**

(*a*) *Reference works*

   (i) *Encyclopedia of Chemical Technology*, R. E. Kirk and D. F. Othmer, Interscience, New York. This multi-volume series is very comprehensive and is usually the first source to consult for information. Publication of the third revised edition has been completed.

  (ii) *Riegel's Handbook of Industrial Chemistry*, 9th edn., J. A. Kent, Van Nostrand Reinhold, New York, 1993. A multiauthor survey of the chemical industry.

 (iii) *The Chemical Process Industries*, 6th revised edn., N. Shreve, McGraw-Hill, New York, 1993. Strong on heavy inorganics and weak on organics.

 (iv) *Chemical Technology*, 1st English edn., F. A. Henglein, Pergamon Press, London, 1969. Very strong on the technology of the German chemical industry.

  (v) *Industrial Organic Chemicals in Perspective*, Vols. I and II, Harold A. Witcoff and Bryan G. Reuben, Wiley–Interscience, New York, 1980. A combined edition was also published in 1985. An excellent account of the production and use of organic chemicals.

 (vi) *Industrial Organic Chemistry*, 2nd English edn., K. Weissernel and H.-J. Arpe, VCH, 1993. An account of organic raw materials and intermediates.

(*b*) *Textbooks*

These are detailed under references 2, 3, 4, and 5 below.

(*c*) *Journals*

A selection are given below, the first four giving a fairly general coverage and the remainder more specific coverage of the chemical industry.

   (i) *European Chemical News*
  (ii) *Chemistry and Industry*
 (iii) *Chemical and Engineering News*
 (iv) *Chemical Age*
  (v) *Chemical Marketing Reporter*
 (vi) *Chemical and Process Engineering*
 (vii) *Hydrocarbon Processing*

(*d*) *Patents*

These are covered in *Chemical Abstracts* plus specialist publications such as those issued by Derwent Publications in the U.K.

References

1. Alan Heaton, *Chem. Brit.*, 1982, **18**, 162.
2. *The Chemical Economy*, B. G. Reuben and M. L. Burstall, Longman, 1973.
3. *Basic Organic Chemistry*, Part V: *Industrial Products*, J. M. Tedder, A. Nechvatel and A. H. Jubb, Wiley, 1975.
4. *Principles of Industrial Chemistry*, Chris. A. Clausen III and Guy Mattson, Wiley–Interscience, 1978.
5. *Industrial Organic Chemicals in Perspective*, Part I. *Raw Materials and Manufacture*, Part II, *Technology, Formulation and Use* Harold A. Witcoff and Bryan G. Reuben, Wiley–Interscience, 1980.

Introduction 1

Alan Heaton

The aim of this first chapter is to give something of an overview of that diverse part of manufacturing industry which is called the chemical industry. In doing so, a number of topics are introduced fairly briefly, but are discussed in more detail later in the book. As well as being a lead in to the other chapters, it should give the reader an idea of what the industry does, which are the major chemical producing countries, the scale of operations, the major products, and briefly discuss environmental issues.

The chemical industry exists to increase wealth, or add value (primarily of the shareholders in the companies which make up the industry), by taking raw materials such as salt, limestone and oil, and turning them into a whole range of chemicals which are then either directly, or indirectly, converted into consumer products. These products arguably improve our lives and lifestyles, and we could not live the way we do without them. Examples such as synthetic fibres being made into garments which drip dry and do not need ironing, and the amazing range of their colours, testify to the achievements of the research chemists, engineers and technologists. Also modern fresh fruit and vegetables are of better quality and remain fresh longer thanks to the products of the agrochemicals sector. However, the latter applications of chemicals plus a number of other areas are not without controversy and these items are addressed in more detail in a number of sections in this book (see section 1.6 particularly). Although some people may question the addition of chemicals to food, for example to hasten ripening of fruit or to extend the life of fresh vegetables, there is no doubt that without the products of the chemical industry less food would be available to us and it would be of inferior quality and have a shorter life time.

We should also draw attention to the many life saving and therapeutic drugs and medicines produced by the pharmaceutical sector of the industry. These have made a major contribution to the dramatically increasing life expectancy rates during the 20th century. These topics are discussed in detail in Volume 2 (throughout this book references to Volume 2 refer to *The Chemical Industry*, 2nd edn. by Alan Heaton, Blackie Academic and Professional), Chapter 8. The few unfortunate tragedies and disasters—thalidomide, Seveso, Bhopal (which are discussed later in this chapter)—should not detract from the remarkable contribution which the chemical industry makes to our lives. Sadly in the past it has, by default, contributed to its own negative image by only speaking out in response to some pollution incident which has occurred. It is surely time that the industry took the initiative to put across to the public all the positive and beneficial things which it does and the remarkable products which it continues to develop. There are welcome signs that this change is starting to take place.

Clearly the chemical industry is part of manufacturing industry and within this it plays a central part even though it is by no means the largest part of the manufacturing sector. Its key position arises from the fact that almost all the other parts of the sector utilize its products. For example, the food industry relies on the chemical industry for its packaging materials; modern automobiles depend heavily on synthetic polymers and plastics, which also play an increasing role in the building industry. Nowadays all manufacturing industry must keep a careful check on the quality of the waste materials and effluents which are produced. This necessitates chemicals for analysis and probably also for treating the waste before discharge or for recovering by-products.

Having emphasized the central role which the chemical industry plays both in our lives and within manufacturing industry, and having been given examples to illustrate this, the reader should now look at section 4.2.1 to appreciate how to define the chemical industry. As you will see it's not as easy as it seems!

Several references to the importance of the chemical industry to our society have already been made and this is further emphasized in a number of places in this book, but particularly in sections 4.2.2 and 4.3.1.

1.1
Characteristics of the industry

In the developed world, Europe, U.S.A. Japan, etc., the chemical industry has now become a mature manufacturing industry, following its explosive growth in the 1960s and early 1970s. However, its rate of growth still exceeds that of most manufacturing industries. In most developed countries the ratio is 1.5–2 to 1 (see p. 81). As expected, growth rates are higher in countries which might be classed as developing countries in chemical terms. Examples are Korea, Mexico and Saudi Arabia. In the latter two cases readily and cheaply available supplies of crude oil and natural gas have created the stimulus for this growth.

One of the main factors responsible for the growth in the developed world is a major characteristic of the chemical industry—its great emphasis on research and development (R&D). This led in the 1950s and 1960s to the introduction of many new products which now command world markets of over 1 million tonnes annually. Examples are synthetic polymers such as polythene, PVC and nylon. Although the number of new products coming forward has declined since those halcyon days, the very high commitment to R&D remains and expenditure as a percentage of sales income is double that of all manufacturing industry. We recognize the importance of R&D by devoting a full chapter (Chapter 3) to it.

The chemical industry is very much a high technology industry with full advantage being taken of advances in electronics and engineering. Thus the use of computers is extremely widespread: from automatic control of chemical plant to automating and/or extending the abilities of analytical instruments. This also partly explains why it is capital rather than labour intensive.

The scale of operations within the industry ranges from quite small plants (a few tonnes per year) in the fine chemical area to the giants (100–500 thousand tonnes per year) of the petrochemical sector. Although the latter take full advantage of the economy of scale effect (section 6.7), if the balance between production capacity and market demand is disturbed the losses due to running at well under design capacity can be extremely high. This is particularly evident when the economy is depressed, and the chemical industry's business tends to follow the cyclical pattern of the economy with periods of full activity followed by those of very low activity.

The major chemical companies are truly multinational and operate their sales and marketing activities in most of the countries of the world, and they also have manufacturing units in a number of countries. For example ICI has sites in 40 countries and sells to over 150 countries. This international outlook for operations, or globalization, is a growing trend within the chemical industry, with companies expanding their activities either by erecting manufacturing units in other countries or by taking over companies which are already operating there. Further discussion of these characteristics can be found in section 4.7.

1.2
Scale of operations

It is important to appreciate that although most of the discussion about the chemical industry tends to revolve around the multinational giants who are household names—Bayer, Ciba-Geigy, DuPont, ICI—the industry is very diverse and includes very many small-sized companies as well. There is a similar diversity in the sizes of chemical plants. By and large these divide according to whether the plant operates in a batch mode or a continuous one. Generally speaking, the batch type are used for the manufacture of relatively small amounts of a chemical, say up to 100 tonnes per annum. They are therefore not dedicated to producing just a single product but are multi-purpose and may be used to produce a number of different chemicals each year. In contrast continuous plants are designed to produce a single product (or a related group of products) and as the name suggests they operate 24 hours a day all the year round. Nowadays they are invariably controlled by computers. They have capacities in the 20 000 to 600 000 tonnes per annum range and are generally used to make key intermediates which are turned into a very wide range of products by downstream processing. Most examples (ethylene, benzene, phenol, vinyl chloride etc.) are to be found in the petrochemicals sector but another well known example is ammonia. These operations are discussed in detail in Chapter 12. Clearly such large and sophisticated plants require a very high capital investment, and this is illustrated by SHOP (Shell Higher Olefines Plant) which cost Shell £100 million to build at Stanlow on Merseyside in the early 1980s. In the late 1980s further investment was made to increase its capacity. In contrast small-scale batch plants are used to manufacture fine chemicals. These are chemicals which are needed in relatively small quantities and high purity. Examples are pharmaceuticals, dyestuffs, pesticides and speciality chemicals such as optical fibre coatings and aerospace advanced materials (speciality polymers).

1.3
Major chemical producing countries

Comparisons between the U.K. chemical industry and those in other countries are made in section 4.3.2. The U.S. chemical industry and other countries are also discussed, in sections 4.4, 4.5, and 4.6. However, it is useful at this stage to indicate which are the important chemical producing countries. The most important by a considerable margin is the United States, whose total production equates roughly with that of Western Europe. Within the latter area West Germany is the largest producer followed in turn by France, U.K., Italy and the Netherlands. Note, however, that the second most important chemicals producer (based on value of sales) is Japan, which has double the output of the third country, West Germany. The U.S.A.'s output is some 50% higher than that of Japan. Although reliable statistics are harder to obtain, the U.S.S.R. and the Eastern Bloc are also important chemical manufacturers.

1.4
Major sectors and their products

The major sectors of the chemical industry are those forming most of the chapter headings in *The Chemical Industry*, 2nd edn. (Volume 2). A similar categorization is shown in Table 1.1, even though this is based more on end uses of the chemicals. Note that here the comparison between the sectors is based on the value added, i.e. roughly the difference between the selling price and the raw material plus processing costs. This means that pharmaceutical products, which sell for very high prices per unit of weight

Table 1.1   Sectors of the U.K. chemical industry (1992) (gross value added)

	% share
Pharmaceuticals	37
Specialized chemical products—industrial/agricultural use	20
Organics	10
Soaps and toilet preparations	10
Synthetic resins, plastics and rubber	7
Paints, varnishes and printing inks	5
Dyestuffs and pigments	3
Inorganics	4
Specialized chemical products—household/office use	3
Fertilizers	1

(up to tens of thousands of pounds per tonne or more), stand out much more than petrochemicals (organics plus synthetic/plastics) which typically sell for several hundreds of pounds per tonne. Even the vastly greater tonnage of the latter does not reverse the positions. Some of the major sectors of the chemical industry are listed below.

Petrochemicals	Chlor-alkali products
Polymers	Sulphuric acid (sulphur industry)
Dyestuffs	Ammonia and fertilizers (nitrogen industry)
Agrochemicals	Phosphoric acid and phosphates (phosphorus
Pharmaceuticals	industry)

The petrochemicals sector provides key intermediates, derived from oil and natural gas, such as ethylene, propylene and benzene. These are then used as raw materials for downstream processing in some of the other sectors listed. It is clearly one of the most important sectors of the industry and forms the subject of Chapter 12.

The polymers sector is the major user of petrochemical intermediates and consumes almost half the total output of organic chemicals which are produced. It covers plastics, synthetic fibres, rubbers, elastomers and adhesives, and it was the tremendous demand for these new materials with their special, and often novel, properties which brought about the explosive growth of the organic chemicals industry between 1950 and 1970.

Although the dyestuffs sector is much smaller than the previous two, it has strong links with them. This arose because the traditional dyestuffs, which were fine for natural fibres like cotton and wool, were totally unsuitable for the new synthetic fibres like nylon, polyesters and acrylics. A great deal of research and technological effort within the sector has resulted in the amazingly wide range of colours in which modern clothing is now available.

Agrochemicals (or pesticides) is an area of immense research effort with demonstrable success in aiding the production of more and better food. Along with pharmaceuticals it is a bluechip sector, i.e. very profitable for those companies which can continue to operate in it.

Pharmaceuticals has been the glamorous sector of the industry for some years now. This arises from an excellent innovative record in producing new products which has led, for many companies, to high levels of profitability. In addition, demand for its products is unaffected by the world's economy and therefore remains high even during recessions. This contrasts with the situation for most other sectors of the chemical industry. Indeed criticism seems to regularly surface that profits are too high, but this must be set against the R&D costs, which exceed £100 million to get one new drug to market launch. The chlor-alkali products sector produces mainly sodium hydroxide and chlorine, both of which are key basic chemicals and are discussed in Chapter 10. This chapter demonstrates nicely the influence of new technology and energy costs on chemicals production.

Sulphuric acid is the most important chemical of all in tonnage terms. Its production can be regarded as having reached maturity some years ago, but even now work is being done to remove the last traces of unreacted SO_2 (for environmental reasons).

Ammonia and fertilizers is a sector in which it has been difficult to achieve a balance between capacity and demand, and this has often led to major cost cutting and losses for many companies. In tonnage terms it is one of the most important sectors and it is based on the Haber process for ammonia. This is very energy demanding (moderately high temperatures and very high pressures) and a fortune is awaiting anyone who can find a viable alternative route. It will not be easy since no one has yet succeeded despite 70 years of intensive research effort!

Various phosphates are produced from phosphoric acid which is made either by adding sulphuric acid to phosphate rock (wet process) or by burning phosphorus in air to give phosphorus pentoxide, which is then hydrated. Major uses of phosphoric acid are the production of phosphate and compound fertilizers, formation of sodium tripolyphosphate (which is used as a builder in detergents where it forms stable water-soluble complexes with calcium and magnesium ions) and the production of organic derivatives like triphenyl and tricresyl phosphate. These are used as plasticizers for synthetic polymers and plastics.

Soaps and detergents represent an interesting and rather different sector. Interesting in that early production of soap, with its demand for alkali, can be viewed as the beginnings of the modern chemical industry. Different from the other sectors in that its products are sold directly to the public and market share probably has more to do with packaging and marketing than the technical properties of the product. Many of its products can be derived from both petrochemical intermediates and from animal and vegetable oils and fats, e.g. alkyl and aryl sulphonates. Chemicals from oils and fats are discussed in section 2.2.4.

1.5
Turning chemicals into useful end products

Although some chemicals, such as organic solvents, are used directly, most require further processing and formulating before they can be put to their end uses. In some cases, where novel materials have been discovered, major technological advances were required before they could be processed and their unique properties utilized. Such a material is polytetrafluoroethylene, which is better known as PTFE or under its trade names Fluon (ICI) and Teflon (DuPont). When this was first made its special properties of great chemical stability, excellent electrical insulation, very low coefficient of friction (hence its non-stick applications) and very wide working temperature range were quickly recognized. However, its use was delayed for several years because it could not be processed by conventional techniques and it had to await the development of powder metallurgy techniques.

In order to appreciate the downstream processing and technology, let us take as an example polyvinyl acetate and one of its applications as a binder in emulsion paint. Here its function is to bind the pigment, e.g. titanium dioxide, such that a homogeneous film is produced on evaporation of the water base. What processing steps are involved in making the polyvinyl acetate and in finally formulating the paint?

The story starts with crude oil or natural gas fractions, e.g. naphtha, which are cracked to give principally ethylene. The ethylene is then reacted with acetic acid and oxygen over a supported palladium catalyst to produce the vinyl acetate (see section 12.7.4). Finally this is polymerized to polyvinyl acetate which is then mixed with the other ingredients to produce the emulsion paint.

The above examples teach us an important lesson; although it is the chemists who make and discover the new chemicals which may have special properties, a considerable input from engineers and technologists may be required before the chemical can be processed and converted into a suitable form in which it can be used. This emphasizes an important aspect of research and technology in the chemical industry, namely the importance of inter-disciplinary teamwork.

1.6 Environment issues

Ever since Rachel Carson drew attention to the adverse environmental effects of some pesticides in her book *Silent Spring* in 1962, there has been a growing concern and awareness of environmental issues. The environmental move-ment has grown very rapidly in recent years and this is evident from the establishment and rapid growth of the Green political parties in countries like Germany and the U.K. It is therefore appropriate to examine the position of the chemical industry with regard to the environment since a number of the problems have been laid at the former's doorstep.

Some of the major problems are much wider than the chemical industry; acid rain and the greenhouse effect are clearly problems created by the energy industry, although in the latter case burning of the rain forests is a significant and worrying contributor. Nuclear waste is also a result of the activities of part of the energy industry. It is interesting to speculate that if all our energy requirements were met by nuclear fission power both the acid rain and greenhouse effect (caused largely by the combustion of fossil fuels) would be considerably reduced, but would this merely replace one set of problems by another? The answer to all this would seem to be the generation of energy by nuclear fusion, but as we all know there are immense technical problems to be overcome before this is a viable process.

Let us now briefly look into several major environmental problems/ disasters which clearly are associated with the chemical industry.

1.6.1 *Flixborough*

This major disaster occurred in 1974 at the Flixborough works of Nypro (U.K.) Ltd. The plant involved was part of the process for producing Nylon 6, and was used for the stage in which cyclohexane is oxidized to cyclohexanol plus cyclohexanone. One of the reactors had been removed for repair and temporarily replaced by a large diameter pipe which was inadequately supported. Cyclohexane began to leak and a very large cloud of it eventually ignited, causing a massive explosion. This resulted in 28 dead, almost 100 injured and damage to nearly 2000 factories, houses and shops in the neighbourhood.

It was the U.K. chemical industry's blackest day but note that it appeared to be caused by human error.

1.6.2 *Minamata Bay (Japan)*

This incident, in 1965, led to almost 50 deaths and to 100 seriously ill people. They displayed symptoms of mercury poisoning and the problem was traced to mercury which had been discharged into the bay by a chemical company. There it was converted into the very toxic dimethyl mercury by micro-organisms at the bottom of the sea. This substance concentrated in fish which were subsequently eaten by the victims.

Incidents such as this led to a major tightening up of pollution laws in Japan and nowadays up to 50% of the capital cost of a new plant can be earmarked for pollution control equipment.

1.6.3 *Thalidomide and drugs*

The thalidomide tragedy in 1961, in which some pregnant women who were prescribed the drug gave birth to grossly malformed babies, is one in which the companies involved (Chemie Grünenthal and Distillers) were criticized for not detecting this problem during testing of the drug before it was marketed. This was unjustified because at that time no one had ever envisaged that a drug could pass from the mother to the foetus and cause such dreadful results. Nowadays of course testing for this—known as tetratogenicity—is routinely carried out with several different mammals for all potential drugs and pesticides. Where the companies were quite rightly criticized was in not withdrawing the product quickly enough and in not compensating the victims (until after litigation taking many years), once the link between the drug and these side effects had been established.

Drug abuse involving e.g. barbiturates and tranquilizers, is an area where the industry sometimes comes in for criticism. This is quite wrong because it is a social problem of our whole society. It is also very important to place these problems in context. These few tragedies must be viewed against the hundreds of millions of lives which have been saved and prolonged by the thousands of new drugs and medicines which the pharmaceutical industry has produced.

1.6.4 *Seveso, Bhopal and pesticides*

Over the years pesticides have probably attracted more adverse comment than any other chemical products. Whilst one should quite rightly discuss the problems associated with products such as Dieldrin, DDT and Agent Orange, and have these replaced by less toxic, more environmentally acceptable products, these difficulties must again be put into context. These few problem pesticides must be viewed against the many thousands which are in regular use and have not caused any difficulties but have helped crop yields increase dramatically. It is an accepted projection that if all pesticides were banned world food production would fall by at least 50%. Those products which caused problems did so because they were not selective enough. However, the selectivity of some recent pesticides is quite remarkable; for example, herbicides now exist which kill wild oats (a weed) growing amongst the oat or barley crop, leaving the crop totally unaffected.

In Seveso in Northern Italy in 1976, a plant used for manufacturing the herbicide 2,4,5-T (2,4,5-trichlorophenoxyacetic acid), which was a component of Agent Orange (used as a defoliant in Vietnam), blew up and released about 2 kg of dioxin. This is one of the most stable and toxic chemicals known and is also teratogenic, like thalidomide. Hundreds of people living nearby were evacuated and the contaminated soil removed and destroyed. Although no one died or appeared to suffer as a result of this accident, it was a potential disaster. Like Flixborough, the cause again appears to be human error. Most of the above points are discussed in more detail in Chapter 7 of Volume 2.

An accident at a plant in Bhopal in India in 1984 accidently released methyl isocyanate onto the surrounding population with terrible consequences—some reports have put the final death toll as high as 3000 and many thousands more were seriously injured. After 4 years of legal wrangling the company owning the plant, Union Carbide, in 1989, agreed to pay $470 million to the victims and in 1994 contributed $20 million for the construction and operation of a hospital. Union Carbide sold its 50.9% share in its Indian affiliate, Union Carbide India Ltd. to the Indian conglomerate McLeod Russel Ltd. This technically and legally severs its connection with the Bhopal incident. Yet again human error appears to have been the cause of the tragedy.

1.6.5 *Hickson and Welch, Castleford*

In 1992 an explosion ocurred in a distillation vessel associated with a mononitrotoluene plant at this works. The vessel had never been cleaned since its installation in the early 1960s and consequently contained a 14″ deep residue of jelly-like sludge. This sludge contained flammable dinitro-toluene and nitrocresols covering one of the unit's steam heating coils.

Plant managers decided to remove the sludge with a metal rake after passing steam through the coil to soften it. However, during the raking, with the heating still on, the sludge ignited, sending a jet of flame over 50 m long shooting through the plant control building into the site's main office block.

Five people died as a result of this and the company was fined £250 000 with £150 000 costs for failing to ensure the safety of its employees and putting them at risk of fire.

The U.K. Health and Safety Executive's report on the incident blames inadequate monitoring, safety and operating procedures and lax plant design for the explosion. So yet again we see that human error was the cause of a tragedy.

1.6.6 CFCs (chlorofluorocarbons)

CFCs are the remarkably inert and non-toxic chlorofluorocarbons such as CFC11, $(CFCl_3)$ and CFC12, (CF_2Cl_2). Their trade names are Freons (DuPont) and Arctons (ICI). They have been widely used for many years as refrigerants, aerosol propellants and polyurethane foam-blowing agents. Ironically it is their very stability which has proved to be their undoing because they rise up into the stratosphere unchanged. There they are decomposed on exposure to the sun's short wave U.V. radiation into Cl· and these radicals cause the breakdown of the ozone layer which screens us on earth from this dangerous U.V. radiation. Most of the world's developed nations, the major users of CFCs, plus some third world countries, signed an agreement known as the Montreal Protocol which is a timetable for reducing and phasing out the use of CFCs. Pressure is being increased to speed up the phase out so that it is complete by the end of this century.

All the major producers, about 14 worldwide, have been working flat out to produce a family of alternatives which are known as HFCs (hydrofluorocarbons) or HFAs (hydrofluoroalkanes). These generally do not contain chlorine and should therefore have a much smaller or even no effect on the ozone layer. Success was achieved in 1993 when ICI and DuPont started up plants to produce the CFC alternative MFA 134a, CF_3CH_2F. This topic is discussed more fully in section 3.5.2.

In concluding this section, although attention has been drawn to some serious pollution problems these must be seen against the background of an industry operating very many potentially very hazardous processes every day. Clearly the vast majority operate safely and efficiently. There does also seem to be a greater concern for the environment by the companies and although one can see that they have to a degree been pushed towards this by organizations like Greenpeace, they can take a good deal of credit themselves for their changing attitudes to safety and the environment. After all, the people who work for the companies live in the same environment as the rest of us.

1.7.1 *Quality*

During the past decade, the recognition of the importance of quality in manufacturing industry has increased enormously. This was initiated by the success of Japanese manufacturing industry which, with its emphasis on quality products, achieved dominance in several sectors, perhaps the best examples being motor cycles and cars, and electrical goods such as television sets. Although this development has been going on for at least two decades, it is only during the last decade, and particularly the latter part of it, that the large chemical firms have acted to educate their staff on the importance of quality and developing a company environment conducive to a quality operation. An important stimulus has been the increasingly competitive nature of the international chemicals business, coupled with the worldwide recession of the late 1980s and early 1990s. Many of the companies have spent over £10 million in putting in place a Total Quality Management System and in educating and training staff to be able to contribute to it and use it. What quality essentially boils down to is 'doing it right, first time'! A detailed discussion of this topic is given in Chapter 2 of Volume 2.

1.7.2 *Safety*

The growing concern in the general public's attitude to the safety of chemicals in the environment has rapidly extended to areas such as chemicals in food, solvents in paint and glue, etc., and the safe operation of chemical plant. Add to this the occasional accident involving a road tanker which resulted in spillage of chemicals on to the highway, which provided another opportunity for the media to criticize the chemical industry, and it is easy to see why public pressure on the companies has increased greatly in recent years. This pressure has resulted in stricter regulations to ensure that plants are operated safely, so that hopefully we will never have another disaster such as Flixborough (section 1.6.1) or Bhopal (section 1.6.4), and even the occasional leak of hazardous chemicals will be less likely to occur. There has also been insistence on much more extensive testing to ensure that compounds are as safe as possible before they can be used as pesticides, drugs, etc. Perusal of Table 7.4 and sections 7.3 and 8.3 of Volume 2 will demonstrate just how comprehensive, time-consuming and expensive this is. Note that we, the general public, ultimately end up paying for the cost of this in the increased price of the chemical which we purchase as a consumer product. Although we all accept the importance and necessity for this, it must be pointed out that however much testing is carried out no compound can ever be proved safe. Water is non-toxic and essential for life but it can still kill people by drowning! It is salutary to note that if we applied the same testing and safety standards to aspirin (and water) that we currently apply to new compounds being developed as drugs they would almost certainly be banned from reaching the market! This testing programme and submission of the results to the appropriate regulatory authorities for approval now take at least seven years, and some of the biologically active compounds start to lose their effectiveness within four to five years of first being marketed because of pest resistance by, for example, insects or bacteria. My personal opinion is that we must now reach a balance and taper off our calls for more and more testing, since the additional tests will probably give us little or no additional safeguard on the safety of the product. There is, however, scope for making the tests more efficient and developing alternatives, for example in the area of animal testing.

Safe working practices and the good health of employees are closely linked. Partly perhaps because of trade-union pressure, triggered by incidents such as some workers on plants using naphthylamines (for dyestuffs) and vinyl chloride monomer (for PVC) developing rare forms of cancer, and partly due to a more enlightened and open approach by the companies, both safety and health and hygiene of workers now command a much higher profile. Cynics might say a lot of this was forced on the companies because of the poor public image of the chemical industry, but this is not justified. As we show in a number of places in this book the public perception of aspects of the chemical industry (usually very negative) can be at variance with the facts. For example, the one or two disasters in the industry (see section 1.6) have led the public to view the industry as a particularly dangerous sector of manufacturing industry. The figures and graph on p. 80 tell a different story and show a very interesting comparison with some other occupations.

It is encouraging that the industry is now taking a much more positive and open approach and bringing the facts on what it is doing, and also often the problems it still faces, to the attention of the public. Schools–industry projects such as the one directed by the author provide factual material for school-children and their teachers and allow them to undertake works visits, enabling them to acquire a much more balanced view of the industry, but still allowing them to make up their own minds about the industry and its operations. Their major concern and interest relate to environmental matters, which are discussed in Chapter 9.

Sources of Chemicals 2

Alan Heaton

The number and diversity of chemical compounds is remarkable: over ten million are now known. Even this vast number pales into insignificance when compared to the number of carbon compounds which is theoretically possible. This is a consequence of catenation, i.e. formation of very long chains of carbon atoms due to the relatively strong carbon–carbon covalent bonds, and isomerism. Most of these compounds are merely laboratory curiosities or are only of academic interest. However, of the remainder there are probably several thousand which are of commercial and practical interest and this text will demonstrate the very wide range of chemical structures which they encompass. It might therefore be expected that there would be a large number of sources of these chemicals. Although this is true for inorganic chemicals, surprisingly most organic chemicals can originate from a single source such as crude oil (petroleum).

Since the term 'inorganic chemicals' covers compounds of all the elements other than carbon, the diversity of origins is not surprising. Some of the more important sources are metallic ores (for important metals like iron and aluminium), and salt or brine (for chlorine, sodium, sodium hydroxide and sodium carbonate). In all these cases at least two different elements are combined together chemically in the form of a stable compound. If therefore the individual element or elements, say the metal, are required then the extraction process must involve chemical treatment in addition to any separative methods of a purely physical nature. Metal ores, or minerals, rarely occur on their own in a pure form and therefore a first step in their processing is usually the separation from unwanted solids, such as clay or sand. Crushing and grinding of the solids followed by sieving may achieve some physical separation because of differing particle size. The next stage depends on the nature and properties of the required ore. For example, iron-bearing ores can often be separated by utilizing their magnetic properties in a magnetic separator. Froth flotation is another widely used technique in which the desired ore, in a fine particulate form, is separated from other solids by a difference in their ability to be wetted by an aqueous solution. Surface active (anti-wetting) agents are added to the solution, and these are typically molecules having a non-polar part, e.g. a long hydrocarbon chain, with a polar part such as an amino group at one end. This polar grouping attracts the ore, forming a loose bond. The hydrocarbon grouping now repels the water, thus preventing the ore being wetted, and it therefore floats. Other solids, in contrast, are readily wetted and therefore sink in the aqueous solution. Stirring or bubbling the liquid to give a froth considerably aids the 'floating' of the agent-coated ore which then overflows from this tank into a collecting vessel, where it can be recovered. The key to success is clearly in the choice of a highly specific surface-active agent for the ore in question.

Chemical treatment depends on the nature of the ore, but for metal oxides and sulphides, reaction with a reducing agent like coke may be employed, as in the blast-furnace where iron oxide is converted into iron. More recent developments have centred on extracting metals from waste-heaps of old mine workings. When these materials (or tailings) were first co-mined they were discarded because the desired metal ore content was too low to make the extraction economically viable. However over recent years the prices of metals have increased and this, coupled with new processing methods, means that the economics have now become favourable. An example is extraction of copper from an aqueous solution of its nitrate using a selective complexing agent or even using a specific micro-organism which concentrates a particular metal.

Atypically, there are a few materials which occur in an elemental form. Perhaps the most notable example is sulphur, which occurs in underground deposits in areas such as Louisiana, Southern Italy and Poland. It can be brought to the surface using the Frasch process in which it is first melted by superheated steam and then forced to the surface by compressed air. This produces sulphur of high purity. Substantial quantities of sulphur are also removed and recovered from natural gas and crude oil (petroleum). This amounted to 24 million tonnes out of a total world sulphur production of 37 million tonnes in 1991, and clearly demonstrates the vast scale on which the oil and petrochemical industries operate since crude oil normally contains between 0.1 and 2.5% of sulphur, depending on its source. Desulphurization of flue gases from some U.K. power stations will be another source of sulphur in the future. Over 80% of all sulphur is converted into sulphuric acid, and approximately half of this is then used in fertilizer manufacture.

A second example of the occurrence in nature of materials in elemental form is air, which may be physically separated into its component gases by liquefaction and fractional distillation. In this way substantial amounts of nitrogen and oxygen, plus small amounts of the inert gases argon, neon, krypton, and xenon are produced. A recent development has been the use of zeolites (p. 323) for carrying out this separation.

In contrast to inorganic chemicals which, as we have already seen, are derived from many different sources, the multitude of commercially important organic compounds are essentially derived from a single source. Nowadays in excess of 90% (by tonnage) of all organic chemicals is obtained from crude oil (petroleum) and natural gas via petrochemical processes. This is a very interesting situation—one which has changed over the years and will change again in the future—because technically these same chemicals could be obtained from other raw materials or sources. Thus aliphatic compounds, in particular, may be produced via ethanol, which is obtained by fermentation of carbohydrates. Aromatic compounds on the other hand are isolated from coal-tar, which is a by-product in the carbonization of coal. Animal and vegetable oils and fats are a more specialized source of a limited number of aliphatic compounds, including long-chain fatty acids such as stearic (octadecanoic) acid, $CH_3(CH_2)_{16}CO_2H$, and long-chain alcohols such as lauryl alcohol (dodecanol), $CH_3(CH_2)_{11}OH$.

The relative importance of these sources of chemicals, or chemical feedstocks, has changed markedly over the past thirty years. In 1950, in the U.K., coal was the source of 60% of all organic chemicals, oil accounted for 9% and carbohydrates the remainder. Since 1970, oil and natural gas have dominated the scene, providing the source for over 90% of chemicals. Coal and carbohydrates complete the total, the latter contributing < 1% of total production. The relative positions in the next century could be quite different because supplies of oil are limited and at the present rates of usage, even allowing for the current discoveries of new oilfields, it is forecast that they will be exhausted some time during the next century. Coal is in a similar situation, although because of its lower rate of use and its vast reserves, the time-scale is greater and is measured in hundreds of years before supplies run out.

Figure 2.1 shows the economically recoverable reserves of fossil fuel, i.e. oil,

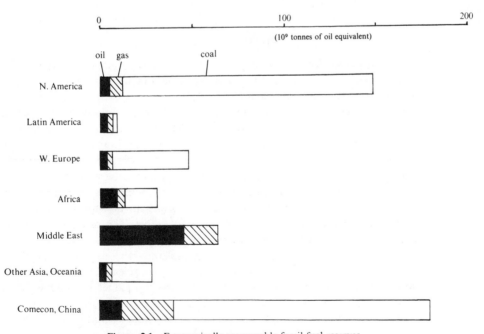

Figure 2.1 Economically recoverable fossil fuel reserves.

gas and coal reserves[1]. The formation of these fossil fuels takes millions of years and once used they cannot be replaced. They are therefore referred to as *non-renewable resources*. This contrasts with carbohydrates which, being derived from plants, can be replaced relatively quickly. A popular source is sugar-cane—once a crop has been harvested and the ground cleared, new material may be planted and harvested, certainly in less than one year. Carbohydrates are therefore described as *renewable resources*. The total *annual* production of dry plant material has been estimated[2] as 2×10^{11} tonnes.

Fossil fuels—natural gas, crude oil and coal—are used primarily as energy sources and not as sources of organic chemicals. For instance various petroleum fractions are used as gas for domestic cooking and heating, petrol or gasoline for automobiles, and heavy fuel oil for heating buildings or generating steam for industrial processes. Typically only around 8% of a barrel of crude oil is used in chemicals manufacture. Thus the price of crude oil is affected by the world supply/demand for energy. The lower-boiling fractions of value as feedstock for the chemical industry have alternative uses as premium fuels. Prices have risen faster than energy prices in general. Thus the increasing popularity of petroleum as a world energy source has had a double impact on the organic-chemical industry, increasing both the price of feedstock and the energy required for chemical conversion and separation processes. The following figures demonstrate why the chemical industry can compete with the fuel- or energy-using industries for the crude oil:

Form of oil	Relative value of oil
Crude oil	1
Fuel	2
Typical petrochemical	10
Typical consumer product	50

This competition for the precious, limited, resource of crude oil will have to be resolved in favour of its use for chemicals manufacture. Alternative energy sources are available such as nuclear and, in certain locations, hydroelectric. Despite varying degrees of opposition from some sections of the public because of the possible hazards associated with the use of radioactive materials, nuclear energy makes a significant contribution to the total energy requirements of most nations in the developed world. A typical figure for Western Europe is around 10%, with France leading the way by producing over 50% of its needs by nuclear fission. If the immense practical problems of bringing about, and controlling, nuclear fusion can be solved then almost unlimited supplies of energy will become available. The process of nuclear fusion is not only a very important natural process but an essential one for life on earth, since the fusion of two hydrogen nuclei together to produce a helium nucleus occurs in the sun and is accompanied by the release of vast amounts of solar radiation energy, which is required to warm our planet and enable plants to carry out photosynthesis and hence grow to provide food. Tremendous worldwide efforts are being made to satisfactorily harness many natural forms of energy, notably solar, wind, tidal and wave. Again, despite the practical problems, the long-term rewards will make a significant diminution in the world use of fossil fuels and nuclear energy. Another energy medium which is attractive from an environmental point of view, since it is clean, is hydrogen. Combustion of this yields only water. There seems, however, to be a psychological barrier to its use since it is thought of as a dangerous material due to its flammability. Solid metallic tanks are required for its storage, but the principal drawback is its currently expensive method of production by electrolysis of water. Development of an efficient process for the photodecomposition of water is the key, and work towards this is progressing slowly. However if some of these barriers can be overcome its attractions are obvious.

To demonstrate and assess the feasibility of hydrogen, a village has been constructed in the United States which uses hydrogen as its only energy medium—for cooking, heating and even powering automobiles and buses. A German company has also developed, and is currently evaluating, a hydrogen-powered bus. Although it is a little early to pronounce judgements on these experiments, nevertheless it would appear that most of the technical problems have been solved and the only difficulties are economic and psychological.

A further reason to discontinue the use of fossil fuels for energy generation is that they produce mainly carbon dioxide and water by complete combustion. Because this has taken place on such an enormous scale over many years, the quantity of carbon dioxide released into the atmosphere has been very large indeed. Considerable concern has been expressed for the consequences of this. It has been suggested that as all this carbon dioxide diffuses into the earth's upper atmosphere it will reduce the amount of screening of the sun's rays, causing the so-called greenhouse effect.

Clearly alternative energy sources to fossil fuels are now available if we have the will to use them, and we can confidently expect other alternatives to become available in the not too distant future. It is therefore essential that we retain our precious oil supplies for chemicals production. The statement that 'the last thing you should do with oil is burn it' becomes more valid every year. It is interesting, and salutory, to note that as early as 1894 Mendeleyev (the Russian chemist who developed the Periodic Table) reported to his government that 'oil was too valuable a resource to be burned and should be preserved as a source of chemicals'.

A further benefit of replacing oil as an energy source is that most of the known reserves—around 370 thousand million barrels out of a free-world total (i.e. excluding the Communist bloc) of about 550 thousand million barrels—are located in the Middle East, which is an area noted for its political instability. Thus security of oil supply and gradual changes in its price cannot be guaranteed. Therefore reducing dependence on oil to a large extent and substituting for it, if possible, an indigenous material like coal not only secures supplies but helps the balance of payments. A country like Germany, which does not have any oil of its own but has large coal reserves, could clearly derive great benefit from such change, provided it were economically viable.

Even a change to nuclear energy would help, since the overall cost of the raw material—the uranium ore—would be smaller due to the much smaller quantities required, and the security of supply should be less of a problem since the ore occurs throughout the world, particularly in North America, Southern Africa, Australia and Sweden. The much higher capital costs for the nuclear power station are an important factor which has to be taken into consideration. Prototype fast-breeder reactors have been operated in the U.K. for some years now and when fully developed they could substantially improve the economics of nuclear energy. This is because they enable more energy to be extracted from 'waste' uranium and in addition utilize the plutonium produced in conventional reactors as fuel.

2.2
Sources of organic chemicals

As we have already seen, these are surprisingly few in number and it is technically possible for each to act as the raw material for the synthesis of the majority of all the organic chemicals of commercial importance. The choice between them is therefore largely a matter of economics, which has been greatly influenced by the scale of operation. The dominant position of oil and natural gas as the source of more than 90% of all organic chemicals is due in considerable measure to the very large scale on which the petrochemical industry operates. This feature is shared with the oil-refining industry from which it developed and which provides its feedstocks. For example, Shell's Stanlow refinery on Merseyside—a typical large modern refinery—has the capacity to process 50 000 tonnes per *day* of crude oil, i.e. 18 million tonnes per annum.

A different job is carried out at the very large on-shore establishments at the U.K. end of the pipelines from North Sea oilfields. They are primarily engaged in removing the dissolved hydrocarbon gases in order to stabilize the oil for export in tankers. Petrochemical plants having capacities between 100 and 500 thousand tonnes per year are commonplace. In times of high production levels this allows the full benefits of the economy-of-scale effect (see Chapter 6) to be realized, resulting in a relatively low (but still profitable) selling price for the chemical. Unfortunately the recession (or period of low demand) at the start of the 1980s showed the other side of the coin, where once production levels on these very large continuous plants fall below something like 70% of capacity considerable losses are incurred, and these rise drastically the further this figure falls. Following an economic resurgence in the late 1980s ethylene capacity increases were planned by existing producers and also new entrants to this business, e.g. Korean companies. Unfortunately the coming on stream of this additional capacity at the start of the 1990s coincided with a downturn in the world's economy. Overcapacity leading to heavy losses was the inevitable result. For example in western Europe in 1991 capacity was 18 million tonnes per annum, with demand estimated at only 14 million tonnes per annum.

The inability of the world chemical industry to match capacity and demand has resulted in some European manufacturers withdrawing from this area, or at least reducing capacity. They prefer to concentrate on downstream products for the reasons outlined in section 3.3, and leave ethylene production to the oil producing countries like Saudi Arabia.

Natural gas is also a major petrochemical feedstock. Because it has a similar chemical nature it will be included with oil (petroleum) in subsequent discussions in this chapter. However it is less versatile than petroleum because carbon–carbon bonds have to be built up. Carbonylation technology is available (see section 12.4) and natural gas is far superior to coal as a source of carbon monoxide. Nevertheless, although coal is much more difficult to extract from the ground and handle than fluid hydrocarbons, it will grow in importance as a source of organic chemicals as petroleum and natural gas stocks decrease because of overall world shortage or political starvation (coal is currently the major source of chemicals and hydrocarbon fuels in South Africa). Carbohydrates (now generally called biomass) also involve solids processing. They are mainly available in tropical countries where there is a much smaller concentration of chemical processing plants.

Each of these chemical feedstocks—oil (and natural gas), coal, carbohydrates, and also animal and vegetable oils and fats—will be considered in turn. Treatment of oil and natural gas will be more limited than its importance merits because they are considered in detail in Chapter 12 on petrochemicals.

2.2.1 *Organic chemicals from oil and natural gas*

The petroleum or crude-oil processing industry dates from the 1920s, and in these early days its operations were confined to separation of the oil into fractions by distillation. These various fractions were then used as energy sources, but the increasing sales of automobiles pushed up demand for the gasoline or petrol fraction. Development of processes such as cracking and reforming was stimulated and by this means higher-boiling petroleum fractions, for which demand was low, were converted into materials suitable for blending as gasoline. Additionally these processes produced olefins or alkenes, and at this particular time there was no outlet for these as petroleum products. Subsequent research and development showed that they were very useful chemical intermediates from which a wide range of organic chemicals could be synthesized. This was the start of the petrochemical industry.

Up to the early 1940s the production of chemicals from petroleum was confined to North America. This was due in no small measure to the policy of locating refineries adjacent to the oilfields. From 1950 this policy was reversed and the crude oil was transported from oilfields (the major ones then being located in the Middle East) to the refineries, now located in the areas occupied by the major users of the end products. This coincided with, or led to, the development of the European petrochemical industry, followed in the 1960s by Japan. Oil-producing countries, such as Saudi Arabia, have continued this trend into chemicals production and in the 1970s and 1980s petro-chemicals production has become a truly worldwide activity.

The meteoric rise of oil as a source of chemicals has already been attributed mainly to economic factors—its remarkably stable price during the 1950–1970 period, when prices generally were rising *each* year due to inflation, and the benefits of the economy-of-scale effect (section 6.7) coupled with process improvement over the years. However, over the past two decades two factors have had a marked effect leading to rising production costs, despite increased competition in petrochemicals. The first of these was the policy of OPEC (Organization of Petroleum Exporting Countries) which led in 1973 to the price of crude oil quadrupling within a very short period. Its effect, not only on chemicals and energy production, was dramatic on the economies of the world, particularly in Europe. The ramifications of the resulting swing of purchasing power from the oil-importing countries to the Arab countries has had a profound influence on world financial dealings. Since 1973 crude oil prices have fluctuated, sometimes quite markedly, in contrast to the period of great stability up to 1970 (see section 12.1). The second factor influencing production costs of petrochemicals and chemicals production in general, has been the increasing concern of the public about the environment and also for the safety of industrial plant. The need, and desirability, of conforming to regulations in attempting to ensure a clean and safe environment have had a significant influence in increasing processing costs. One extreme example of this quotes 70% of the total capital cost of a plant constructed in Japan being attributed to pollution-control equipment. These aspects are discussed in detail in Chapter 9.

Crude oil or petroleum has been discovered in many locations throughout the world, important reserves occurring in the Middle East, North America (including Alaska), the North Sea (U.K. and Norwegian) and North Africa (Algeria and Libya). The precise composition of the oil varies from location to location, and this may be apparent even from its odour and its physical appearance, particularly with reference to viscosity, since oils vary from those with very high viscosities, almost treacle-like in consistency, to much more fluid liquids. Indeed analysis of oil slicks at sea, by gas-liquid chromatography, has been used to identify the country of origin, and this, together with knowledge of tanker movements, has been used in the prosecution of skippers for illegal washing-out of empty crude-oil tanks at sea.

Crude oil consists principally of a complex mixture of saturated hydro-carbons—mainly alkanes (paraffins) and cycloalkanes (naphthenes) with smaller amounts of alkenes and aromatics—plus small amounts ($< 5\%$ in total) of compounds containing nitrogen, oxygen or sulphur. The presence of the latter is undesirable since many sulphur-containing compounds, e.g. mercaptans, have rather unpleasant odours and also, more importantly, are catalyst poisons and can therefore have disastrous effects on some refinery operations and downstream chemical processes. In addition, their combustion may cause formation of the air pollutant sulphur dioxide. They are therefore removed at an early stage in the refining of the crude oil (or else they tend to concentrate in the heavy fuel oil fraction). One way of achieving this is by hydrodesulphurization in which the hot oil plus hydrogen is passed over a suitable catalyst. Sulphur is converted into hydrogen sulphide, which is separated off and recovered.

The complex mixture of hydrocarbons constituting crude oil must first be separated into a series of less complex mixtures or fractions. Since the components are chemically similar, being largely alkanes or cycloalkanes, and ranging from very volatile to fairly involatile materials, they are readily separated into these fractions by continuous distillation. The separation is based on the boiling point and therefore accords largely with the number of carbon atoms in the molecule. Table 2.1 shows the typical fractions which are obtained, together with an indication of their boiling range, composition and proportion of the starting crude oil.

In terms of producing chemicals it is the lower-boiling fractions which are of importance—particularly the gases and the naphtha fractions. However consideration of the nature of the components of these fractions (they are largely alkanes and cycloakanes) suggests that they will not be suitable for chemical synthesis as they stand. Alkanes are well known for their lack of chemical reactivity—indeed their old name, paraffins, is derived from two Latin words, *parum* and *affinis*, meaning, 'little affinity'. They need to be converted into more reactive molecules and this is achieved by chemical reactions which produce unsaturated hydrocarbons such as alkenes and aromatics. As expected, because of the alkanes' lack of reactivity, the reaction conditions are very vigorous, and high temperatures are required. Alkenes are produced by a process known as cracking, which may be represented in

Table 2.1 Distillation of crude oil

Fraction	Boiling range (°C) (at atmospheric pressure)	Number of carbon atoms in molecule	Approximate % by volume
→GASES	< 20	1–4	1–2
→LIGHT GASOLINES OR LIGHT NAPHTHA	20–70	5–6	20–40
→NAPHTHA (MID-RANGE)	70–170	6–10	
→KEROSENE	170–250	10–14	10–15
→GAS OIL	250–340	14–19	15–20
→DISTILLATE FEEDSTOCKS for LUBRICATING OIL and WAXES, or HEAVY FUEL OILS	340–500	19–35	40–50
→ BITUMEN	> 500 i.e. Residue	> 35	

CRUDE OIL

very simple terms as

$$2C_6H_{14} \xrightarrow[\text{catalyst}]{800-1000°C} CH_4 + 3C_2H_4 + C_2H_6 + C_3H_6$$

In practice the feedstock, being a crude-oil fraction such as gases or naphtha, is a mixture and therefore the product consists of a number of unsaturated and saturated compounds (cf. p. 358). Cracking is used to break down the longer-chain alkanes, which are found in (say) the gas-oil fraction, producing a product akin to a naphtha (gasoline) fraction. This process was developed because of the great demand for naphtha and the relatively low demand for gas oil, particularly in the U.S.

Aromatics are made from alkanes and cycloalkanes by a process aptly named reforming, which may be represented, again in over-simplified terms, as

$$C_6H_{14} \xrightarrow[\text{catalyst, e.g. Pt metal}]{\text{heat}} \bigcirc + 4H_2$$

As in cracking, the feedstock is a mixture of compounds. A substantial conversion (c. 50%) to aromatic compounds is achievable. The principal components, benzene, toluene and the xylenes, are separated for further processing.

Separation and purification of products is a major cost item in industrial chemical processes. It is important not only to isolate the desired product but also to recover the by-products. The economic success of many processes involves finding uses for co-produced materials. Greater selectivity in the reaction will minimize by-product formation and hence reduce the purification requirements. It is often economically desirable to run a reaction at a lower conversion level in order to increase selectivity even though this increases the amount of recycling of reactants.

Natural gas is found in the same sort of geological areas as crude oil, and may occur with it or separately. It consists mainly of methane plus some ethane, propane and small amounts of higher alkanes, and has long been a very important chemical feedstock for ethylene (ethene) production in the U.S.A., although its importance has started and will continue to decrease as supplies dwindle. In Europe natural gas was first discovered in the province of Groningen in Holland in 1959. This turned out to be the largest natural gas field in the world, containing an estimated 60 million million cubic feet of gas and extended out under the North Sea. The gas consists almost entirely of methane, with very few other alkanes being present. This discovery stimulated searches in the other areas of the southern North Sea and was rewarded in 1965 when British Petroleum found natural gas. Further discoveries soon followed and this gas has been used as a fuel for heating (both domestic and industrial) and cooking. Drilling in the more exposed northern areas of the North Sea has resulted, during the 1970s, in the discovery of several other important oil and gas fields in areas east of the Shetland Isles. This has made natural gas readily available in the United Kingdom as a chemical feedstock. Steam reforming of natural gas (discussed in detail in section 12.4) is a very large scale and important reaction for producing synthesis gas (syngas), which is a mixture of carbon monoxide and hydrogen, viz.

$$CH_4 + H_2O \xrightarrow[850°C]{\text{Ni catalyst}} CO + 3H_2$$
$$\underset{\text{Natural gas}}{} \qquad\qquad \underset{\text{syngas}}{}$$

The importance of syngas as an intermediate for the production of a variety of organic chemicals is demonstrated later (section 12.4). Large quantities of hydrogen, produced as indicated above, are used in ammonia synthesis using the high-temperature and high-pressure Haber process, viz.

$$N_2 + 3H_2 \rightleftharpoons 2NH_3 \quad \text{(see section 12.4.1.2)}$$

2.2.2 *Organic chemicals from coal*

Coal, like crude oil (petroleum), is a fossil fuel which forms over a period of millions of years from the fossilized remains of plants. It is therefore also a non-renewable resource. However, reserves of coal are several times greater than those of petroleum[1], and, in contrast to petroleum, most European countries have deposits of coal varying from significant to very large quantities. The United States also has large reserves of coal. Extraction and handling of the coal is more difficult, and expensive, than for oil.

Although the precise nature of coal varies somewhat with its source (like crude oil), analysis of a representative sample shows it to be very different from oil. Firstly, its H:C ratio is on average about 0.85:1 (the corresponding figure for oil being around 1.70:1). Secondly, it consists of macro, or giant, molecules having molecular weights up to 1000. Thirdly, a significant proportion of heteroatoms—particularly oxygen and sulphur—are present[1]. The complexities of coal suggest that coal chemistry will develop along different lines to that of oil[3], although the wealth of knowledge and experience gained with the latter will provide considerable and valuable help in this development.

2.2.2.1 *Carbonization of coal.* Traditionally, and even today to some extent, chemicals have been obtained from coal via its carbonization. This is brought about by heating the coal in the absence of air at a temperature of between 800 and 1200°C, viz.

$$\text{coal} \xrightarrow{800-1200°C} \text{coke + town gas + crude benzole + coal-tar}$$

The major product by far is the coke, followed by the town gas (a mixture of largely carbon monoxide and hydrogen) with only small amounts of crude benzole and coal-tar ($\sim 50\,kg$ per $1000\,kg$ of coal carbonized) being formed, but it is from these that chemicals are obtained. It is therefore clear that for the carbonization process to be viable there must be a market for the coke produced. The steel industry is the main outlet for this, and therefore demand for coal carbonization is closely linked to the fortunes of this industry. Until some twenty years ago demand was high, certainly in Europe, not only for coke but also for the town gas which was widely used for cooking and heating. With its replacement by natural gas there is now no demand at all for town gas in the United Kingdom. Coupled with this is a much lower demand for coke by the steel industry due to a marked decrease in the level of output of steel plus increases in the efficiency of the process which has reduced the quantity of coke required per tonne of steel produced. Thus the demand for coal carbonization has fallen considerably over the past few decades and it is not economically feasible to carry out this operation merely to obtain the crude benzole, coal-tar, and chemicals derived from them.

Coal-tar is a complex mixture of compounds (over 350 have been identified) which are largely aromatic hydrocarbons plus smaller amounts of phenols. The most abundant individual compound is usually naphthalene, but this only comprises about 8% of the coal-tar. Typical weights of major chemicals

Table 2.2 Distillation of coal-tar

	Fraction	Boiling range (°C) (at atmospheric pressure)	Approximate % by volume	Main components
COAL-TAR	→ AMMONIACAL LIQUOR			
	→ LIGHT OIL	up to 180		benzene, toluene, xylenes (dimethyl benzenes), pyridine, picolines (methyl-pyridines)
	→ TAR ACIDS (carbolic oil)	180–230	8	phenols, cresols (methylphenols), naphthalene
	→ ABSORBING OIL (creosote oil)	230–270	17	methylnaphthalenes, quinolines, lutidines (dimethylpyridines)
	→ ANTHRACENE OIL	270–350	12	anthracene, phenanthrene, acenaphthene
	→ RESIDUE (pitch)		60	

obtainable from coal (isolated from coal-tar plus crude benzole) are

$$\text{coal} \xrightarrow{\hspace{1cm}} \text{benzene} + \text{naphthalene} + \text{phenol}$$

coal	benzene	naphthalene	phenol
1000 kg	5.3 kg	2.9 kg	0.4 kg

The initial step in the isolation of individual chemicals from coal-tar is continuous fractional distillation (cf. oil) which yields the fractions shown in Table 2.2

Each of these fractions still consists of a mixture of compounds, albeit a much simpler and less complex mixture than the coal-tar itself. Some are used directly, e.g. absorbing oil is used for absorbing benzene produced during carbonization of coal and it is also used—under the name creosote—for preserving timber. More usually they are subjected to further processing to produce individual compounds. Thus the light-oil fraction is washed with mineral acid (to remove organic bases such as pyridine, plus thiophene), then with alkali (to remove tar acids, i.e. phenols). The remaining neutral fraction is subjected to fractional distillation which separates benzene, toluene and the xylenes.

Figure 2.2 shows some of the main uses of primary products which are obtained from crude oil and from coal.

2.2.2.2 *Gasification and liquefaction of coal.* In view of the difficulties of obtaining chemicals from coal via carbonization and coal-tar (discussed above), it is not surprising that alternative routes starting from this source have been under very active investigation for some years now. These routes have all reached at least the pilot-plant stage of development and indeed some have achieved full commercialization. They may be grouped under two headings: (a) gasification and (b) liquefaction. Before considering representative examples of these methods it is important to realize that, as in the case of oil, the same products are also used as fuels, i.e. as energy sources. Although the technological problems in these coal conversion processes are being overcome fairly rapidly, the swing back from oil to coal as a source of organic chemicals will tend to be gradual. Even by the end of this century some estimates suggest that only between 10 and 30% of organic chemicals will be derived from coal. Reasons for this include the following.

(*a*) *Capital required.* Very large capital sums will be required to expand mining operations and transportation facilities. Due to the considerable differences in the nature of oil on the one hand, and coal and its conversion products on the other, much of the existing petrochemical plant may not be suitable, and in any case it may require replacing by the end of the century. This clearly requires a vast amount of capital expenditure.

(*b*) *Commercial availability.* As indicated, this varies from process to process, and although some are already available many are still only undergoing pilot-

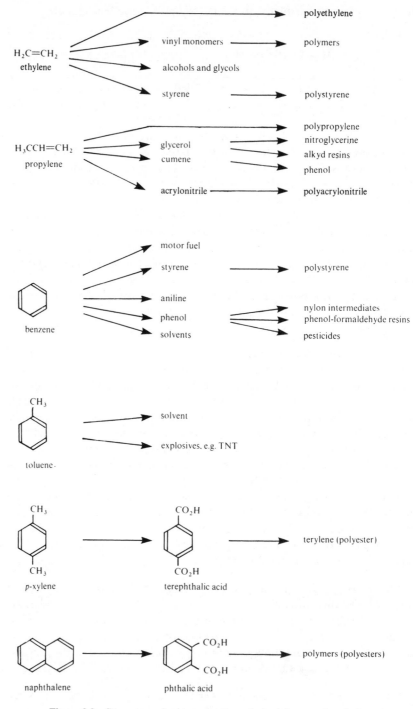

Figure 2.2 Some uses of primary products derived from coal and oil.

plant trials and it may be some years before they are suitable for full-scale manufacturing.

(c) *Incentives for change*. The principal advantage of oil is that (i) it contains molecules with linked carbon atoms, and (ii) it contains hydrogen already attached to carbon atoms. Other attractions of oil as a source of chemicals are its mobility due to its liquid nature and the scale and efficiency of the established processes for its conversion. Coal is less attractive because, as a solid, it is much more difficult to handle and transport. Recent events have reduced the momentum of the drive from oil to coal. Many of the oil-producing countries have maintained or even increased production levels (although Saudi Arabia is a notable exception) since the oil revenues are the cornerstone of their development and industrialization, but consumption of oil has been reduced due to energy conservation measures, and this has led to a current surplus. The present world recession is another major contributor to reduced oil usage. Stabilized, or reduced, prices resulted, maintaining the advantage of oil over coal. Certainly continued lower levels of consumption will prolong the time-scale for the availability of oil, postponing the swing back to coal. Any major switch away from oil to alternative sources of energy will have the same effect although the net effect would be much more dramatic.

There is no doubt about the switch from oil to coal as a source of organic chemicals; the only uncertainty concerns the time-scale and rate at which it happens.

Coal-conversion processes under development are directed towards producing either gaseous or liquid feedstocks which approximate in composition to petroleum-derived feedstocks. They can then be utilized directly in existing petrochemical plant and processes. To achieve this, however, two problems must be overcome, which are a consequence of the differing natures of coal and oil. Firstly, the H:C ratios are different for coal and for petroleum-derived liquid feedstocks. Secondly, significant amounts of heteroatoms are present in coal, particularly sulphur which may reach levels as high as 3%. The sulphur has to be removed for two reasons: (i) on combustion it will form the atmospheric pollutant SO_2, and (ii) it is a potent catalyst poison, and most of the downstream petrochemical processes are catalytic. However, its removal from coal is difficult and it is therefore removed from the conversion products instead.

Consider now the two types of coal conversion processes, (a) liquefaction and (b) gasification.

(a) *Liquefaction*. Liquefaction of coal, via hydrogenation, is quite an old process which was operated commercially in Germany during World War II when external fuel supplies were cut off. Tens of millions of tonnes of gasoline were produced in this way during this period. Interestingly, Germany is today playing a leading part in the development of more efficient processes. For the resulting liquids to be suitable chemical feedstocks the H:C ratio must be improved in favour of more hydrogen. Clearly this can be achieved in two ways—either by adding hydrogen or by removing carbon. Although many of the new processes are only at the pilot-plant stage of development, their superiority over the old methods is due to increased sophistication of the chemical engineering employed plus improvements in the catalysts available. The basic problems, however, remain the same—poor selectivity in producing the desired fractions and a relatively high rate of consumption of hydrogen.

The fundamental processes involved are pyrolysis, solvent extraction and hydrogenation. Differences between the techniques being developed lie in how these fundamental processes are combined. Thus the *Solvent Refined Coal Process* (Gulf Oil) uses the minerals in the coal as the catalyst, and hydrogenation with hydrogen is effected in the liquefaction reactor at 450°C and 140 atm. pressure. In contrast the *Exxon Donor Solvent* process uses tetralin (1, 2, 3, 4-tetrahydronaphthalene) as the source of hydrogen in the liquefaction reactor, which also operates at 450°C and 140 atm. pressure. Further hydrogenation of the product liquids is carried out in a separate reactor, and the overall yield is about 0·4 tonnes of liquids per tonne of coal used. The National Coal Board in the U.K. is developing a supercritical direct solvent extraction of the coal process. This process uses toluene (the best solvent) at 350–450°C and 100–200 atm pressure, and it is claimed that separation of liquids from the solid material remaining is easier and the solvent can be recycled. Hydrogenation is effected in an additional step. Note that in the above processes the main product is a highly carbonaceous solid material known as char. This is either burnt to provide process heat or reacted with water and oxygen to produce hydrogen. Table 2.3 summarizes some typical coal liquefaction processes.[1]

The liquids produced by coal liquefaction are similar to fractions obtained by distillation of crude oil (although they are much richer in aromatics), and therefore require further treatment, e.g. cracking, before being used for synthesis.

(b) *Gasification.* Coal gasification is commercially proven—for example in the SASOL plants in South Africa—and may be directed towards producing either high-energy fuel gas which is rich in methane, or synthesis gas (syngas) for chemicals production. Some well-known processes are summarized in Table 2.4[1]. In these processes the coal, in a suitable form, plus steam and oxygen enter the gasification reactor where they undergo a complex series of reactions, the balance between these depending on the temperature employed. Thus at high temperatures, i.e. 1000°C,

$$2C + 2H_2O \longrightarrow 2CO + 2H_2$$

dominates.

At lower temperatures competition with

$$CO + H_2O \longrightarrow CO_2 + H_2$$

Table 2.3 Typical coal liquefaction processes

Process	Typical catalyst	Conditions		Hydrogenation		Comments
		$T(^\circ C)$	P (atm)	Prime source	Method	
Bergius (original process developed early 1900s)	Iron oxide	465	200	H_2	in liquefaction reactor	Most severe conditions; catalyst discarded
Solvent Refined Coal (Gulf Oil)	Minerals in coal (none added)	450	140	H_2	in liquefaction reactor	Recycle of portion of product liquid to reactor; lack of hydrogenation specificity
H-Coal® (Hydrocarbon Research Inc.)	CoO—MoO₃/Al₂O₃	450	200	H_2	in liquefaction reactor	Catalyst ages rapidly
Exxon Donor Solvent	Minerals in coal in liquefaction reactor CoO—MoO₃/Al₂O₃ in separate hydrogenation reactor	450	140	Tetralin in liquefaction reactor. Recycled after hydrogenation in separate reactor		Further hydrogenation of product liquids in separate reactor—catalyst deactivation slow. Typical product yields 0·3 to 0·4 te liquid/te coal feed
National Coal Board supercritical extraction	In separate hydrogenation reactor	350–450	100–200	H_2 in separate reactor		Supercritical gas extraction of portion of coal with PhMe as solvent

Table 2.4 Gasification processes (commercially proven)*

Process	Conditions	Typical products (vol. %)				Comments
		CH_4	H_2	CO	CO_2	
Lurgi	Fixed bed reactor ~1000°C; 30 atm	12	37	18	32	Production of by-product heavy tar ($\sim 1\%$) restricts coal to 'non-caking' types. 'Slagging Gasifier' under development by British Gas Corporation to enable the difficult 'caking' coals to be handled
Koppers–Totzek	Entrained bed reactor ~1800°C; 1 atm	—	34	51	12	Can handle all coals; high temperatures destroy heavy organic tars. 'Shell–Koppers' pressurized version (15–30 atm) under development
Winkler	Fluidized bed reactor ~900°C; 1 atm	3	42	36	18	Higher pressure process (15 atm) under development
Texaco	Entrained bed reactor ~1200°C; 20–80 atm					Commercially successful process for partial oxidation of fuel oil to synthesis gas being developed to handle coal as coal/water or coal/oil slurries

*The field of coal gasification is in a very active state of development. Nearly 20 other processes at various stages of development have been described; see A. Verma, 1978, *Chemtech*, 372 and 626.

and

$$CO + 3H_2 \longrightarrow CH_4 + H_2O$$

take place.

Hence higher temperatures are used to produce predominantly mixtures of carbon monoxide and hydrogen, or syngas. The Texaco process operated by Ruhrkohle in Germany produces a gas consisting mainly of carbon monoxide and hydrogen and only 1% methane. The production and value of syngas as a chemical and fuel feedstock is summarized in Fig. 2.3. Syngas, ammonia and methanol are considered in more detail in section 12.4. Note that small amounts of alcohols and esters are produced in the Fischer–Tropsch process. Synthesis of specific oxygenated products is described in Chapter 12.

Although the Fischer–Tropsch process (based on coal) has been un-economic for many years compared to oil-based routes, the peculiar political situation of South Africa allied with its large reserves of cheap coal led it to continue operating this process. Indeed, the experience gained led in the 1970s to the building of a second-generation plant, (the scheme for this SASOL II plant is shown in Fig. 2.4[4]) and a SASOL III plant is now on stream.

As indicated in Fig. 2.3, methanol is readily obtained from syngas, and its conversion to hydrocarbons over a metal-oxide catalyst is well known. However, this gives a wide range of hydrocarbons, and the catalyst's lifetime is relatively short. Mobil gained a breakthrough with their discovery in the 1970s of the ZSM-5 zeolite catalysts, which are much more selective, have longer lifetimes and can produce ethylene as well. Their selectivity is towards lower-molecular-weight hydrocarbons ($< C_{10}$). Also, the product is rich in aromatics ($\sim 25\%$) and the alkanes present show a high degree of branched chains (both

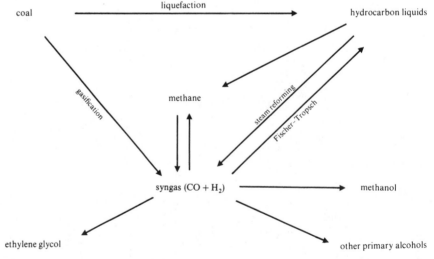

Figure 2.3   Uses of syngas.

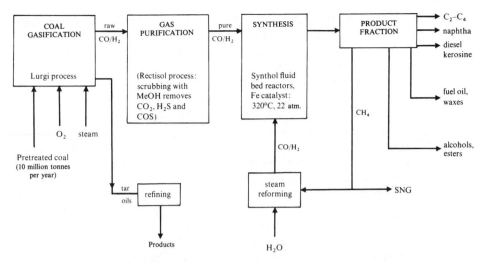

Figure 2.4 Fischer–Tropsch process—Sasol II plant.

of these contrast with the Fischer–Tropsch product). The product therefore has a high octane rating and is very suitable for use as gasoline. The sharp cut-off at C_{10} products is associated with the small pore size (5–6 Å) of the catalyst. A commercial plant utilizing this process came on stream in 1985 in New Zealand.

As indicated, the process can be directed towards ethylene production and hence chemical synthesis. The use of ZSM-5 catalysts for direct conversion of syngas into hydrocarbons (i.e., without the need to produce methanol first) and selective preparation of benzene, toluene, and xylene aromatics only are already being actively investigated.

2.2.3 Organic chemicals from carbohydrates (biomass)

The main constituents of plants are carbohydrates which comprise the structural parts of the plant. They are polysaccharides such as cellulose and starch. Starch occurs in the plant kingdom in large quantities in foods such as cereals, rice and potatoes; cellulose is the primary substance from which the walls of plant cells are constructed and therefore occurs very widely and may be obtained from wood, cotton, etc. Total dry biomass (i.e. plant material) production has been estimated at 2×10^{11} tonnes *annually*[2]. In some regions of the world—particularly in underdeveloped countries—biomass in the form of wood is the sole energy source. Also in all parts of the world much biomass is grown and harvested for food, and the material remaining when the food has been extracted can be utilized for chemicals production. An example is molasses which is left after the sugar (sucrose) has been extracted from sugar cane. Thus, not only is the potential for chemicals production from biomass considerable, as the figure above demonstrates, but the feedstock is renewable. The major route from biomass to chemicals is via fermentation processes. However these processes cannot utilize polysaccharides like cellulose and starch, and so the latter must first be subjected to acidic or enzymic hydrolysis to form the simpler sugars (the mono- or disaccharides, e.g. sucrose) which are suitable starting materials.

Fermentation processes utilize single-cell micro-organisms—typically yeasts, fungi, bacteria or moulds—to produce particular chemicals. Some of these process have been used in the domestic situation for many thousands of years, the best-known example being fermentation of grains to produce alcoholic beverages. Indeed up until about 1950 this was the most popular route to aliphatic organic chemicals, since the ethanol produced could be dehydrated to give ethylene, which is the key intermediate for the synthesis of a whole range of aliphatic compounds. This is illustrated in Fig. 12.2. Although chemicals production in this way has been declining there is a lot of interest in producing automobile fuel in this way. This has been led by Brazil, which has immense resources of biomass, e.g. from the Amazon jungle, to utilize in alcohol production, and thus reducing its dependence on imported oil. Since 1930 it has been mandatory to use 5% of alcohol in gasoline or petrol in Brazil, and under the 'Proalcool' programme commenced in 1975 this has been increased to 15%. More recently 96%-alcohol gasoline has become available for engines which have been specially modified to run on this fuel. This seems likely to be followed by other developing nations, particularly if the price of crude oil continues to rise; Kenya's first gasohol plant, having a capacity of 18 million litres per year, has also come on stream.

The current low level of production of chemicals from carbohydrates, or biomass, is largely a consequence of the unfavourable economics *vis à vis* oil. Disadvantages reflected in this can be divided into two parts (a) raw materials (b) the fermentation process. Raw-material costs are higher than that of crude oil, because biomass is an agricultural material and therefore in comparison its production and harvesting is very labour-intensive. Also, being a solid material transportation is more difficult and expensive. Major disadvantages of fermentation compared with petrochemical processes are, firstly, the time scale, which is usually of the order of days compared to literally seconds for some catalytic petrochemical reactions, and secondly, the fact that the product is usually obtained as a dilute aqueous solution ($< 10\%$ concentration). The separation and purification costs are therefore very high indeed. Since the micro-organism is a living system, little variation in process conditions is permitted. Even a relatively small increase in temperature to increase the reaction rate may result in death of the micro-organism and termination of the process.

On the other hand particular advantages of fermentation methods are that they are very selective and that some chemicals which are structurally very complex, and therefore extremely difficult to synthesize, and/or require a multi-stage synthesis, are easily made. Notable examples are various antibiotics, e.g. penicillins, cephalosporins and streptomycins. In addition to the antibiotics, citric acid (2-hydroxypropane-1, 2, 3-tricarboxylic acid) is a good example of this:

$$C_{12}H_{22}O_{41} \xrightarrow[\text{6–12 days}]{\textit{Aspergillus niger}} \begin{array}{c} CH_2CO_2H \\ | \\ HO-C-CO_2H \\ | \\ CH_2CO_2H \end{array}$$

sucrose   citric acid

The feedstock for these fermentations is usually a carbohydrate, but hydrocarbon fractions, or methanol derived from petroleum sources, have also been used in fermentation processes to produce single-cell protein. There is a shortage of protein for human consumption in many of the developing nations and fermentation protein could make a valuable contribution as a foodstuff supplement in these cases. However concern over safety, particularly with regard to even low concentrations of hydrocarbon residues, means that none of this protein has yet been approved for human consumption. There is an important psychological barrier to be overcome in these considerations and the difficulties have already forced BP to dismantle its 100 000 tonne per annum hydrocarbon-based plant, which used yeast as the micro-organism, in Sardinia. In contrast ICI have been operating a plant since 1980 at their Billingham works, using a methanol feedstock and the micro-organism *Methylophilus methylotropus* to produce their protein 'Pruteen'[5], which is sold as an animal feed. Since methanol is readily soluble in water, any that is not converted does not contaminate the 'Pruteen'. This plant's capacity is 60 000 tonne per annum, but this plant has closed down because it could not compete with the cheaper protein obtained from soyabean. The USSR is reported to have several alkane-based plants operating with capacities up to 200 000 tonne per annum. In contrast there has been little interest in single-cell protein in the U.S.A because it is the world's leading soyabean-protein producer.

Provided that the immense practical problems associated with the rapidly developing field of genetic engineering, where micro-organisms such as bacteria are 'tailor-made' to produce the required chemical, can be overcome, then the interest in fermentation methods will be very considerable. Eli Lilly is one of the companies showing the way, and has started producing insulin at its Indianapolis and Liverpool plants via recombinant DNA techniques in which the genes to produce chain A of the insulin are inserted into one batch of the bacterium *E. coli*. A second batch of the bacterium receives the genes to produce chain B of the structure. After isolation and purification the two chains are then joined chemically to give the insulin. In tests this has proved identical to human insulin and its production will end total dependence on animal-derived insulin. This will be of considerable benefit to diabetes sufferers since it should reduce any allergic or side-effects, and eliminate, or at least minimize, long-term effects such as blindness. However it seems unlikely that bulk chemicals, i.e. those required in very large quantities such as ethylene and benzene, will be produced in this way in the foreseeable future because of the slow reaction rate and the very high product separation costs.

2.2.4 *Organic chemicals from animal and vegetable oils and fats*

Animal and vegetable oils and fats—commonly known as lipids—are composed of mixtures of glycerides, which are esters of the trihydric alcohol, glycerol (propane-1, 2, 3-triol). They have the general structure

$$CH_2-O-\overset{\overset{\displaystyle O}{\|}}{C}-R$$
$$CH-O-\overset{\overset{\displaystyle O}{\|}}{C}-R^1$$
$$CH_2-O-\underset{\underset{\displaystyle O}{\|}}{C}-R^2$$

The groups R, R^1, R^2, which may be similar or different, are straight-chain aliphatic hydrocarbon groupings containing between 10 and 20 carbon atoms. They may be saturated or unsaturated, e.g.

$$CH_2-O-\overset{\overset{\displaystyle O}{\|}}{C}-(CH_2)_7-CH=CH-(CH_2)_7CH_3$$
$$CH-O-\overset{\overset{\displaystyle O}{\|}}{C}-(CH_2)_{16}-CH_3$$
$$CH_2-O-\underset{\underset{\displaystyle O}{\|}}{C}-(CH_2)_{16}-CH_3$$

There are many different sources of these oils and the nature and proportion of the R groups varies with the source. Some popular sources are soya, corn, palm-kernel, rapeseed and olive, animal fats and even sperm whales. Table 2.5 gives an indication of the distribution of R groups. The oils are isolated by solvent extraction and considerable quantities are used in the food industries as cooking oils and fats, and for production of butter, margarine and various other foodstuffs such as ice-cream. There is still controversy about the effect of the R-groups in these foodstuffs on human health, particularly on high cholesterol levels in blood which may lead to high blood pressure and heart disease. Opinion now seems to favour a high proportion of unsaturated groups as being beneficial in lowering cholesterol levels and reducing the risk of heart attacks. This has led to a trend away from cooking fats and ordinary butter or margarine (which are all rich in saturated R groups) to cooking oils and the use of margarines rich in polyunsaturates.

Table 2.5 Distribution of R groups in various oils

R group	(corresponding fatty acid)	Palm oil	Corn oil	Soybean oil	Cod-liver oil
Saturated groups	C_{14} (myristic)	1·0	1·0	0·5	4·0
	C_{16} (palmitic)	47·5	9·0	9·0	10·5
	C_{18} (stearic)	5·0	2·5	3·5	0·5
Unsaturated groups	C_{18} (oleic) with one $C=C$ bond	38·0	40·0	28·0	28·0
	C_{18} (linoleic) with two $C=C$ bonds	8·5	45·0	55·0	0[†]

[†] cod-liver oil contains a large proportion of C_{20} and $> C_{20}$ R groups

Being esters, the use of lipids for chemicals production starts with hydrolysis. Although this can be either acid- or alkali-catalysed, the latter is preferred since it is an irreversible reaction, and under these conditions the process is known as saponification, viz.:

stearin   glycerol   sodium stearate

Salts of long-chain carboxylic, or fatty, acids such as sodium stearate, are the basic ingredients of soaps. The long hydrocarbon chain dissolves oily or greasy dirt, whereas the polar part of the molecule—the carboxylate anion—dissolves in the water enabling the dirt to be washed away.

The hydrolysis can be effected without the aid of a catalyst, but as expected requires more vigorous reaction conditions, i.e. heating strongly under pressure with steam. This procedure is known as 'fat splitting':

stearin   glycerol   stearic acid

A similar result can be obtained under milder conditions using acidic catalysts, e.g. sulphuric acid. Long-chain fatty acids, such as stearic acid, are difficult to synthesize and this is therefore the route for their manufacture.

Hydrogenolysis of either glycerides themselves or the corresponding methyl esters (easily prepared by an ester interchange reaction with methanol) produces long-chain or 'fatty' alcohols e.g.

$$CH_3(CH_2)_{10}CO_2CH_3 \xrightarrow{\text{H}_2} CH_3(CH_2)_{10}CH_2OH$$

methyl laurate   lauryl alcohol

This reaction is brought about by reaction with hydrogen over a copper chromite catalyst at a pressure of about 250 atm. Under these conditions carbon–carbon double bonds in the glycerides or methyl esters are also reduced by the hydrogen. These alcohols can also be made from ethylene but the vegetable-oil-based route remains competitive at the present time.

It is important to note that for the sake of simplicity the examples given above of saponification, hydrolysis (fat splitting) and hydrogenolysis reactions have each used a single glyceride (or methyl ester). In practice, the vegetable oil which is used is a mixture of various glycerides and the product is therefore a mixture which requires separating.

2.3
Sources of inorganic chemicals

The diversity of sources of inorganic chemicals was touched on in the introduction to this chapter and the book entitled *The Modern Inorganic Chemicals Industry*[6] discusses, in detail, the production of a number of important inorganic chemicals. Comment here therefore will be restricted to a summary of world consumption (Table 2.6) and major uses of the more important inorganic raw materials, plus a brief study of the changing raw-material usage for sulphuric acid production and the reasons for this. Between 1960 and 1975 four alternatives were utilized in sulphuric acid production. They are sulphur, anhydrite (a calcium sulphate mineral), zinc concentrates

Table 2.6   Major sources of inorganic chemicals[7]

Source	World chemical consumption 1975 (millions of tonnes)	Examples of uses
Phosphate rock	120·0	Fertilizers, detergents
Salt	120·0	Chlorine, alkali production
Limestone	60·0	Soda ash, lime, calcium carbide
Sulphur	50·0	Sulphuric acid production
Potassium compounds	25·0	Caustic potash, fertilizers
Bauxite	8·0	Aluminium salts
Sodium carbonate	8·0	Caustic soda, cleaning formulations
Titanium compounds	4·4	Titanium dioxide pigments, lightweight alloys
Magnesite	3·0	Magnesium salts
Borates	2·4	Borax, boric acid, glazes
Fluorite	1·5	Aluminium fluoride, organofluorine compounds

(largely ZnS), and pyrites (largely FeS). Nowadays almost all sulphuric acid production is based on sulphur as the raw material.

The raw material for sulphuric acid production via the lead-chamber process or the contact process is sulphur dioxide. However, whereas the former could accept impure sulphur dioxide, in the contact process (which is the only process now operated) this is not possible since the impurities would poison the vanadium pentoxide catalyst used in the conversion to sulphur trioxide. Hence the disappearance of zinc concentrates and pyrites as sources today. In contrast, elemental sulphur, extracted from the earth or obtained from petroleum sources, is very pure and is ideal for burning to sulphur dioxide. Additionally it is favoured on economic grounds over anhydrite. Burning sulphur to form sulphur dioxide is a highly exothermic process and this makes the overall process for manufacturing sulphuric acid energy-producing. This excess energy is used for steam generation. Sulphuric acid plant managers are probably the only people (excepting the oil producers) who do not like to see the price of oil fall—this lowers energy costs and reduces their net credit for the sale of their energy! This results in an increase of their manufacturing costs for the sulphuric acid.

Due to its many uses sulphuric acid can be used as a measure of a country's industrial development, since consumption of the former increases with the latter. Also, because production responds rapidly to changes in consumption it is also a barometer of industrial activity, even in highly developed countries such as our own.

2.4
Recycling of materials

The finite nature of most of our sources of chemicals, e.g. oil, coal, metallic ores, suggests that every effort should be made to conserve these valuable resources.

Limiting demand for them is clearly one way of approaching this problem but it may not be practicable if competition, which stimulates demand, is intense. Since many materials can be recycled, this is an alternative, or indeed complementary, approach. Some materials have been recycled for many years, ferrous metals being perhaps the most notable examples. Thus at the end of their useful life automobiles are crushed and the (rusted) metal returned as part of the feed to the blast furnace where it is reconverted into iron and steel. Paper and cardboard are other well-known examples.

More recently many municipal authorities, in conjunction with glass-making companies, have started to provide 'bottle banks'. Benefits to the community accrue in two ways as a result. Firstly, the municipal authority receives a cash payment related to the amount of glass recovered. Secondly, despoilment of the environment due to unsatisfactory disposal of the bottles is eliminated, or at least considerably reduced.

Some municipal authorities recover organic waste from refuse, and although it is not actually recycled, it is combusted and the heat utilized in central heating systems for complete housing estates. McCauliffe of

Manchester, U.K., has developed a method of converting organic refuse into high-quality crude oil, and in collaboration with the local authority is operating a pilot plant to demonstrate its commercial viability.

Smaller-scale operations are the recycling of automobile engine oil and also plastics. The latter scheme is made very difficult because the diversity of plastics makes their identification and separation troublesome.

References

1. R. Pearce and M. V. Twigg, in *Catalysis and Chemical Processes* (eds. Ronald Pearce and William R. Patterson), Leonard Hill, 1981, chapter 6.
2. August Vlitos, in *The Chemical Industry* (eds. D. Sharp and T. F. West), Ellis Horwood, 1982, p. 315.
3. Joseph Haggin, *C&EN*, 1982 (Aug. 9), pp. 17–22.
4. As ref 3, p. 167.
5. *C & I*, 1980 (24), 921.
6. *The Modern Inorganic Chemicals Industry*, ed. R. Thompson, The Chemical Society, 1977, pp. 21–25.
7. B. M. Coope, *Industrial Minerals for Inorganic Chemicals*, ECN Large Plants Supplement, 1976.
8. Annual Report of the National Sulphuric Acid Association Limited, 1982.

Bibliography

Industrial Organic Chemicals in Perspective, Part 1, *Raw Materials and Manufacture*, Harold A. Wittcoff and Bryan G. Reuben, Wiley–Interscience, 1980, chapters 2 and 3.
An Introduction to Industrial Organic Chemistry, Peter Wiseman, Applied Science, 1972, pp. 5–20.
The Chemical Industry (eds. D. Sharp and T. F. West), Ellis Horwood, 1982, chapters 20, 21, 25 and 26.
Catalysis and Chemical Processes (eds. Ronald Pearce and William R. Patterson), Leonard Hill, 1981, chapters 5 and 6.
The Modern Inorganic Chemicals Industry, 2nd Edition, The Royal Society of Chemistry, 1995 R. Thompson.
Introduction to Industrial Chemistry, Howard L. White, Wiley, 1986, Chapters 2 and 4.

Research and Development 3
Alan Heaton

No one can fail to be impressed, even amazed, by the staggering technological changes that have occurred during this century. Who could have imagined at the time of the Wright Brothers' first flight in 1903 that this would lead to Concorde and modern 'fly by wire', computer controlled, aircraft that can cross the world in less than one day? Other 'dreams' which have been realized include landing a man on the moon, harnessing atomic energy and exploring the deepest parts of our oceans. As this century has unfolded the pace of technological change has been ever increasing. In what areas has this been most apparent to us, the general public?

Undoubtedly in the areas of computers and information technology. The amount of information which a modest personal computer (PC) can hold and process already exceeds that which only large, state of the art, research machines could 25 years ago. Remarkably, this tremendous technological advance has been accompanied by a considerable REDUCTION in cost. The latter is the main factor in accounting for the enormous growth in home PCs. For example in the U.K. around 70% of homes have one PC and possibly 20% two PCs. Introduction of CD-ROM Multi-Media versions will continue to stimulate demand. Other interesting developments already being introduced include the use of the PC as an international communications vehicle to, for example, make phone calls—at significant cost savings over conventional networks—and perusing and ordering goods without leaving your armchair, and sending and receiving faxes!

Two threads link all of these staggering advances—the chemical industry and research and development. The chemical industry provides either the raw materials or even finished products which form the hardware. Just think of a modern aircraft such as an Airbus 300 or a Boeing 747 Jumbo Jet and the chemical industry's contribution to their development and manufacture. How can such an enormous structure take off as well as withstand the enormous air pressures and stresses and strains of many hours, and indeed years, of flight? The answer lies in the light, yet very strong, metal alloys developed to make the air frame and the turbine blades in the jet engines. Inside the aircraft the synthetic polymers and plastics made by the chemical industry are apparent everywhere, not to mention the hidden miles of electrical wiring with its coating of synthetic polymer insulation. Even the air conditioning system will utilize CFCs (or their alternatives). The numerous flight microprocessors and computers would not be available without the production of ultra-pure silicon chips and inert solvents to clean (i.e. degrease) electronic circuits. Space exploration also would not have been possible without the contribution of the chemical industry. Examples are the fuels to provide the enormous thrust which the launch vehicle requires, and the synthetic materials for the spacesuits. The U.S. Space Shuttle depends on the specially developed ceramic tiles, and powerful adhesive which holds them to its frame, to enable it to withstand the very high temperatures to which it is exposed on re-entry to the earth's atmosphere.

The second interconnecting thread is research and development, or R&D as it is commonly referred to. This is an activity which is carried out by all sectors of manufacturing industry but its extent varies considerably, as we will see shortly. Let us first understand, or at least get a feel for, what the terms mean. Although the distinction between research and development is not always clear-cut, and there is often considerable overlap, we will attempt to separate them. In simple terms research can be thought of as the activity which produces new ideas and knowledge whereas development is putting those ideas into practice as new processes and products. To illustrate this with an example, predicting the structure of a new molecule which would have a specific biological activity and synthesizing it could be seen as research, whereas testing it and developing it to the point where it could be marketed as a new drug could be described as the development part (see section 8.3 in Volume 2).

3.2
Research and
development activities

3.2.1 *Introduction*

In industry the primary reason for carrying out R&D is economic and is to strengthen and improve the company's position and profitability. The purpose of R&D is to generate and provide information and knowledge to reduce uncertainty, solve problems and to provide better data on which management can base decisions. Specific projects cover a wide range of activities and time scales, from a few months to 20 years.

We can pick out a number of areas of R&D activity in the following paragraphs but if we were to start with those which were to spring to the mind of the academic, rather than the industrial, chemist then these would be basic, fundamental (background) or exploratory research and the synthesis of new compounds. This is also labelled 'blue skies' research.

Fundamental research is typically associated with university research. It may be carried out for its own intrinsic interest and it will add to the total knowledge base but no immediate applications of it in the 'real world' will be apparent. Note that it will provide a valuable training in defining and solving problems, i.e. research methodology for the research student who carries it out, under supervision. However, later 'spin offs' from such work can lead to useful applications. Thus physicists claim that but for the study and development of quantum theory we might not have had computers and nuclear power. However, to take a specifically chemical example, general studies on a broad area such as hydrocarbon oxidation might provide information which would be useful in more specific areas such as cyclohexane oxidation for the production of nylon intermediates.

Aspects of synthesis could involve either developing new, more specific reagents for controlling particular functional group interconversions i.e. developing synthetic methodology or complete synthesis of an entirely new molecule which is biologically active. Although the former is clearly fundamental the latter encompasses both this and applied aspects. This term 'applied' has traditionally been more associated with research carried out in industrial laboratories, since this is more focused or targeted. It is a consequence of the work being business driven.

Note, however, that there has been a major change in recent years as academic institutions have increasingly turned to industry for research funding, with the result that much more of their research effort is now devoted to more applied research. Even so, in academia the emphasis generally is very much on the research rather than the development.

3.2.2 Types of industrial research and development

The applied or more targeted type of research and development commonly carried out in industry can be of several types and we will briefly consider each. They are: (a) product development, (b) process development, (c) process improvement and (d) applications development. Even under these headings there are a multitude of aspects so only a typical example can be quoted in each case. The emphasis on each of these will vary considerably within the different sectors of the chemical industry (see section 3.2.3).

3.2.2.1 *Product development.* Product development includes not only the discovery and development of a new drug, referred to above, but also, for example, providing a new longer-acting anti-oxidant additive to an automobile engine oil. Developments such as this have enabled servicing intervals to increase during the last decade from 3000 to 6000 to 9000 and now to 12 000 miles. Note that most purchasers of chemicals acquire them for the effects that they produce, i.e. a specific use. Thus 2,4-dichlorophenoxyacetic acid is purchased because it acts as a selective weedkiller (or herbicide) in destroying broad-leaved weeds such as dandelions and daisies growing amongst grass or cereal crops. Teflon^{TM}, or polytetrafluoroethylene (PTFE), may be purchased because it imparts a non-stick surface to cooking pots and pans, thereby making them easier to clean.

3.2.2.2 *Process development*. Process development covers not only developing a manufacturing process for an entirely new product but also a new process or route for an existing product. The push for the latter may originate for one or more of the following reasons: availability of new technology, change in the availability and/or cost of raw materials. Manufacture of vinyl chloride monomer (discussed in section 3.5.1) is an example of this. Its manufacturing route has changed several times owing to changing economics, technology and raw materials. Another stimulus is a marked increase in demand and hence sales volume which can have a major effect on the economics of the process. The early days of penicillin manufacture afford a good example of this.

The ability of penicillin to prevent the onset of septicaemia in battle wounds during the Second World War (1939–1945) resulted in an enormous demand for it to be produced in quantity. Up until then it had only been produced in small amounts on the surface of the fermentation broth in milk bottles! An enormous R&D effort jointly in the U.S. and the U.K. resulted in two major improvements to the process. Firstly a different strain of the mould (*Penicillium chrysogenum*) gave much better yields than the original *Penicillium notatum*. Secondly the major process development was the introduction of the deep submerged fermentation process. Here the fermentation takes place throughout the broth, provided sterile air is constantly, and vigorously, blown through it. This has enabled the process to be scaled up enormously to modern stainless steel fermenters having a capacity in excess of 50 000 litres. It is salutary to note that in the First World War (1914–1919) more soldiers died from septicaemia of their wounds than were actually killed outright on the battlefield!

Process development for a new product depends on things such as the scale on which it is to be manufactured, the by-products formed and their removal/recovery, and required purity. Data will be acquired during this development stage using semi-technical plant (up to 100 litres capacity) which will be invaluable in the design of the actual manufacturing plant. If the plant is to be a very large capacity, continuously operating one, e.g. petrochemical or ammonia, then a pilot plant will first be built and operated to test out the process and acquire more data. These semi-technical or pilot plants will also supply increased quantities, e.g.10–100 kg of the product which will be required for testing, e.g. a pesticide, or customer evaluation, e.g. a new polymer.

Note that by-products can have a major influence on the economics of a chemical process. Phenol manufacture provides a striking example of this. The original route, the benzenesulphonic acid route, has become obsolete because demand for its by-product sodium sulphite (2.2 tonnes/1 tonne phenol) has dried up. Its recovery and disposal will therefore be an additional charge on the process, thus increasing the cost of the phenol. In contrast the cumene route owes its economic advantage over all the other routes to the strong demand for the by-product acetone (0.6 tonnes/1 tonne phenol). The sale of this therefore reduces the net cost of the phenol.

A major part of the process development activity for a new plant is to minimize, or ideally prevent by designing out, waste production and hence possible pollution. The economic and environmental advantages of this are obvious.

Finally it should be noted that process development requires a big team effort between chemists, chemical engineers, and electrical and mechanical engineers to be successful.

3.2.2.3 *Process improvement*. Process improvement relates to processes which are already operating. It may be a problem that has arisen and stopped production. In this situation there is a lot of pressure to find a solution as soon as possible so that production can restart, since 'down time' costs money.

More commonly, however, process improvement will be directed at improving the profitability of the process. This might be achieved in a number of ways. For example, improving the yield by optimizing the process, increasing the capacity by introducing a new catalyst, or lowering the energy requirements of the process. An example of the latter was the introduction of turbo compressors in the production of ammonia by the Haber process. This reduced utility costs (mainly electricity) from $6.66 to $0.56 per tonne of ammonia produced. Improving the quality of the product, by process modification, may lead to new markets for the product.

In recent years, however, the most important process improvement activity has been to reduce the environmental impact of the process, i.e. to prevent the process causing any pollution. Clearly there have been two inter-linked driving forces for this. Firstly, the public's concern about the safety of chemicals and their effect on the environment, and the legislation which has followed as a result of this (see section 9.5). Secondly the cost to the manufacturer of having to treat waste (i.e. material which cannot be recovered and used or sold) so that it can be safely disposed of, say by pumping into a river. This obviously represents a charge on the process which will increase the cost of the chemical being made. The potential for improvement by reducing the amount of waste is self-evident.

Note, however, with a plant which has already been built and is operating there are usually only very limited physical changes which can be made to the plant to achieve the above aims. Hence the importance, already mentioned, of eliminating waste production at the design stage of a new plant. Conserving energy and thus reducing energy cost has been another major preoccupation in recent years (see section 8.4).

3.2.2.4 *Applications development*. Clearly the discovery of new applications or uses for a product can increase or prolong its profitability. Not only does this generate more income but the resulting increased scale of production can lead to lower unit costs and increased profit. An example is PVC whose early uses included records and plastic raincoats. Applications which came later included plastic bags and particularly engineering uses in pipes and guttering.

Emphasis has already been placed on the fact that chemicals are usually purchased for the effect, or particular use, or application which they have. This often means that there will be close liaison between the chemical companies' technical sales representatives and the customer, and the level of technical support for the customer can be a major factor in winning sales. Research and development chemists provide the support for these applications developments. An example is HFA 134a, CF_3CH_2F. This is the first of the CFC replacements and has been developed as a refrigerant gas. However, it has recently been found that it has special properties as a solvent for extracting natural products from plant materials. In no way was this envisaged when the compound was first being made for use as a refrigerant gas, but it clearly is an example of applications development.

3.2.3 *Variations in research and development activities across the chemical industry*

Both the nature and amount of R&D carried out varies significantly across the various sectors of the chemical industry. In sectors which involve large-scale production of basic chemicals and where the chemistry, products and technology change only slowly because the processes are mature, R&D expenditure is at the lower end of the range for the chemical industry. Most of this will be devoted to process improvement and effluent treatment. Examples include ammonia, fertilizers and chloralkali production (section 10.5) from the inorganic side, and basic petrochemical intermediates such as ethylene (sections 12.3.1–3) from the organic side.

At the other end of the scale lie pharmaceuticals and pesticides (or plant protection products). Here there are immense and continuous efforts to synthesize new molecules which exert the desired, specific biological effect. A single company may generate 10 000 new compounds for screening each year. Little wonder that some individual pharmaceutical company's ANNUAL R&D expenditure is now approaching $1000 million! Expressing this in a different way they spend in excess of 14% of SALES INCOME (note not profits) on R&D. This makes the pharmaceutical sector of the chemical industry the most research intensive area of U.K. manufacturing industry, save for, perhaps, aerospace and electronics. Full details of these activities are given in sections 7.3 and 8.3 of Volume 2.

3.3
The importance of research and development

The importance of R&D in the chemical industry will already have become apparent in the preceding sections, but it is worth while re-emphasizing and expanding on some of the points already made in this section. In addition other general points relating to the need for R&D will be made. Indeed let us start with the latter.

We live in a constantly changing world yet we in the U.K. are much happier with things as they are and we tend to be suspicious of anything 'new'. This contrasts with the situation in the U.S.A. and Japan where newness and novelty are welcomed. It is essential for any organization in the modern, intensely competitive world to be forward looking and progressive. If not they stagnate, are overtaken by their competitors and may even go out of business. This is even truer for the high technology industries like the chemical industry where change is very rapid. These points are well made by Sir John Harvey-Jones, the former I.C.I. Chairman, in his book *Making It Happen*. He also makes the valid point that a new plant is obsolete by the time it starts production because competitors will already be developing alternative processes and plants for the same product.

An example of the failure to recognize the need for change and the (almost) disastrous consequences is afforded by the watch-making industry. The Swiss had almost reached perfection in the manufacture of mechanical and automatic watches and not surprisingly dominated the industry. They ignored the development of the electronic watch in Japan and almost paid the ultimate price, but fortunately realized the error of their ways and adapted in the nick of time. Examples from the chemical industry include the changing routes to vinyl chloride monomer (section 3.5.1) and phenol manufacture.

Nations whose chemical industries can still be regarded as being at the developing stage, e.g. Mexico, Saudi Arabia and South Korea, in contrast to the mature industries of Japan, Western Europe and the U.S.A., tend to concentrate on basic chemicals and intermediates for which the manufacturing processes are well established and mature. Examples are the manufacture of basic petrochemical intermediates such as ethylene and benzene, and ammonia. With these they often have an economic advantage over the 'developed' countries because of the availability of abundant and cheap supplies of the raw materials crude oil and natural gas. Since the alkene and aromatic products are only a few processing steps downstream from these raw materials, the latter's price represents a significant proportion of the cost of making the products. They can therefore make ethylene more cheaply than U.K. or U.S. firms. As a result many chemical companies in the developed countries are switching resources to higher tech. chemicals. These are several, to many, steps downstream from the crude oil raw material. This enables them to exploit their greater experience and expertise in research and development over the developing countries. Many of these high tech. chemicals are described as speciality chemicals. As such they have special properties and applications, resulting in premium prices. Note though that only small tonnages are normally required. Examples are Aramid Fibres, e.g. Kevlar which commands tens of thousands of pounds per tonne, and the world's largest selling drug Zantac, whose price equates to about 3 million pounds sterling per tonne. Clearly, however, one tonne of the latter will be transformed into millions of anti-ulcer pills! For comparison the selling price of the basic petrochemicals is around several hundred pounds per tonne. An additional advantage of these speciality chemicals is that, because they are so many steps removed from the crude oil starting material, they make a very much smaller contribution to the selling price. Thus cheap oil would offer little advantage.

Attention has already been drawn to the enormous and continuous demand for new products in some sectors of the chemical industry, notably pharmaceuticals and agrochemicals. Indeed the discovery and development of new products is regarded as the life blood of these sectors. Clearly there is a need for drugs to combat new types of illness or disease such as AIDS. There is also a requirement for improved drugs which eliminate the undesirable side effects of some current drugs. An additional driving force for new products, particularly those having a novel mode of action, which both drugs and pesticides share, is the development of resistance to existing products by bacteria, fungi, insects and weeds. This considerably reduces their effectiveness and in time means that it is essential to replace them. An example is penicillin where some bacteria have been able to produce the enzymes, β-lactamases, which are able to cleave the β-lactam ring of the penicillin structure and hence destroy its antibiotic activity. Even worse this can lead to cross-resistance where the bacteria (or pests) also become resistant to other types of antibiotic (or pesticide) which have the same mode of action. Hence the drive to develop new products which have novel modes of action. An example of this was glyphosate, sold under the trade name Roundup by Monsanto, which had both a novel mode of action and a totally new type of structure for a pesticide when introduced to the market in 1971. As a result it rapidly became a major revenue earner, and not surprisingly was christened the first 'million dollar' pesticide. Other areas where the demand for new products is high include aerospace, where alloys and polymers which are lighter but stronger (and in the latter case thermally and oxidatively stable to ever higher temperatures) are needed, and replacement parts for the human body, e.g. hip joints. The development of catalytic converters on new automobiles has reduced the emission of pollutants such as nitrogen oxides.

The stimulus to carry out research and development, as exemplified above, is often attributed to either demand pull or technology push, although these represent the two extreme ends of the spectrum. Demand pull, as the term suggests, means identifying a market need or demand for a product. In some areas considerable market research will be necessary first in order to do this. Examples here are biodegradable detergents and plastics, which will reduce environmental pollution. Biodegradable detergents were produced by replacing the branched alkyl chain of alkylbenzenesulphonates, which the bacteria in sewage plants could not break down, to linear (or straight chain) ones which were readily biodegradable. Another market demand, which despite billions of pounds worth of research work has still not been met, is a drug to combat HIV.

In contrast technology push is where a product has been made or developed and you are looking for a use for it. The early days of the petrochemical industry afford us an example of this. When ethylene was made by cracking naphtha some propylene was co-produced. At the same time there was no use for this and it was merely burned as a fuel. However, here were quantities of a cheap, readily available chemical. The success of the research chemists and chemical engineers in exploiting this as a feedstock for making other chemicals is apparent from Figure 12.2

Another example is polytetrafluorethylene or Teflon^{TM}. For some time after it was first made uses for its remarkable properties, e.g. non-stick, electrical insulation, chemical inertness could not be developed and exploited because the material could not be processed using the processes then available. The problem was solved by applying newly developed powder metallurgy techniques.

Finally the importance of research and development as an activity within the chemical industry is confirmed by the enormous financial resources which are devoted to it. As indicated in section 3.2.3 the industry as a whole devotes approximately 7% of sales INCOME to R&D, with this doubling for the most research intensive sector, pharmaceuticals. For comparison the figure for the engineering sector in the U.K. would be around 1%, and all manufacturing 2%.

3.4
Differences between academic and industrial research

Some aspects of this have already been discussed in section 3.2.1., and it may be apparent from this that the differences between these two types have narrowed during the last decade. This is largely a consequence of the universities having to change to attract more industrial sponsorship for their research projects. Prior to this, academic research, by and large, concentrated very much on fundamental research and its purpose was essentially to increase the knowledge base. It could be very exploratory and tested new ideas. Examples would include confirming postulated reaction mechanisms, discovering new catalysts and developing novel reagents which are highly selective for carrying out specific functional group interconversions and, perhaps, giving stereochemical control also. Specific applications for the results of the research would not normally be apparent at this stage. The topic will often be a follow-on from work which the supervisor carried out when he/she in turn was a postgraduate research student.

The project will be carried out within a research group of, say, up to 10 people who are all working on related topics, but overall within a fairly narrow area of chemistry. Normally all the group will be chemists and will often identify themselves with their specialist area, e.g. colloid chemists or fluorine chemists.

Although the trend has been towards larger units, or groups, in recent years, because this makes attracting research funding easier, there are still some individual research students and supervisors. These research projects do provide a valuable training for the student in research methodology and problem solving.

Having gained a higher degree (M.Sc. or Ph.D.) the person joining the R&D department of a very large chemical company will find things rather different. Firstly, they will find that they are joining a team within what may be a large group of up to 100 scientists, and they will need good inter-personal skills, as well as chemical ability, in order to integrate and work well within the team. In industry a research problem is just that; it is something to be solved and is not inorganic, organic or physical (as it would be in a university), just a chemical problem. Its solution may involve ideas and methods from all these areas and beyond, and help and interactions with not only other chemists but with other scientists and engineers also. This is largely a consequence of the research being business driven. Little fundamental or 'blue skies' research is carried out in industry. It is either left to universities alone or else the company may sponsor it in a university. This long-term more speculative research is inevitably the first to be cut during the regular cyclical recessions which hit the chemical industry.

Thus industrial research is much more specifically targeted and applied generally than university research work. Also, in contrast to industry, academia usually has little interest in the development part of R&D. It should be emphasized that the above comments represent generalizations, so there will inevitably be the odd exceptions to them. It is also worth repeating that universities are moving more towards applied research in order to attract funding from industry.

3.5
Research and development case studies

This chapter is rounded off with two case studies to illustrate research and development activities. It is not possible to present here a really detailed study, indeed to a large extent only the results/outcomes of the work are presented and discussed. However, it is hoped that this will give the reader the flavour and the excitement and challenge of R&D. On a personal level I can well remember the moment during the first year of research for my Ph.D., when I had confirmation that the compound I had made had the proposed structure. To realize that I was the first person in the world ever to have made this was immensely exciting and satisfying!

The first case study relates to vinyl chloride monomer (VCM) production and provides an excellent example of how changing raw materials, economics and technology stimulated research and development to develop new routes to this compound.

The second outlines a remarkable success story on two counts. It was the first worldwide political agreement on solving a global pollution problem and this stimulated the research and development to find alternative chemical compounds to those which were destroying the ozone layer in the stratosphere. As we will see later this has been achieved on a remarkably short time scale due to the will to achieve it and the unprecedented co-operation which occurred.

3.5.1 *Vinyl chloride monomer*

Vinyl chloride monomer (VCM) is a typical high tonnage chemical and most of the production is converted into polymers, particularly polyvinyl chloride (PVC). The processes and routes of manufacture of VCM have changed significantly over the years and these are summarized in section 12.3.3.2. However, we will look here at the pre-petrochemical days and also in more detail at the processes and the reasons why they changed.

The story starts in the early part of this century, in the era before the development of the petrochemical industry, when the major source of organic chemicals was coal. The starting point for many aliphatic compounds then was acetylene (ethyne) whose production utilized coke which was produced from coal. The reactions involved were as follows:

$$CaCO_3 \xrightarrow{\text{roast}} CaO + CO_2$$
$$\text{Limestone}$$

$$CaO + 3C + HEAT \longrightarrow CaC_2 + CO$$
$$\text{Calcium carbide}$$

$$CaC_2 + 2H_2O \longrightarrow H{-}C{\equiv}C{-}H + Ca(OH)_2$$
$$\text{Acetylene}$$

The acetylene was then reacted with HCl:

$$H{-}C{\equiv}C{-}H + HCl \longrightarrow H_2C{=}CH{-}Cl + HEAT$$
$$\text{VCM}$$

Yield 80–90%

This seemed to be the perfect chemical process for the following reasons:

(1) It uses HCl which is cheaply and readily available from chlorination reactions in which it is often formed as a by-product.
(2) It requires little energy because once started the reaction is exothermic.
(3) No by-products are formed.
(4) It is a single-step reaction.

With all these advantages how did the process ever come to be replaced? The answer lies in the one serious disadvantage—the high cost of acetylene—which eventually came to outweigh all the above advantages. The first two steps in its synthesis are endothermic. However, it is the second which is the killer blow, since this reduction requires a temperature in excess of 2000°C because it is so endothermic. This can only be achieved by using an electric arc furnace, and electricity is a very expensive form of energy.

The start of the petrochemical era made ethylene readily available at a low (and falling) price when compared to that of acetylene. Research chemists and chemical engineers therefore developed the first ethylene based route as follows:

$$H_2C{=}CH_2 + Cl_2 \longrightarrow Cl{-}CH_2CH_2{-}Cl$$
Ethylene dichloride
(EDC)

$$\Big\downarrow 500°C$$

$$H_2C{=}CH{-}Cl + HCl$$
VCM

Although this used the much cheaper ethylene as the hydrocarbon feedstock, it also needed the more expensive chlorine and was a two stage process. Even worse half the chlorine ended up as HCl instead of in the VCM. The next development overcame this latter problem and was called the Balanced Process. It linked the two methods together, viz.

$$H_2C{=}CH_2 + Cl_2 \longrightarrow Cl{-}CH_2CH_2{-}Cl$$

$$\downarrow$$

$$HCl + H_2C{=}CH{-}Cl$$

$$\downarrow$$

$$H{-}C{\equiv}C{-}H + HCl \longrightarrow H_2C{=}CH{-}Cl$$

Thus all the chlorine ends up as VCM. There are several disadvantages though, including the fact that the process has three stages and therefore would have increased capital and operating costs, and half the feedstock is the expensive acetylene. Despite these it held sway for a while. A later improvement was to produce a mixture of ethylene and acetylene by cracking naphtha (see section 12.3.1) and using this as the hydrocarbon feedstock.

The most recent development in this story brings us to the currently used route, the oxychlorination route. The start of this part of the story provides a valuable lesson for all chemists and particularly research chemists. This is the importance of searching that immense treasure house, the chemical literature. As a result it was discovered that disposal of by-product HCl (from substitutive chlorination reactions) was nothing new, since Deacons in 1868 had overcome this by patenting a process for converting it into the much more valuable chlorine by oxidation with air over a copper chloride catalyst at 450°C. However, just adding this on to an existing VCM process would add an extra stage and therefore increase capital and operating costs. Research workers not only managed to overcome this problem, by carrying out the oxidation in the presence of ethylene, but they also improved the catalyst and enabled the reaction to take place at a lower temperature, thereby reducing energy costs. This OXYCHLORINATION process is summarized as follows:

$$H_2C{=}CH_2 + 2HCl \xrightarrow[250\text{--}350°C]{O_2\,|\,CuCl_2\,|\,KCl} Cl{-}CH_2CH_2{-}Cl + H_2O$$

$$\Big\downarrow 500°C$$

$$HCl + CH_2{=}CH{-}Cl$$

Although this is a two step vapour phase process, a liquid phase version for the first step has also been developed. Yields of over 90% at 95% conversion are achieved.

What is the final improvement that further research might ultimately achieve? This would be to produce VCM in a single step. Although this has been claimed, by carrying out the oxychlorination at 500°C, it has never been commercialized. This case study has demonstrated how the changing raw materials availability and costs have stimulated research and development to produce new and more economic routes to a particular product, VCM.

3.5.2 *CFC replacements*

CFCs (chlorofluorocarbons) were hailed as wonder products, following Midgley's discovery of the first one, CF_2Cl_2 (known as CFC 12), in 1930. Their chemical inertness, non-flammability and non-toxic nature rapidly led to their large-scale use as safe refrigerants, aerosol propellants and foam blowing agents in the production of polyurethane foams.

In 1974, however, Roland and Molina first suggested that because of their great stability they could rise, unchanged, into the stratosphere where they would be broken down by the short wavelength UV-B radiation to form $Cl^{\bullet}$ radicals. These would then attack and destroy ozone molecules, viz.

$$CF_2Cl_2 \xrightarrow[\text{light}]{\text{U.V.-B}} CF_2Cl^{\bullet} + Cl^{\bullet}$$

$$Cl^{\bullet} + O_3 \longrightarrow ClO^{\bullet} + O_2$$

$$ClO^{\bullet} + O^{\bullet} \longrightarrow Cl^{\bullet} + O_2$$

Note that $Cl^{\bullet}$ is regenerated and can therefore go on and destroy many more ozone molecules. To prove this link was clearly going to be extremely difficult, and was to take another 20 years of extensive research and investigation. The consequence of the thinning of the ozone layer would be that it could no longer prevent the dangerous short wavelength UV radiation reaching the Earth's surface. This would result in a considerable increase in the number of cases of skin cancers.

So great was the concern about this that it led to the signing of the first ever international agreement on an environmental issue, the Montreal Protocol in 1987. More than 50 nations, who between them produced more than 80% of the world's CFCs, agreed to cap production and consumption at 1986 levels, and work towards phasing them out by early in the next century. The band wagon was now really rolling on this issue and at the 1992 United Nations Ozone Layer Conference in Copenhagen some 80 nations essentially agreed to complete the total phasing out of CFCs by 1996, although there were some essential-use exemptions.

However, because of the important uses of CFCs the above would only be possible if suitable replacements were available to take their place. Two points to note about this are that a whole family, not just one or two compounds, of replacements for the family of CFCs would be needed and that the time-scale would be incredibly short for such a major R&D project. Normally a time scale of 10–20 years would be envisaged.

How did the chemical industry view this situation in the late 1980s? As a disaster with the loss of the billion dollar CFC market? No, they regarded it more of a challenge to produce these new products and the market opportunities that they might generate. The company that was first to have a plant up and running and making the very first replacement would clearly steal a march on its competitors.

The very short time scale dictated that enormous resources, both financial and manpower, would have to be devoted to this research and development. Also the approach would have to be different to the traditional one. Normally the chemical and other engineers, who would design and commision the plant to produce a chemical, do not tend to be brought into the project until the research chemists have largely finished their work. Here they were brought in to work with the chemists almost from the start of the project.

Where did the research start? The first thing to do was to identify compounds that would match, as far as possible, the properties of the CFCs. Clearly, chemically similar compounds, but without chlorine preferably, would be the starting point. All the companies involved in this research quickly came up with pretty much the same list of compounds. They are HFCs (hydrofluorocarbons) or HCFCs (hydrochlorofluorocarbons) having 1–3 carbons in the molecule, although the latter would only be interim replacements. Interestingly, the industry prefers to call the former HFAs (hydrofluoroalkanes) because the alternative name of HFCs is too close to that of CFCs and their bad image! Early research focused on checking their properties and initiating toxicity and environmental impact (or hopefully lack of this) studies. It also concentrated on refrigerants and foam blowing agents since alternative 'ozone friendly' aerosol propellants were already on the market. Examples are butane and dimethyl ether but, unlike CFCs, they are flammable!

Conscious of the very tight timetable plus the fact that they were working on the same compounds, the 14 major companies agreed to jointly fund independent toxicity and environmental studies, another innovation. They concentrated their research scientists on discovering and developing the best and most economic route(s) for making the compound(s). Early efforts concentrated on HFA 134a and their extent can be gauged by the fact that one company had over 100 scientists and engineers working on the project and spent a total of over £100 million in getting as far as designing the plant to produce this compound.

Routes investigated for the synthesis were mainly based on using CFCs since the companies already had well established processes for these. ICI's process that was developed and implemented for HFA 134a is as follows:

$$Cl_2C{=}CClH + 3HF \xrightarrow{\text{catalyst}} CF_3CH_2Cl + 2HCl$$

$$CF_3CH_2Cl + HF \xrightarrow{\text{catalyst}} \underset{\text{HFA 134a}}{CF_3CH_2F} + HCl$$

Two stages are required to improve the yield by 'cheating' the thermodynamics. An alternative process uses a final hydrogenation step:

$$CF_3CCl_2F + 2H_2 \xrightarrow[\text{catalyst}]{\text{Pd/C}} CF_3CH_2F + 2HCl$$

A lot of the research work involved the catalyst development and ensuring it had an acceptable lifetime.

On the development/application side some unexpected problems were encountered during the testing of HFA 134a. Firstly, unlike CFC 12 (CF_2Cl_2), it was immiscible with the mineral oil lubricant used in the refrigeration compressors. This necessitated a further programme of research to find lubricants which were compatible. Polyalkylene glycols were found to meet this requirement, resulting in a new outlet for them! Secondly, HFA 134a caused the gaskets to shrink causing leaks, whereas with CFC 12 they had expanded to give a tight seal. Thus, another problem had to be overcome.

The success of this enormous R&D programme can be judged by the fact that two of the major CFC producers, DuPont and ICI, had plants producing HFA 134a in 1993, less than 5 years from the commencement of the R&D work. Indeed ICI now have plants in operation in Japan and the U.S.A. in addition to the first plant at Runcorn, U.K.

Why are these HCFCs and HFAs acceptable alternatives to the CFCs? They all contain one or more hydrogens in the molecule; CFCs do not contain any. This results in them having much shorter atmospheric lifetimes

Table 3.1 Atmospheric lifetimes of some CFCs and their replacements

Structure	Code	Atmospheric lifetime (years)
$CFCl_3$	CFC 11	55
CF_2Cl_2	CFC 12	116
$ClCF_2CF_2Cl$	CFC 113	110
$CHClF_2$	HCFC 22	16
CF_3CHCl_2	HCFC 123	2
CH_2F_2	HFA 32	7
CF_3CH_2F	HFA 134a	16

Table 3.2 Ozone depletion potentials (ODPs) of some CFCs and their replacements

Structure	Code	ODP
$CFCl_3$	CFC 11	1·00
CF_2Cl_2	CFC 12	1·00
$ClCF_2CF_2Cl$	CFC 113	0·80
$CHClF_2$	HCFC 22	0·06
CF_3CHCl_2	HCFC 123	0·02
CH_2F_2	HFA 32	0·00
CF_3CH_2F	HFA 134a	0·00

than CFCs (see Table 3.1). They therefore break down in the atmosphere and hence even if they eventually got up to the stratosphere they could not cause the problems which CFCs do. HFAs do not contain chlorine and therefore will not be able to attack ozone molecules. They therefore have an ODP (Ozone Depletion Potential) of zero. Table 3.2 shows ODP values for some CFCs and their replacements. R&D continues to produce further replacements.

The development of CFC replacements such as HFA 134a in such a short time represents a major research and development success story. Note, however, the innovative approaches of: (1) involving engineers at a very early stage; and (2) the inter-company collaboration on joint toxicity and environmental studies.

**3.6
Conclusions**

This chapter has demonstrated the very large commitment of the chemical industry to research and development, with about 7% of SALES INCOME typically being invested in these activities. However, considerable variation across the different sectors was noted, ranging from a much lower figure for fertilizers to as much as 14% in the pharmaceutical sector, where individual companies' investment now approaches $1 million per annum, e.g. Glaxo, Merck. They have a large (approx. 25% of total personnel) and highly educated work force in the research and development function and a high proportion of these will have degrees and higher degrees. Indeed in the drug and pesticide areas R&D was quite rightly described as the lifeblood of the industry.

The reasons for, and the importance of, research and development activities were discussed, particularly in section 3.3. This followed consideration of the various types of research and development in section 3.2.2. Differences between academic and industrial research were considered in section 3.4. University research is often basic or fundamental and may be carried out for its own intrinsic interest to add to the total knowledge base or pool. The results are usually freely available following publication in scientific journals. It is also often confined to a fairly narrow area of chemistry.

Industrial research, being business driven, tends to be much more applied with a lot of emphasis on the development part of R&D. The ability to work with a range of other chemists, scientists, and even engineers, in a team to solve a common problem was emphasized. The results are usually patented giving the company exclusive rights to exploit the discovery/invention for a certain period of time.

Finally, two case studies were presented briefly to give a better feel for what research and development activities are all about. Vinyl chloride monomer production showed how the availability of cheap raw materials (e.g. ethylene) stimulated the development of processes to utilize these, and how continuing research and development led to new, even better processes. It also emphasized the importance of reading the chemical literature. Development and production of CFC replacements demonstrated what an enormous R&D effort can achieve in such a short time. Great emphasis on research and development is a key characteristic of high technology industries like the chemical industry.

Bibliography

The Chemical Industry, 2nd edn, Alan Heaton (ed), Blackie A&P, Glasgow, 1994. Sections 3.5, 7.3, 8.3, and 9.7–9.10.
CFCs—The Search for Alternatives, STEAM (ICI Science Teachers Magazine), No. 12, January 1990, 14.

4 The World's Major Chemical Industries

Alan Heaton

4.1
History and development
of the chemical industry

4.1.1 *Origins of the chemical industry*

The use of chemicals dates back to the ancient civilizations. For example, many chemicals were known and used by the ancient Egyptians—they used soda (known to them as 'natron') mixed with animal fats as soap to wash corpses, and on its own in the mummifying process which followed. Glass objects and glazed pottery, which were buried with the mummies for their use in their assumed after-life, were made from soda and sand.

Evolution of an actual chemical industry is much more recent, and came about, as with many other industries, during the industrial revolution, which occurred in the U.K. around 1800 and rather later in other countries. Its initial development was stimulated by the demand of a few other industries for particular chemicals. Thus soapmaking required alkali for saponification of animal and vegetable oils and fats; the cotton industry required bleaching powder; and glassmaking required sand (silica) and soda (sodium carbonate). A key advance had been made by Roebuck and Gardner in Birmingham in 1746 when they substituted lead chambers for glass reaction vessels, which had been used until that time, because the construction in glass had previously limited the scale of vitriol (sulphuric acid) manufacture. The vitriol was required in the Leblanc process for making soda from salt, the chemical reactions being

$$2NaCl + H_2SO_4 \rightarrow Na_2SO_4 + 2HCl$$
$$Na_2SO_4 + CaCO_3 + 2C \rightarrow Na_2CO_3 + CaS + 2CO_2$$

In addition to the uses mentioned above, and as washing soda, sodium carbonate was also a ready source of sodium hydroxide, viz.

$$Na_2CO_3 + Ca(OH)_2 \rightarrow 2NaOH + CaCO_3$$

In the early days of the process, condensing the hydrochloric acid fumes proved difficult (or was ignored!) as this reference to the Liverpool Works of James Muspratt (one of the founding fathers of the chemical industry in the

North West of England) shows[1]. It is from a letter, published in the *Liverpool Mercury* dated 5th October 1827, which stated that Muspratt's works poured out 'such volumes of sulphureous smoke as to darken the whole atmosphere in the neighbourhood; so much so that the church of St. Martin-in-the-Fields now erecting cannot be seen from the houses at about one hundred yards distance, the stones of which are already turned a dark colour from the same cause. The scent is almost insufferable, as well as injurious to the health of persons residing in that neighbourhood.' Public outcry against this serious atmospheric pollution—which also rotted curtains and clothes—led to the passing of the Alkali Act in 1863, the first legislation in the world concerned with emission standards. Clearly environmental pollution is not merely a problem of the 20th century. Thus at the beginning of the 19th century the chemicals industry produced a small number of inorganic chemicals, with the Leblanc process for making soda from salt (summarized above) at its heart.

Many of these early processes had evolved by trial and error, and there was a considerable art in getting and keeping them working. Advances in scientific theories and knowledge during the 1800s—from Dalton's atomic theory onwards—provided a much more solid foundation from which the industry could develop a clearer understanding of the basis of a process and advances in technology were the result. Recognition of the importance and application of scientific principles needed men of great foresight, and one such person was John Hutchinson, who may be considered a pioneer of modern chemicals manufacture. In 1847 he established a chemical works on Merseyside, not far from Liverpool. However, he had the vision to recognize the importance of scale and scientific control to his manufacturing processes. He also proved to be an excellent manager, certainly in his judgement of men. Not only did he engage J. W. Towers to develop analytical methods, but also J. T. Brunner as office manager and Ludwig Mond from Germany to assist in the scientific development of the manufacturing processes. Friendship between the latter two led to the formation of the Brunner Mond Company in 1872. This in turn became a founder member of the well-known chemical giant ICI, Imperial Chemical Industries. Mond appreciated at a very early stage the importance of the challenge of the new Solvay ammonia–soda process to the Leblanc process. This not only overcame to a large degree the considerable effluent problems of the Leblanc process, but was cheaper in terms of raw materials and labour, although the capital cost of the plant was considerably higher. The first Solvay plant in Britain started production for Brunner Mond in 1874, and due to its economic advantages had captured 20% of the country's alkali production within 10 years. It went from strength to strength, and eventually in 1914 had 90% of the U.K.'s total production.

The process itself may be summarized as

$$NH_3 + H_2O + CO_2 \rightarrow NH_4HCO_3$$
$$NaCl + NH_4HCO_3 \rightarrow NaHCO_{3\downarrow} + NH_4Cl$$

and
$$2NaHCO_3 \xrightarrow{heat} Na_2CO_3 + H_2O + CO_2$$

Treatment of the ammonium chloride liquor with calcium hydroxide regenerates ammonia for the use in the first reaction:

$$2NH_4Cl + Ca(OH)_2 \rightarrow CaCl_2 + 2NH_3 + 2H_2O$$

The calcium hydroxide is obtained from limestone:

$$\underset{\text{limestone}}{CaCO_3} \overset{\text{roast}}{\longrightarrow} CaO + CO_2$$

$$CaO + H_2O \rightarrow Ca(OH)_2$$

There was no organic chemicals side of the industry to speak of before 1856, since up to that time any organic materials were obtained from natural sources. Examples are animal and vegetable oils, fats and colouring matter, and natural fibres such as cotton and wool. In 1856, as has happened many times since (sometimes leading to major advances in the subject!), some planned research work did not give the results expected. William Henry Perkin was trying to synthesize the antimalarial drug quinine, a naturally-occurring compound found in the bark of *Cinchona* trees. Using sodium dichromate he was attempting to oxidize aniline (phenylamine) sulphate to quinine, but instead obtained a black precipitate. Rather than just rejecting this reaction which had obviously failed in its purpose, he extracted the precipitate and obtained a purple compound. It showed great promise as a dye and 'mauve', as it was named, became the first synthetic dyestuff. Rather ironically, the mauve was formed because Perkin's aniline was impure—he had obtained it in the standard way by nitrating benzene and then reducing the product, but his benzene had contained significant amounts of toluene. Although only 18, and a student in London, he had the confidence to terminate his studies in order to manufacture mauve, which he did very successfully. The synthetic dyestuffs industry grew rapidly from this beginning and was dominated by Britain into the 1870s. However, chemical research in Britain by this time tended to be very academic, whereas in Germany the emphasis was much more on the applied aspects of the subject. This enabled the Germans to forge ahead in the discovery of new dyes, early successes being alizarin and the azo-dyes. So successful were they that by the outbreak of World War I in 1914 they dominated world production, capturing over 75% of the market.

The very sound base which their success in dyestuffs provided for the large German companies—BASF, Bayer, Hoechst—particularly in terms of large financial resources and scientific research expertise and skills, enabled them to diversify into and develop new areas of the chemical industry. By the early years of the 20th century advances into synthetic pharmaceuticals had been made. Early successes were Salvarsan, an organoarsenical for treating syphilis, and aspirin.

BASF concentrated on inorganic chemistry and achieved notable successes in developing the contact process for sulphuric acid and the Haber process for ammonia production. The latter particularly was a major technological breakthrough, since it required novel, very specialized plant to handle gases at high temperatures and pressures. Thus by 1914 Germany dominated the world scene and was well ahead in both applied chemistry and technological achievement.

However, the outbreak of World War I changed this situation dramatically. Firstly, in both Britain and Germany stimulus was given to those parts of the industry producing chemicals required for explosives manufacture, e.g. nitric acid. Secondly, Germany was particularly isolated from its raw material supplies and could not of course export its products, such as dyestuffs, which led to shortages in Britain and the U.S.A. The result was a rapid expansion in the manufacture of dyestuffs in these countries. In contrast Germany had to rapidly develop production of nitrogen-based compounds—particularly nitric acid for explosives manufacture and ammonium ·salts for use in fertilizers—since its major source, sodium nitrate imported from Chile, had been removed. Commercialization of the Haber process (catalytic oxidation of ammonia leading to nitric acid) had come at just the right time to make this possible.

Overall, therefore, the effect of World War I on the chemical industries of the world was to stimulate home-based production.

4.1.2 Inter-war years, 1918–1939

The war had alerted governments to the importance of the chemical industry, and the immediate post-war years were boom years for the British and American industries. An important factor aiding this was the 'protectionist' policy operated by their governments at the time. Thus Britain brought in a heavy import duty on most synthetic organics, and the introduction of a tariff system—the American Selling Price—effectively closed the American market to most exporters. During the war and immediately afterwards most countries had, largely out of necessity, expanded their chemical industry. By the early 1920s considerable overcapacity was the result and international competition became correspondingly fierce. With its large and expanding home market and lack of competition from imports the American chemical industry grew rapidly during the 1920s. This was in complete contrast to the situation in most other countries at this time. Germany continued to dominate the international scene, and in 1925 an event occurred which was to not only support this position, but to set a trend which was to characterize the inter-war period, and remain with us up to the present time. This was the amalgamation of the major German dyestuff companies to form one giant company—I. G. Farben. This immediately became the largest chemical company in the world, with a very broad financial base, enormous applied scientific and technological expertise, and dynamic management. Certainly no European, and few American firms could hope to compete with it.

The consequences of this were soon appreciated in Britain and in the following year, 1926, amalgamation of Brunner Mond, United Alkali

Company, British Dyestuffs Corporation and Nobel Industries led to the formation of another chemical giant—Imperial Chemical Industries (ICI). Although the integration and rationalization process was not without its difficulties, the result was a much more sound company. This was demonstrated in the early 1930s, during the very severe world recession, when although the chemical industry also suffered severely (from overcapacity caused by the big fall in demand) the effect on ICI was reduced because it had been able to participate in market-sharing agreements (or cartels) with other large companies. Within the continuing protection policies which it enjoyed the British industry grew steadily up to 1939, becoming virtually self-sufficient in the most important chemicals and utilizing and gaining expertise in new technology such as high-pressure gas reactions, e.g. the Haber process. It also became more innovative, introducing several important new dyestuffs.

As indicated previously, during this time the American industry was growing steadily, although largely in isolation. Up until 1940 it was also the only country to have petrochemical plants. The availability for some years of their own supplies of crude oil, and refineries, led to interest in the use of petroleum fractions for the synthesis of organic compounds and started in the 1920s with the manufacture of iso-propanol from the refinery off-gas propylene. However up until 1940 this synthetic scope was limited to producing oxygenated solvents.

Apart from the start of the trend towards very large companies, and bigger manufacturing units, the inter-war period will be remembered as marking the beginning of two areas which have had profound effects on the industry's subsequent development, growth, and profitability. The first, petrochemicals, has already been referred to above. Its all-pervading influence is apparent from the text and the devotion of the final chapter of this volume to it. The second, the synthetic polymer sector, has over the last few decades owed much to the development of petrochemicals. As we shall see later this has grown into the most important sector of the whole chemical industry.

The chemical industry's interest in polymers dates back to the 19th century. In those days it was a case of synthetically modifying natural polymers with chemical reagents to either improve their properties or produce new materials with desirable characteristics. Notable examples were nitration of cellulose giving the explosive nitrocellulose, production of regenerated cellulose (rayon or artificial silk) via its xanthate derivative, and vulcanization of rubber by heating with sulphur. Manufacture of acetylated cellulose (cellulose acetate or acetate rayon) developed rapidly from 1914 onwards with its use both as a semi-synthetic fibre and as a thermoplastic material for extrusion as a film.

The first fully synthetic polymer to be introduced (1909) was a phenol-formaldehyde resin known as bakelite. Although the reaction had been discovered some twenty years previously, it was only after a careful and systematic study of it that it was properly controlled to give a useful thermosetting resin. In the late 1920s two other types of thermosetting resins followed, namely urea-formaldehyde, and alkyd resins.

Assisted by advances in high-pressure technology, a more scientific approach to the study of synthetic polymers (macromolecules) led to the discovery and production of several important polymers in the 1930s and early 1940s. These plastics, elastomers and synthetic fibres have had a profound effect not only on the chemical industry but also on the quality of all our lives.

Germany, in the guise of I. G. Farben, had the foresight to realize the importance of this area, and they dominated the early research and development work. Their commercial successes ranged from rubber-like materials—the co-polymers poly(styrene-butadiene) and poly(acrylonitrile-butadiene)—to thermoplastics like poly(vinyl chloride) and polystyrene.

In the U.K., Gibson of ICI was studying high-pressure reactions of alkenes in autoclaves. His reaction with ethylene appeared to have failed since he obtained a small amount of a waxy white solid. Fortunately his curiosity led him to investigate this. It proved to be polyethylene, first produced commercially by ICI in 1938.

America's interest was led by the Dupont Company and its major success was the commercial introduction of the first nylon in 1941. This certainly was a reward for perseverance and effort since from Carother's first researches on the subject to commercial production took 12 years and reportedly cost $12 million

4.1.3 *Second World War period, 1939–1945*

In general terms the impact on the chemical industry was similar to that of the First World War. Thus Germany especially was cut off from its raw-material supplies and therefore relied entirely on synthetic materials, e.g. poly(styrene-butadiene) rubber, and gasoline produced from coal. Britain and America were not affected to quite the same extent but demand for polymers like nylon and polyethylene for parachutes and electrical insulation was high. By the end of the war facilities for synthetic polymer production had expanded considerably in all three countries.

4.1.4 *Post-1945 period*

The Second World War had two major effects on the German chemical industry (apart from the consequences of the political separation into East and West Germany). Firstly, much of the chemical plant was either destroyed or damaged by bombing and process details were freely acquired by the British and Americans. Secondly, the Allies broke up I. G. Farben into a number of smaller companies such as Bayer, Hoechst and BASF, and therefore eliminated its competitiveness. As we shall see later, this turned out to be merely a temporary setback.

The major changes in the world-wide chemical industry since 1945 have been concerned with organic chemicals, in particular the raw materials used to produce key intermediates such as ethylene (ethene), propylene (propene), benzene and toluene. In the U.K. up to and including 1949 coal was the major source of raw materials, followed by carbohydrates, with oil accounting for under 10% of total production. However, the situation was starting to change due to events in the oil industry. Worldwide demand for petroleum products was increasing, this being most apparent for transport. Increasing ownership of automobiles and rapid development of air transport raised demand for gasoline. Since this demand was in the developed countries, oil refineries were now located in them, e.g. in Europe. Processes were developed in the refineries to convert low-demand fractions, e.g. gas oil, into high-demand lower-boiling fractions, e.g. gasoline. Alkenes such as ethylene and propylene were produced as by-products. Thus they were available for chemical synthesis—and at a very cheap price largely because of the vast scale of operation. Note that until this time ethylene had been obtained by dehydration of ethanol which in turn had been made by fermentation of carbohydrates. The other important source of aliphatics was acetylene (ethyne) which was obtained from calcium carbide. Although there were no technical difficulties in these routes, the ethylene and acetylene were both rather expensive to produce in this way. By 1959 oil was on a par with coal as the major source of chemicals in Europe, mirroring the position of the American industry some two decades previously. During this time the price of oil had hardly increased whereas that of coal was steadily increasing because its extraction is labour-intensive. Also, demand for the carbonization of coal was starting to decrease with falling demand for coke and town gas. This meant less benzole, and by-product coal-tar, from which the chemicals were obtained, was likely to be available. Although the processing costs of the oil tended to increase this was counterbalanced by the ever-increasing size of the operations (hence the benefits of the economy-of-scale effect) and the improved efficiency of the processes. Thus the economic advantages were swinging very much in oil's direction, and by 1969, oil dominated the scene, and this it has continued to do ever since. Petrochemicals had truly come of age.

Let us consider further the reasons for the explosive growth of the petro-chemicals industry. As indicated, oil-refinery processes such as cracking supplied key chemical intermediates—ethylene and propylene—at low prices compared with traditional methods for their preparation. In real terms these prices were more or less maintained for a considerable period. However this factor alone cannot account completely for the vast increase in the tonnage of organic chemicals produced from petroleum sources. Much of the credit may be placed with research chemists, process-development chemists and chemical engineers. Once it was realized that abundant quanties of ethylene and propylene were available, research chemists had the incentive to develop processes for the production of many other compounds. Success in the laboratory led to process development and eventually construction of manufacturing units. Chapter 12 demonstrates the versatility of the alkenes, and the achievements of the scientists. Consideration of the chemical reagents and reactants used in many of these processes shows the importance and influence on the development of the petrochemical industry of one particular area of chemistry, namely catalysis. This influence has been a two-way process: research into catalysis has been stimulated by the needs of the industry and, in turn, better, more efficient catalysts have been discovered which have improved process efficiency and economics, and may even have rendered an existing process obsolete by permitting introduction of an entirely new, more economic route. Introduction of these catalysts has made possible not only entirely new routes and interconversions of compounds, but also immensely shorter reaction times (sometimes now just a few seconds) compared with the corresponding non-catalytic route. Another major advantage has been that catalysts have introduced better control of the reaction, particularly improved selectivity towards the desired product. This is a vital aim in chemicals production, since very few processes yield only the desired product and by-product formation represents material and efficiency loss. Furthermore separation of product from by-products can be a major cost item in the overall process economics. A notable example is the introduction of catalytic cracking units in place of purely thermal cracking units for ethylene production. Evolution of manufacturing processes utilizing these new catalysts has often necessitated technological developments. Examples are design and construction of equipment to handle very high-pressure gas reactions, facilities for rapid quenching of products (as in cracking to cheat the kinetics and obtain predominantly alkenes rather than alkynes when the process has to be operated under conditions favouring the latter) and development of fluidized-bed reactors. We emphasize the importance of catalysts and catalysis by devoting Chapter 11 to this topic.

The major industrial developments in organic chemicals initiated in the 1930–1940 period have continued since that time. Most important of all is the introduction of purely synthetic polymers. Table 4.1 shows the growth in importance of these materials over the period 1950 to 1988. Although much of this growth was due to the increased demand for the polymers introduced in the 1930s and 1940s—urea-formaldehyde resins, nylon, polyethylene (low-density), poly(vinyl chloride), and butadiene co-polymers—new polymers

Table 4.1 World consumption of synthetic polymers (millions of tonnes)

	1960	1970	1978	1986	1994e
Plastics and resins	6·0	25·0	42·0	58·5	95·5
Fibres (excluding cellulosics)	1·5	5·0	12·0	11·4	16·0
Rubbers	1·0	5·0	4·0	5·4	8·8
Total	3·5	35·0	58·0	75·3	120·3

e = estimated.

have also made a significant contribution. The most familiar examples (together with their year of commercial introduction and the company responsible) are polyacrylonitrile (1948, Du Pont); Terylene (1949, ICI); epoxy resins (1955, Du Pont); and polypropylene (1956, Montecatini). Their special properties and economic advantage have enabled them to displace traditional materials. Thus plastics such as poly(vinyl chloride) have replaced wood in window frames and metal drainpipes, since they are unaffected by weathering, lighter in weight, and maintenance-free. Synthetic fibres like nylon and terylene have enabled non-drip, machine-washable, and crease-resistant clothing to be introduced, and we are made aware many times each day of the widespread usage of synthetic plastics in all facets of the packaging industry. Indeed, so profound is the influence of plastics and polymers on our lives (something we tend to take for granted) that it is difficult to imagine our existence without them. Not only are all these polymers well established now, but so are the processes for making them. It must also be mentioned in passing that the actual use of these polymers required major technological advances in developing processes to get the material in the correct form for its particular applications. Techniques such as injection-moulding and blow-moulding are the result. Existing dyestuffs would not adhere to these synthetic materials and so new ones were in turn discovered and developed. Polymer science and technology has grown into a large field of study.

These 'standard' polymers are well established nowadays and are produced by many companies; this is therefore a very competitive area of the chemical industry. However, the overall market for these products is still expanding, as the figures for 1986 and 1994 clearly show.

A major advance in polymer chemistry was provided by the work of Karl Ziegler and Giulio Natta, which led in 1955 to the introduction of some revolutionary catalysts which bear their name. The great significance of this event was highlighted by them being awarded a Nobel Prize in 1963 for their work. Ziegler–Natta catalysts are mixtures of a trialkyl aluminium plus a titanium salt and they bring about the polymerization by a coordination mechanism in which the monomer is inserted between the catalyst and the growing polymer chain. Industrial interest in this centred on the fact that it gave more control over the polymerization process. Significantly, it allowed the polymerization to occur under milder conditions *and* produced a very stereoregular polymer. Taking polypropylene as an example, the Ziegler–Natta catalyst gives the isotactic product whereas a free-radical process leads to the atactic stereochemical form. These are shown below:

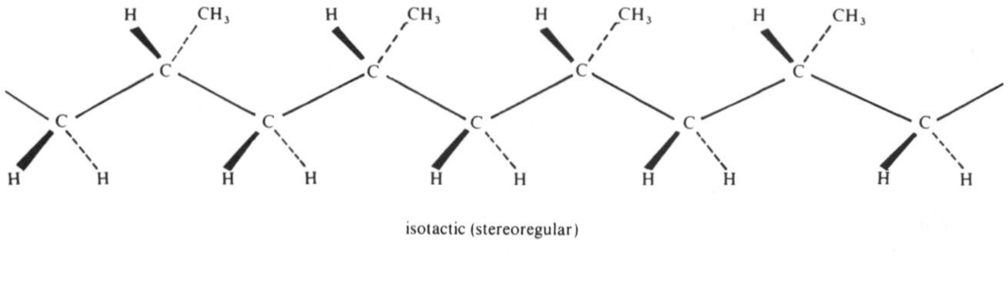

isotactic (stereoregular)

atactic (stereochemistry random)

These stereochemical forms show differences in their properties; for example, the isotactic polymer chains can pack together better and this form has a higher degree of crystallinity and hence a higher softening point than the atactic variety. Ziegler–Natta polymerization of ethylene gives high-density polyethylene (HDPE) which is different in some properties from the low-density form (LDPE) produced by free-radical polymerization, as is shown in Table 4.2.

During the late 1970s production of linear low-density polyethylene (LLDPE) was commercialized. Here the polymer chain is much more linear containing only short branching chains, in contrast to conventional LDPE. The polymer has greater strength and toughness, particularly in film applications, than ordinary low-density polyethylene.

Over recent years in addition to continuing production of the bulk polymers described above many companies have developed certain speciality polymers. These can be thought of as polymers whose structure has been tailor-made to yield certain specific properties, and therefore have limited, very specialized applications. For example, polymers which are exceedingly thermally and oxidatively stable are required for space capsules which will re-enter the earth's atmosphere. To develop these requires particular knowledge and skills. There is therefore little competition in producing these polymers; the market for them, although small, is assured and the product therefore commands a high price by comparison with 'standard' polymers. The polymer sector is the subject of Chapter 4 in Volume 2.

A very important trend during the post-Second World War period has been the increasing size of plants, and nowhere is this more apparent than in the petrochemicals sector, where capacities of 100 000 tonnes per annum are commonplace. Large integrated chemical complexes have evolved, due to this increase in scale, and the need to locate petrochemical plants adjacent to refineries, so as to minimize transportation of vast quantities of chemicals.

Table 4.2 Some properties of LDPE and HDPE

Property	LDPE	HDPE
Specific gravity/g cm^{-3}	0·920	0·955
% Crystallinity	55	80
Softening point/K	360	400
Tensile strength/atm.	85–136	204–313

Many of the downstream processes, which utilize petrochemical inter-mediates, are located on the same site, and this arrangement can lead to reduced utility costs. For example, a large complex of units can justify its own power station to produce electricity and steam. In the U.K., in times of low demand surplus electricity may be fed into the national grid and this generates a small payment to the company from the Electricity Generating Companies. In contrast, when demand is high electricity is taken from the national grid—but at a much higher cost.

This trend to larger-sized individual chemical plants and large complexes of chemical plants has been matched by movement to bigger chemical companies by mergers, or takeovers, or both, as we have already seen happening in the 1920s with the formation of ICI and I. G. Farben. Nowadays the multinational giants dominate the international chemical scene. Interestingly, in 1993 ICI split into two parts—ICI and Zeneca, with the latter being largely the pharmaceutical and speciality part of the original company.

The chemical industry has been to the fore during the last decade or two in utilizing the tremendous advances in electronics. Thus complete automation (even full computer control) of large continuous plants is commonplace. An additional advantage is the automatic data collection of throughput, tempera-tures, pressures, etc. This can be invaluable when subsequent analysis of the data may suggest small adjustments to the plant to improve its efficiency and hence reduce product costs and improve profitability. The rapid developments in microelectronics which have led to reduced selling prices for micropro-cessors is making their introduction for fairly small plants an economical and practical proposition.

Control of pollution, i.e. effluent control and treatment, is another important factor which has markedly influenced the industry over the past decade. Indeed there are legal requirements to be met in control of the emission or discharge of effluent. Although, as indicated previously, utilization of by-products is a feature of the industry, nevertheless some effluent is generated and due to the very large scale of operation the quantity can be substantial. Its treatment and disposal can be a considerable cost item of the process. Equipment for the prevention of loss of untreated wastes can also add significantly to the capital and running costs of a plant. Unfortunately the legislation controlling treatment varies quite a lot from country to country and those with strict requirements may place companies which manufacture there at a disadvantage, in terms of production costs, compared with companies in other countries. Chapter 9 takes up these aspects in detail.

Its widening international base has also been a recent feature of the chemical industry. As in so many other areas of manufacturing industry the shining star has been Japan. From a position of insignificance in 1945 it has grown to be the second most important chemical industry in the world today. Its growth is even more remarkable when one realizes that it has no oil of its own—it must all be imported. In more recent years countries having oil reserves, such as the Arab countries, have moved from merely exporting the oil to refining it, and now into processing it to produce petrochemicals (downstream operations). Though small at present these industries are growing rapidly.

Since the rapid rise in the price of oil in 1973, energy costs have been a major concern of chemical producers. The chemical industry is a major energy consumer and therefore even small percentage savings in energy costs can mean tens or hundreds of thousands of dollars or pounds, and the difference between profit and loss. It is no surprise to find therefore that immense efforts are now being made to assess and analyse the energy usage of plants to see where savings can be made. We have emphasized this importance in the text: energy is a recurring theme and we have devoted a complete chapter (Chapter 8) to it.

The most exciting area of development at the start of the 1980s, arising from the immense amount of research and development work which the chemical and related industries carry out each year, is biotechnology. In biotechnology genes are manipulated to ensure that the micro-organism—often *E. coli*—will synthesize the desired specific chemical. Although the risks and practical problems are considerable, the rewards for their solution are immense. The technology is still in its infancy and its effect on the chemical industry as a whole is difficult to predict. Clearly it will make a major contribution to pharmaceuticals—as the successful development of insulin has already shown—and production of interferon and antibiotics are on the near horizon. In contrast it is difficult to see biotechnology being a competitive route to bulk chemical intermediates like ethylene and benzene, because of its slowness and the high product separation costs. However, it is an area whose development chemists are watching with interest and fascination. This topic is discussed in more detail in Chapter 9 of Volume 2.

4.2.1 *Definition of the chemical industry*

At the turn of the century there would have been little difficulty in defining what constituted the chemical industry since, as shown earlier in this chapter, only a very limited range of products was manufactured and these were clearly chemicals, e.g. alkali, sulphuric acid. At present, however, many thousands of chemicals are produced, from raw materials like crude oil through (in some cases) many intermediates to products which may be used directly as consumer goods, or readily converted into them. The difficulty comes in deciding at which point in this sequence the particular operation ceases to be part of the chemical industry's sphere of activities. To consider a specific example to illustrate this dilemma, emulsion paints may contain poly(vinyl chloride)/poly(vinyl acetate). Clearly, synthesis of vinyl chloride (or acetate) and its polymerization are chemical activities. However, if formulation and mixing of the paint, including the polymer, is carried out by a branch of the multinational chemical company which manufactured the ingredients, is this still part of the chemical industry or does it now belong in the decorating industry?

4.2
The chemical industry today

It is therefore apparent that, because of its diversity of operations and close links in many areas with other industries, there is no simple definition of the chemical industry. Instead each official body which collects and publishes statistics on manufacturing industry will have its definition as to which operations are classified as 'the chemical industry'. It is important to bear this in mind when comparing statistical information which is derived from several sources. Perhaps the best known international definition for chemicals is that contained in Section 5 of the United Nations Standard International Trade Classification. Individual countries' definitions will differ from this to varying degrees, as will the companies' trade organizations—such as the Chemical Industries Association (CIA) in the U.K.—in each country.

4.2.2 *The need for a chemical industry*

As indicated in Chapter 1, the chemical industry is concerned with converting raw materials, such as crude oil, firstly into chemical intermediates, and then into a tremendous variety of other chemicals. These are then used to produce consumer products, which make our lives more comfortable or, in some cases such as pharmaceutical products, help to maintain our wellbeing or even life itself. At each stage of these operations value is added to the product and provided this added value exceeds the raw material plus processing costs then a profit will be made on the operation. It is the aim of chemical industry to achieve this.

It may seem strange in a textbook like this one to pose the question 'do we need a chemical industry?' However, trying to answer this question will provide (i) an indication of the range of the chemical industry's activities, (ii) its influence on our lives in everyday terms, and (iii) how great is society's need for a chemical industry. Our approach in answering the question will be to consider the industry's contribution to meeting and satisfying our major needs. What are these? Clearly food (and drink) and health are paramount. Others which we shall consider in their turn are clothing, and (briefly) shelter, leisure and transport.

(*a*) *Food.*   The chemical industry makes a major contribution to food production in at least three ways. Firstly, by making available large quantities of artificial fertilizers which are used to replace the elements (mainly nitrogen, phosphorus and potassium) which are removed as nutrients by the growing crops during modern intensive farming. Secondly, by manufacturing crop protection chemicals, i.e. pesticides, which markedly reduce the proportion of the crops consumed by pests. Thirdly, by producing veterinary products which protect livestock from disease or cure their infections.

These and related topics are discussed in Chapter 7 of Volume 2.

(*b*) *Health.*   We are all aware of the major contribution which the pharmaceutical sector of the industry has made to help keep us all healthy, e.g. by curing bacterial infections with antibiotics, and even extending life itself, e.g. β-blockers to lower blood pressure.

This topic is discussed more fully in Chapter 8 of Volume 2.

(c) *Clothing.* The improvement in properties of modern synthetic fibres over the traditional clothing materials (e.g. cotton and wool) has been quite remarkable. Thus shirts, dresses and suits made from polyesters like Terylene and polyamides like Nylon are crease-resistant, machine-washable, and drip-dry or non-iron. They are also cheaper than natural materials.

Parallel developments in the discovery of modern synthetic dyes and the technology to 'bond' them to the fibre has resulted in a tremendous increase in the variety of colours available to the fashion designer. Indeed they now span almost every colour and hue of the visible spectrum. Indeed if a suitable shade is not available, structural modification of an existing dye to achieve this can readily be carried out, provided there is a satisfactory market for the product.

Other major advances in this sphere have been in colour-fastness, i.e. resistance to the dye being washed out when the garment is cleaned. For example, the Procion dyes, developed by ICI, actually chemically bond to cotton, rather than attaching by the more usual physical type of adherence to the fibre. This clearly leads to greater colour-fastness.

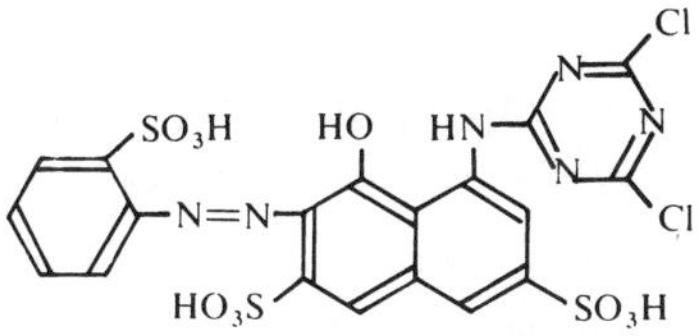

Figure 4.1   A typical Procion dye.

The chlorine atoms on the triazine are very reactive and can be displaced by even a weak nucleophile like the —O—H groups in cotton, giving

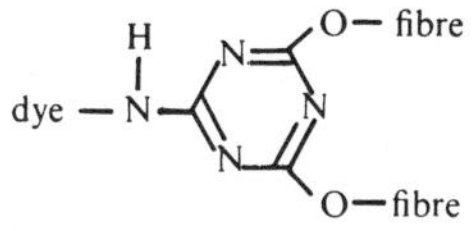

See also page 167 of Volume 2.

(d) *Shelter, leisure and transport.* In terms of shelter the contribution of modern synthetic polymers has been substantial. Plastics are tending to replace traditional building materials like wood because they are lighter, maintenance-free (i.e., they are resistant to weathering and do not need painting). Other polymers, e.g. urea-formaldehyde and polyurethanes, are important insulating materials for reducing heat losses and hence reducing energy usage.

Plastics and polymers have made a considerable impact on leisure activities with applications ranging from all-weather artificial surfaces for athletic tracks, football pitches and tennis courts to nylon strings for racquets and items like golf balls and footballs made entirely from synthetic materials.

Likewise the chemical industry's contribution to transport over the years has led to major improvements. Thus development of improved additives like anti-oxidants and viscosity index improvers for engine oil has enabled routine servicing intervals to increase from 3000 to 6000 to 12 000 miles. Research and development work has also resulted in improved lubricating oils and greases, and better brake fluids. Yet again the contribution of polymers and plastics has been very striking with the proportion of the total automobile derived from these materials—dashboard, steering wheel, seat padding and covering etc.—now exceeding 40%.

So it is quite apparent even from a brief look at the chemical industry's contribution to meeting our major needs that life in the developed world would be very different without the products of the industry. Indeed the level of a country's development may be judged by the production level and sophistication of its chemical industry.

4.2.3   *The major chemicals*

Table 4.3 shows the 20 most important chemicals produced in the U.S.A. in 1993 with the 1988 figures for comparison. Ranking order will be similar for

Table 4.3   Most important chemicals in the U.S.A. (thousands of tonnes)

Position (1993)	Chemical	1993	1988
1	Sulphuric acid	36 422	38 803
2	Nitrogen	29 610	23 628
3	Oxygen	21 098	16 821
4	Ethylene (ethene)	18 707	16 581
5	Lime	16 689	14 667
6	Ammonia	15 646	15 370
7	Sodium hydroxide	11 660	10 871
8	Chlorine	10 912	10 277
9	Methy *t*-butyl ether	10 907	—
10	Phosphoric acid	10 449	10 626
11	Propylene (propene)	10 159	9 057
12	Sodium carbonate	8 980	8 662
13	Ethylene dichloride (1,2-dichloroethane)	8 141	6 190
14	Nitric acid	7 741	7 157
15	Ammonium nitrate	7 615	6 522
16	Urea (carbamide)	7 102	7 147
17	Vinyl chloride (chloroethene)	6 236	4 109
18	Benzene	5 587	5 370
19	Ethylbenzene	5 333	4 508
20	Carbon dioxide	4 848	4 254

other major chemical-producing countries. As in 1988 inorganic chemicals occupy eight of the ten places.

With the notable exception of sulphuric acid, the most important chemical of all (and to a much smaller degree phosphoric acid and urea), which suffered a significant decrease, all the other chemicals achieved an increase in 1993 compared with 1988. The most spectacular increase goes to methyl *t*-butyl ether (MTBE) which was way down the list in 1988 but is now 9th and climbing. Demand for this is increasing enormously because it is used instead of tetraethyl lead to increase the octane rating of unleaded petrol.

**4.3
The United Kingdom chemical industry**

The origins and development of the U.K. chemical industry have already been dealt with in the early parts of this chapter. This section will therefore be devoted to a comparision with other U.K. manufacturing industries and with other countries' chemical industries. Major manufacturing locations will be considered, as will each of the major companies. This will necessarily include quite a lot of statistical data and in view of the difficulties in obtaining figures which are firstly up-to-date and secondly are truly comparable (particularly international sales figures which require currency conversion), it is better to regard them as relative rather than absolute. In any case it is the analysis of the figures that is important to us rather than individual statistics.

Two publications are extremely valuable sources of information. Firstly, the leaflet *U.K. Chemical Industry Facts* which is published annually by (and is obtained from) the Chemical Industries Association. Although small it contains a wealth of statistical and graphical information which has been obtained from many sources. A number of tables and graphs from the July 1994 issue are published, with permission, on the following pages. Secondly, another annual publication, *Chemicals Information Handbook* from the Shell International Chemical Company Ltd., London, is more international in outlook and also deals with important individual chemicals which are derived from crude oil. This is not surprising since the Royal Dutch Shell companies' major activities are exploration, extraction, and refining of crude oil plus chemical processing of suitable fractions.

4.3.1 *Comparison with other U.K. manufacturing industries*

(i) *Growth rates*. Figure 4.2 shows the growth rates for various industries over the period 1984–1994. It is readily apparent that the chemical industry expanded much more rapidly than manufacturing industry generally—in fact almost twice as fast. As will become clear later, this is generally true throughout the world. Indeed chemicals growth rate exceeded most others in this list over the period 1984–1994.

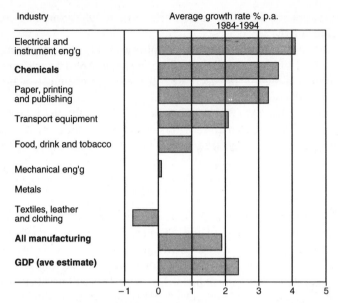

Figure 4.2   U.K. industrial growth rate comparisons 1984–1994.

(ii) *Total sales*. In 1994 the value of exports of chemicals exceeded that of imports by £4400 million. Royalty payments received in 1993 totalled £963 million whereas payments made came to only £208 million. For the U.K. manufacturing industry as a whole royalty payments exceeded income.

Table 4.4   Total sales (gross output) of U.K. chemical industry

Year	Sales (£ millions, current prices)
1970	3 448
1980	13 661
1991	28 309
1992	29 465
1993(e)	32 600
1994(e)	35 500

Table 4.5   Number of employees in the U.K. chemical industry

Year	Thousands
1972	426
1980	402
1988	346
1992	310
1993	303
1994	284

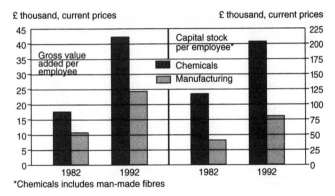

Figure 4.3 Gross value added and capital stock per employee comparisons, U.K.

(iii) *Number of employees.* Steadily increasing output has been achieved with a steadily declining labour force (Table 4.5). Although it employs only 7% of the manpower, it is responsible for 10% of the value of output of all manufacturing industry.

Linking employees and sales leads to

(iv) *Output per employee.* Figure 4.3 shows in value terms the much higher output by each employee in chemicals compared to all manufacturing. A significant part of this difference is attributed to the greater capital investment in the chemical industry—as Figure 4.3 shows.

(v) *Capital stock per employee.* See Figure 4.3.

(vi) *Safety.* The chemical industry operates many hundreds of potentially very hazardous processes—particularly those which require extremely high pressures of several hundred atmospheres. Although a zero level is the only acceptable accident level, it is testimony to the good practices employed that serious accidents are relatively rare. However, it is the occasional disaster which attracts the headlines in the newspapers and on television. The most serious such disasters of the recent past involved the caprolactam (the intermediate for Nylon-6 production) plant owned by Nypro (U.K.) Ltd., at Flixborough, in 1974 (see section 1.6.1), and also the tragedy at Hickson Welch's Castleford works in 1992 (see section 1.6.5).

However, let us place this in its proper context by looking at the industry's performance in two ways. Firstly, let us consider the incidence of reported accidents for chemicals versus all manufacturing industry (Figure 4.4).

The figures show a commendable trend of a decrease in the number of accidents and the 1993–1994 figure of 1.20 fatal and major injuries per thousand employees in the chemical industry is better than the equivalent figure of 1.25 in manufacturing industry in general.

Secondly, let us compare the chemical industry to some specific high-risk

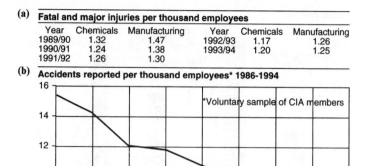

(a) **Fatal and major injuries per thousand employees**

Year	Chemicals	Manufacturing	Year	Chemicals	Manufacturing
1989/90	1.32	1.47	1992/93	1.17	1.26
1990/91	1.24	1.38	1993/94	1.20	1.25
1991/92	1.26	1.30			

(b) **Accidents reported per thousand employees* 1986–1994**

*Voluntary sample of CIA members

Figure 4.4   (a) Fatal and major injuries per thousand employees. (b) Accidents reported per thousand employees among CIA members (voluntary sample).

Table 4.6   Risk of death for various activities

Industry/activity	Deaths per thousand in a 40-year working life
chemical	3·5
all industry	4·0
fishing	35·0
coal-mining	40·0
staying at home	1·0
travelling (a) by train	5·0
(b) by automobile	57·0
(c) by motor-cycle	660·0

industries, and also everyday activities, by showing the risk of death. These figures are calculated ones, and are not as reliable as those above, but are certainly acceptable in relative terms (Table 4.6). They were generated in 1975. Comparing these everyday activities certainly shows that the risk of death from working in the chemical industry is far lower than the qualitative impression which the public obtains from the media.

(vii) *Cost and price indices.* The chemical industry has been quite successful at keeping down the increases in selling or output prices—certainly this has been well below the general rises in selling prices of goods (details are given in Figure 4.5).

4.3.2 *International comparisons in the chemical industry*

(i)   *Growth rates.* It has already been shown that in the U.K. the chemical industry has grown, since 1984, at approximately twice the rate of all

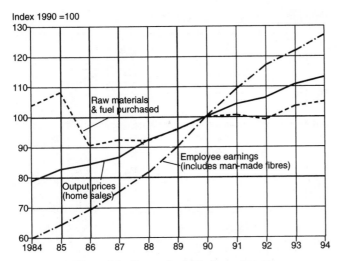

Figure 4.5   Cost and price indices, 1984–1994.

Table 4.7   International growth rate comparisons 1984–1994

| Country | Average growth % p.a. | | Ratio of chemicals to all industry |
	Chemicals	Manufacturing industry	
U.K.	3·6	1·9	1·9
Germany (11 BL)	2·1	2·0	1·1
France	3·5	1·1	3·2
Italy	2·0	1·8	1·1
Total EEC	2·9	1·7	1·7
U.S.A.	3·1	3·0	1·0
Japan	4·0	1·8	2·2

manufacturing industry. Table 4.7 gives the corresponding figures for all the major chemical-producing countries. The general pattern of chemicals growing at a much faster rate than industry in general is very clear with France providing the most striking example.

(ii) *Chemical sales*. Figure 4.6 shows the value of chemical sales in 1994. Note firstly that there is a shortfall in certain figures because they do not include sales of man-made fibres. Secondly, all countries have achieved substantial increases in sales between 1988 and 1994. The U.S. industry dominates the scene and its sales were approximately the same as the combined efforts of the Western European or EEC (European Economic Community) countries.

Most countries increased sales by a factor of 1.0 to 1.4 times between 1988 and 1994. Although Japan's sales were approximately 10% higher than

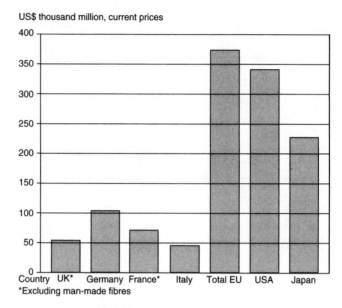

US$ thousand million, current prices

Figure 4.6   Chemical sales, international comparisons.

Germany's in 1988, by 1994 the differential had expanded to almost 225%. As we will see later, when looking at the world's major chemical companies, this has been achieved by a chemical industry whose structure is very different to that of most other countries.

Table 4.8   U.K. external trade in chemicals

Annual growth rate in volume of chemical exports, 1984–1994	5·8%
Annual growth rate in volume of chemical imports, 1984–1994	7·5%
Chemical exports as a percentage of all U.K. visible exports, 1994	14·0%
Chemical imports as a percentage of all U.K. visible imports, 1994	9·8%

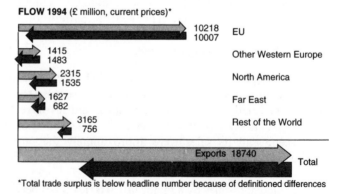

Figure 4.7   U.K. export and import flows, 1994.

Table 4.9 Total new capital invested by U.K. and EEC chemical industry 1983–1994 plus sales proportions

Year	U.K. capital investment (current prices, £m)	U.K. capital investment as % of total EEC	U.K. sales as % of total EEC
1983	1060	19·0	14·5
1987	1520	15·5	13·5
1990	2100	14·8	13·3
1993	1920	14·2	14·5
1994	1880	14·3	14·3

(iii) *U.K. external trade in chemicals.* Table 4.8 and Figure 4.7 provide data on the U.K.'s international trading in chemicals.

Trade with fellow EEC countries accounts for almost half the total but note the negative trade balance here. In contrast there is a positive trade balance with all the remaining areas and this leads to an overall favourable balance.

(iv) *Capital investment by EEC countries.* Attention has already been drawn to the highly capital-intensive nature of the chemical industry, particularly the petrochemicals sector, and this important point will again be emphasized later in this chapter. Table 4.9 supports this with some actual figures for the U.K. industry. It also expresses those as a percentage of all EEC capital investment and shows our percentage of total sales. This concludes our comparisons of the U.K. chemical industry with other U.K. manufacturing industries, and also with the chemical industry in other countries.

However, to complete our picture of the U.K. chemical industry the major manufacturing locations, plus a brief indication of the major U.K. chemical companies and their interests, will be considered.

4.3.3 Major locations of the U.K. chemical industry

The locations in which the chemical industry has grown and expanded have been governed by several factors. During the very early days the manufacturing location was dictated by the source of the raw materials plus the proximity of major users, since transportation of bulk chemicals was extremely difficult; at this time railways had only recently begun to develop. Thus the alkali industry grew up around the river Mersey in Lancashire and Cheshire, where large deposits of salt occur. Close by were the major users of alkali, the textile and oils and fats (particularly soap-making) industries. These factors are equally important now for chemicals which are produced in large quantities. Our best-known examples are again petrochemicals, and the large complexes for their manufacture have grown up next to the oil refineries which are the source of their raw materials. Also, processing plants to use their output, e.g. polymerization units, have been constructed either as part of the same complex or else nearby. This has also meant that transport facilities for products have been good. The complexes have invariably developed on river estuaries, since the large tankers bringing in crude oil need to get close to the refineries to discharge their very large cargo. Hence bulk chemicals transport by sea is readily available. Also, the road and rail network is good and well-developed.

There are several additional advantageous factors in this arrangement. Firstly the chemical industry, as we have already seen, is a very big energy user. Much of this is oil-derived, so that a location near to the refinery eases transportation of large volumes of fluids. Secondly, being located on rivers means that the vast quantities of cooling water which are needed by the chemical plants are readily available. Rivers, flowing a short distance into the sea, are useful for the discharge of treated effluent.

Almost all the areas in the U.K. where the chemical industry is concentrated are situated on river estuaries and have a very strong petrochemical link. Examples are Fawley (near Southampton), the Thames Estuary, Baglan Bay (south-west Wales), Merseyside, Humberside, Teesside and Grangemouth (near Edinburgh). Plans are also well advanced for petrochemical developments on the east coast of Scotland, well north of Aberdeen, where oil and gas from the North Sea oilfields are pumped ashore.

Avonmouth (near Bristol) is one of the few concentrations which does not have a strong petrochemical base. It is noted for fertilizer manufacture and other inorganic chemicals.

The above comments apply to the bulk of the chemical industry's operations. However, there are a few exceptions, and these are all concerned with products which are manufactured in small quantities, typically 10–100 tonnes per annum, but have a very high value per unit weight. For some products this might mean tens of thousands of pounds per tonne. Agrochemicals and pharmaceuticals are the best-known products of this type. Due to their high value and small bulk, transportation (and its costs) is no problem, and manufacturing units may therefore be sited anywhere. Quite often this means that they are located in attractive rural areas and many miles from the coast. Examples are Barnard Castle (County Durham), Loughborough, Macclesfield, and Ulverston (Cumbria).

4.3.4 *Some major U.K. chemical companies*

Although all the major multinational chemical giants have manufacturing facilities in the U.K., either directly or via subsidiary companies, we shall confine our attention here to those which are incorporated, or have their headquarters, in the U.K. They are Imperial Chemical Industries (ICI), Zeneca, Shell Chemicals, B.P. Chemicals, Croda International, Smith Kline Beecham, Glaxo, Fisons, Albright and Wilson, Laporte, Rio Tinto Zinc, and Unilever. Let us now briefly consider each of these in turn.

ICI. For many years this has been the U.K.'s largest chemical company and indeed was always placed in the top five in the world. In 1993 the decision was taken to split the organization into two separate companies, ICI and Zeneca.

ICI retained the bulk chemical businesses such as petrochemicals, industrial chemicals and also materials and paints. Profits in 1993 were £290 million from sales of £7.6 billion. This still made it the fifth largest chemical producer in the world, but down from the third position which it occupied prior to Zeneca splitting off. Although it has manufacturing plants in many locations, its main plants are found on Merseyside and Teesside.

Zeneca was formed in 1993 from the agrochemical, pharmaceutical and speciality chemicals parts of ICI. It therefore concentrates its activities on fine chemicals, particularly those that are biologically active. Note that it also encompasses the areas which are the most research intensive. Its main manufacturing locations include Fernhurst, Grangemouth, Huddersfield and Macclesfield.

Shell Chemicals. The Royal Dutch/Shell company is known best as one of the 'seven sisters' that dominate the oil industry. Developments into chemicals led to the setting up of the Shell Chemical Company in 1929. As one might expect, its major interests are petrochemicals and derived products. One of the latter, the liquid detergent Teepol, was the first petroleum-based chemical to be produced in Western Europe, in 1942. Their product range now covers several hundred chemicals, largely industrial organics, but also polymers (plastics, resins, synthetic rubbers). A small but growing area is speciality products, e.g. speciality rubbers and plastics. After the difficult period at the beginning of the 1980s, petrochemicals production is much healthier. Major locations are to be found at Carrington, Stanlow, Shell Haven and Mossmorran.

B.P. Chemicals. Like Shell, it is part of an oil giant—the British Petroleum Company. The move into chemicals manufacture in the 1950s and 60s was achieved by setting up several joint companies with established chemical manufacturers, and particularly with the Distillers Co. Ltd. This culminated with the takeover of almost all of Distillers' chemical interests in 1967, including, logically, the Chemicals and Plastics Divisions. Incidentally, Distillers' main interest, as their name suggests, is production of whisky. Like Shell its major interests are petrochemicals and derived products e.g. plastics. Major manufacturing locations include Hull, Grangemouth and Baglan Bay.

Croda International. This group of companies started with the manufacture of lanolin from wool greases, and oleochemicals, i.e. glycerides, fatty acids and compounds derived from them, remain an area of major importance. Other areas of interest include coal-tar chemicals; heterocyclic compounds; polymers used in adhesives, inks, paints and resins; and foodstuffs. Manufacturing locations are centred on Humberside, Wolverhampton and Leek in Staffordshire.

Smith Kline Beecham. This group of companies produces proprietary and ethical pharmaceuticals, and was formed by a merger of Beechams with the U.S. company Smith Kline French. It also has extensive interests in related areas such as food, health beverages, soft drinks, confectionery, and toiletries. Many of its products are well known to the general public by their brand names, including Beecham's pills and powders (for counteracting the effects of colds, headaches, etc.) on which the company was founded in 1842. In terms of chemical achievement their development and introduction in 1963 of the first semi-synthetic penicillin antibiotics was notable. They have developed quite a number of these over the years, and they represent a substantial part of the company's earnings. Manufacturing tends to be concentrated in and around London (and also in the U.S.A.).

Glaxo. This group is in many ways similar to the Beecham group, and in fact the latter made a takeover bid for Glaxo in 1971, but this was turned down by the Monopolies Commission. Since that time the company has gone from strength to strength with rapid growth and good profitability and it is now the second largest pharmaceutical company in the world. This is due in considerable measure to a big research effort leading to new products coming through and onto the market. Examples are the steroid Betnovate (used in treating skin conditions and some allergies), cephalosporin antibiotics and Zantac (an anti-ulcer drug) which has become the world's top selling drug. Many vaccines and veterinary products are also produced, and baby and health foods are also important group products. Manufacturing plants include Annan, Ulverston, Barnard Castle, and Greenford. In 1995 government approval was given for them to take over Wellcome.

Fisons. Ironically this company was founded on fertilizer manufacture and for many years this was a strength of the company. However, losses resulted in the whole fertilizer division being sold to Norsk-Hydro in 1982. This helped profits for the company to rise from £9.3 million in 1981 to £21.1 million in 1982. Its major interests are pharmaceuticals—where its most important success has been the anti-asthmatic drug Intal—and agrochemicals. Diversification into making scientific apparatus and laboratory chemicals, and horticultural materials like composts and peat, were tried, but in the 1990s these were sold to leave the company to concentrate on pharmaceuticals. In 1995 two thirds of its pharmaceutical R&D activities were sold to Astra Pharmaceuticals.

The company's main location is at Loughborough.

Albright and Wilson. This company started by making phosphorus for matches in 1844. In the late 1960s a decision was taken to relocate phosphorus manufacture in a new plant in Newfoundland. However this was plagued with technical and pollution problems, which had such a serious effect on the company's finances that major help from the U.S. company Tenneco was needed and as a result it has become a subsidiary of the U.S. company. Its predominant interests are built on phosphorus and silicon chemistry but include other areas as well. Examples are detergents, shampoos, plasticizers, silicones, and flavours and essences.

Locations include the London area and Whitehaven.

Laporte. This company started by producing hydrogen peroxide in Yorkshire in 1888 for the textile industry. Hydrogen peroxide, other peroxides, and perborates continue to be a major interest today. They also produce a variety of predominantly inorganic chemicals, e.g. titanium dioxide, fuller's earth.

Major manufacturing plant is located at Widnes near Liverpool.

Rio Tinto Zinc. This is primarily a mining company and its chemical interests relate largely to the ores which it mines. It is therefore concerned with inorganic compounds of metals like copper, aluminium, iron, lead and zinc. The takeover of Borax Consolidated in 1968 took it into the area of boron compounds.

Avonmouth is the location of its lead and zinc smelting plant.

Unilever. Like Shell this is a joint U.K./Dutch company. It has major interests in the food industry (ice-cream, sausages, frozen foods, margarine), many of its subsiduaries being household names, and soaps and detergents. Its raw materials are largely animal and vegetable oils and fats and these are extracted and processed largely at its many works on Merseyside.

4.4
The U.S. chemical industry

The U.S. industry is by far the largest national chemical industry, being more than twice the size of its nearest rival Japan and approximately the same size as that of all the EEC countries combined. It has the advantage of a very large home market, and tariff barriers plus its physical remoteness from its major competitors combine to moderate external competition. In fact these restrictions virtually mean that any foreign company wishing to achieve a reasonable market penetration in the U.S. has to set up manufacturing facilities there. Another traditional advantage of U.S. chemical firms, certainly compared with their European counterparts, has been the ready availability of cheap raw materials, particularly natural gas and oil. This has been beneficial twice over. Firstly it has meant energy has been relatively cheap, and as we have already noted the chemical industry is a big energy user. Secondly, raw materials for bulk chemicals and intermediates such as petrochemicals and ammonia are cheap. However these advantages are gradually being eroded as natural gas supplies dwindle in the U.S.A. at the time that countries like the U.K. are making large discoveries of oil and natural gas around their shores. At the present time the U.S.A. still retains an advantage.

With their very solid home and financial base, U.S. companies have not found exporting to be too difficult despite the geographical remoteness of the major markets. They have tended to set up manufacturing units and subsidiary companies in Europe, and particularly in the U.K. In 1993 U.S. chemical exports were valued at $45.07 billion and with imports at $29.17 billion there was a positive trade balance of $15.90 billion.

Several aspects of the U.S. chemical scene have been commented on in the earlier section of this chapter concerned with international comparisons with the U.K. chemical industry (section 4.3.2). These were items such as sales, growth rates and trade with the U.K. The remainder of this chapter will therefore be devoted to a brief consideration of some of the larger U.S. chemical companies. Bear in mind that, as with any very large chemical company, they are all multinationals, and their activities cover practically all the important sectors of chemicals manufacture. Comments on their interests are therefore confined to just a few areas that the company is particularly strong in, or noted for. Another point to remember is that, partly due to the physical size of the country, important manufacturing locations are numerous and, unlike the U.K., it is not reasonable to try and list them.

Du Pont. This has been the largest American chemical company for a number of years, and is noted for its high level of expenditure on research and development. It is the world's largest producer of man-made fibres. Its Teflon (polytetrafluoroethylene) is well known to the general public as the coating for non-stick cooking utensils. Since its takeover of the giant Conoco oil company in 1981, which has doubled total sales, the whole nature of the company has altered.

Dow Chemical. Another giant of the industry which produces a very wide range of both inorganic and organic chemicals. Basic inorganic and organic chemicals plus plastics account for 70% of sales, and pesticides and pharmaceuticals account for a further 10%. It is well known for its chlorinated products, such as herbicides like 2, 4-D and the much discussed 2, 4, 5-T, which was marketed under the trade name Agent Orange. (2, 4-D and 2, 4, 5-T are 2, 4-dichloro- and 2, 4, 5-trichlorophenoxyacetic acid.) Approximately half of total sales are made outside the U.S.A. It has many associated and subsidiary companies, and major manufacturing sites in Europe at Stade in West Germany and Terneuzen in the Netherlands. Rapid expansion in recent years has more than doubled its sales since 1976.

Union Carbide. As its name suggests, this company started by making calcium carbide. It ranks third in the U.S.A. for chemical sales, but is a diverse company with only about 40% of its revenue coming from chemical sales. Areas of particular interest are agricultural chemicals, industrial gases, and plastics. Although it has many overseas subsidiaries, it sold most of its European interests to BP in 1978.

Exxon. This is part of the giant Exxon Corporation, which is perhaps still better known by its former name of Standard Oil of New Jersey in the U.S.A.

In the U.K. the company is well known for its petrol sales under the Esso name. As one might expect, therefore, its interests are primarily in petrochemicals and derived products. The majority of its sales are outside the U.S.A.

Monsanto. Another large company with a diversity of interests. It is known to the general public through its acrilan (polyacrylic) wear-dated fabrics. The company has experienced severe difficulties with its activities in the synthetic polymer field, and as a result it has abandoned nylon, polyester fibres and polystyrene in Europe during the last few years. Nevertheless one-third of its income is generated overseas. It also withdrew from the polyester fibre market in the U.S.A. in 1980. It has shown its interest in biotechnology by putting almost £5 million into the £9·5 million Advent Eurofund. This capital will be used to back new and established companies involved particularly in the biotechnology and genetics areas by acquiring a minority holding for between £100 000 and £500 000. Its pesticide 'Roundup' is a very important revenue earner.

Allied Chemicals. In order to reflect better its wide range of interests the company has dropped the word 'chemical' from its title, becoming just Allied Corporation. It was originally (in the 1920s) a major alkali and dyestuffs manufacturer. Currently its interests include basic inorganic and organic chemicals, plastics, fibres, fertilizers and pesticides. A very active and growing involvement in the energy field has been a feature of recent years and this area now accounts for one quarter of total sales. Diversification continues with the acquisition of a scientific instruments company.

American Cyanamid. Although its main post-war growth has been in pharmaceuticals it has interests also in pesticides and fertilizers, speciality chemicals (e.g. acrylic fibres and pigments) and consumer products (e.g. laminates). It is noted for its part (with Pfizer) in the introduction of the tetracycline antibiotics. Overseas sales account for one-third of total sales.

Hoechst Celanese. This is one of the world's largest producers of cellulosic and synthetic fibres, which represent more than half of total company sales. Other interests are in a variety of organic chemicals.

Pfizer Inc. This is a more specialized company which concentrates on pharmaceuticals. Its specialities are fine chemicals and fermentation products, particularly penicillin and streptomycin antibiotics.

The U.S. chemical industry and the country in general tend to show a greater entrepreneurial flair, and are more willing to back risky projects, than their European counterparts. A good example is the field of biotechnology. Several companies, such as Biotech and Celltech, have been set up and share issues in them were oversubscribed many, many times—despite the fact that success, although potentially extremely rewarding, lies many years ahead owing to the formidable practical problems to be overcome.

4.5
Other chemical industries

From the international comparisons of the chemical industry made earlier in this chapter, it can be seen that apart from the U.K. and the U.S.A., the other major producers in order of importance are Japan, Germany, France Italy and the Netherlands. Several aspects of the chemical industry in these countries are already apparent from the figures and tables which were used in these comparisons.

4.5.1 *Japan*

Although Japan's chemical industry is second only to that of the U.S.A., most of its production is for home use. Even the Netherlands, whose chemicals production is far below that of Japan, outranks it both for exports and imports. Another feature of the Japanese chemical industry is its organizational structure. Although it took some time for a Japanese company to feature in the top twenty chemical companies, Table 4.10 shows that four are now included, and all of these improved their positions in 1993. However, there are still none amongst the top 10. Contrast this with Germany where 4 companies are also listed, but with 3 of these in the top 4 and they dominate the German chemical industry. The Japanese chemical industry must therefore consist of a large number of medium- and small-sized companies.

4.5.2 *Germany*

Like Japan, Germany has no indigenous oil or natural gas, and the performance of its chemical industries is therefore even better than appears at first sight. The German industry is dominated by the three giants Hoechst, BASF, and Bayer who in 1993 occupied first, second and fourth places in the world's top 20 chemical companies. As might be expected they are multinationals in every sense.

4.5.3 *France*

As we have already seen, the French chemical industry is similar in size to that of the U.K. Its two major companies are Rhône–Poulenc and the oil company Elf–Aquitaine, which has extensive chemical interests.

4.5.4 *Italy*

The Italian industry is approximately two-thirds of the size of the U.K. industry. It has only one really big chemical company—Enimont—which was formed at the beginning of 1989 by merging Enichem with 60% of Montedison.

Table 4.10   World's largest chemical companies (by sales of chemicals only)

1993	Position (1992)	[1982]	Company	Country	1993 Sales ($ millions)
1	(2)	[2]	Hoechst	Germany	16 682
2	(1)	[4]	BASF	Germany	16 082
3	(5)	[1]	DuPont	U.S.	15 603
4	(4)	[3]	Bayer	Germany	15 012
5	(3)	[5]	ICI	U.K.	13 162
6	(6)	[6]	Dow Chemical	U.S.	12 524
7	(7)	[12]	Exxon	U.S.	10 024
8	(8)	[9]	Shell	U.K./Netherlands	9 883
9	(10)	[8]	Ciba–Geigy	Switzerland	9 072
10	(12)	[20]	Elf–Aquitaine	France	8 675
11	(13)	[—]	Asahi Chemical	Japan	8 557
12	(9)	[13]	Rhône–Poulenc	France	8 276
13	(14)	[14]	Akzo	Netherlands	7 032
14	(11)	[10]	Enimont	Italy	6 762
15	(17)	[18]	Mitsubishi Kasei	Japan	6 363
16	(20)	[—]	Sekusui Chemical	Japan	6 280
17	(16)	[—]	Veba	Germany	6 103
18	(19)	[11]	Monsanto	U.S.	5 651
19	(30)	[—]	Takeda Chemical	Japan	5 131
20	(27)	[—]	General Electric	U.S.	5 042

4.5.5 *Netherlands*

Although it is only about one-third of the size of the U.K. industry the Dutch industry boasted $1\frac{1}{2}$ of the top 20 companies in 1993, with Royal Dutch Shell in 8th position and AKZO in 13th position.

4.6
World's major chemical companies

Table 4.10[7] lists the world's top 20 companies based on their 1993 sales figures. The 1992 and 1982 figures are also given. (Caution must be exercised over the detailed positions since currency conversions were carried out in arriving at some of the figures—this may favour some companies to a certain extent and disadvantage others.) The conversions were carried out at the end of the appropriate year. Analysis shows that of these 20 companies, 5 are based in the U.S.A., 4 in Germany, 4 in Japan, 2 in France, $1\frac{1}{2}$ each in the U.K. and Netherlands, and 1 each in Italy and Switzerland.

4.7
General characteristics and future of the chemical industry

4.7.1 *General characteristics*

This section collects together important points which have previously been made in this chapter. Comments on the general characteristics of the chemical industry apply to a large extent to any individual national industry.

The first point to make is that, as described in Chapter 3, the industry is very research-intensive, the research and development effort being devoted to (i) development of entirely new products, (ii) better and more economical routes to existing products, and (iii) improving the efficiency of existing processes and developing new applications for existing procedures. In the 1950s and 1960s the accent was very much on (i), and many new polymers, drugs and pesticides were developed. As the need to reduce costs is now more acute than ever, research and development directed towards (iii) is becoming increasingly important. An example, already encountered, is the increasing effort being directed towards energy conservation.

The range of chemicals produced and of scale of operation is very wide. Despite this, the various processing plants have many similarities and have been designed on the basis of a number of common unit operations, examples of which are discussed in Chapter 7.

It is important to visualize a typically high-technology, very capital-intensive industry, where an individual plant and its control equipment can cost hundreds of millions of dollars. All large plants run on a continuous basis, and complete automatic control by computer or microprocessor is quite common. Some of these giants (mainly ethylene crackers) have capacities in excess of 500 000 tonnes per annum. Development of more cost-effective routes and advances in technology have quite often meant that existing plant and processes have rapidly become redundant. For example, development of the cumene route to phenol meant that the benzenesulphonic acid route became obsolete some years ago.

The industry plays a very important part in every highly industrialized country and, excepting Japan, a limited number of giant multinational chemical companies play a key role throughout the world. Even Japanese companies are now appreciating the necessity for globalization of their activities. All countries which are in the early stages of real industrialization are developing chemical industries, e.g. Saudia Arabia, Mexico. Be prepared also, over the next few decades, for China, with its massive home market, to become a major chemicals producer.

Rapid growth has been a dominant feature of the post-war chemical industry, with growth rates typically twice those of manufacturing industry generally. However, in countries where the industry may be regarded as mature, i.e. Western Europe and North America, growth rates are falling and are likely to be much lower during the next decade.

4.7.2 *The future*

Looking ahead towards the turn of the century it is logical to divide the world's chemical industries into two categories, as indicated above; the well-established, mature ones such as those in North America and Western Europe, and those in countries just starting, or continuing their industrialization.

Following the major recession at the beginning of the 1980s and 1990s, the world chemical industry is now more buoyant with a high level of demand for its products being accompanied by reasonable profits. This is the reward for the hard decisions which had to be taken during the recession to trim staff and rationalize businesses. For the latter the guiding principle was to strengthen the activities that you were good at and move out of activities that you were not so good at.

Product and plant swaps with BP chemicals in the early 1980s enabled ICI to concentrate on PVC production and BP to strengthen its low-density polyethylene activities. There are many other recent European examples of inter-company agreements, and other cost-cutting moves such as moving company headquarters out of major cities[8]. Globalization of activities by takeovers has been a recent trend. Thus the structure of the chemical industry is changing and may be significantly different by the end of the century. A study by Dow Chemical Europe on the likely situation in the year 2000 has been presented by Frank P. Popoff, its president[9]. They predict that as many as 10 of today's top 30 firms will not be around as we currently know them—due to mergers, acquisitions, rationalization and nationalization. Further it is suggested that only 10 of the top 20 chemical companies will then be from Europe and the U.S.A. At present 19 are from these areas. The move by ICI in the opposite direction by demerging into ICI and Zeneca in 1993 will continue to be followed with interest over the next few years.

New chemical industries, although small at present, are growing rapidly in the industrializing countries which have an abundance of the raw materials for chemical feedstocks, namely crude oil. Saudi Arabia, for example, through its Saudi Basic Industries Corporation, is pressing ahead with a massive downstream petrochemical development in the Al Jubail area[10]. Large oil discoveries in Mexico will mean it will also follow Saudi Arabia's lead in developing a major chemical industry of its own. Other rapidly developing nations in this sphere are Korea and China, which, with its massive home market, will be a major chemical producer by the end of the century. These new developments have been considerably aided by the fact that many petrochemical processes are now so well established that the plant and process technology can readily be purchased, almost off the shelf.

Due to the difficulty in competing with these developing countries in terms of raw material costs we can expect European, Japanese and U.S. companies to continue to move further downstream in petrochemicals, where they can use their greater technical skill, expertise, and experience. This is already evident in polymers, where the only profitable area is speciality polymers which require the above attributes for their production. Finally, the availability of feedstocks for organic chemicals as oil reserves diminish will also have an influence, but this is difficult to predict. The influence of the communist block is similarly difficult to predict but the construction of a pipeline to transport natural gas from Russia to Germany has been completed. Reports that Russia may become a major oil exporter (as its large reserves of natural gas are tapped) are being viewed with concern by OPEC. This apparent abundance of feedstocks will lead to Russia becoming a significant exporter of basic organic intermediates, alongside Mexico and Saudi Arabia.

As ever, the difficulty for the industry is to balance capacity with expected demand, although the latter is extremely difficult to predict even 5 years ahead because it is so strongly affected by the world's economy which is very cyclical with a boom inevitably followed by a slump. The diversity of experts' views clearly illustrates the difficulties. Views on ethylene in the mid 1990s range from predictions of significant over-capacity in Europe to a worldwide shortage. Clearly the actual outcome depends on how the market for this product expands (or contracts).

References

1. *The Liverpool Section of the Society of Chemical Industry* 1881–1981: *A Centennial History*, D. Broad, Society of Chemical Industry, 1983.
2. *Reigel's Handbook of Industrial Chemistry*, 8th edn, ed. James A. Kent. Van Nostrand Reinhold, 1983, p. 363.
3. *Annual Review of the Chemical Industry* 1979, by the Economic Commission for Europe (ECE/CHEM/34), Table 86.
4. Derived from: *C&EN*, 1978, (May 1), 33, and *C&EN*, 1982, (Dec. 20), 48.
5. *C&EN*, 1982, (Dec. 20), 50.
6. *C&EN*, 1983, (Feb. 14), 15.
7. Taken from *Chemical Insight*.
8. See for example ref. 4, 54, 55.
9. *C&EN*, 1982, (Oct. 11), 6.
10. *European Chemical News*, 1982, (Dec. 20/27), 16.

Bibliography

A History of the Modern British Chemical Industry, D. W. F. Hardie and J. Davidson Pratt, Pergamon Press, 1966.
The Chemical Economy, B. G. Reuben and M. L. Burstall, Longmans, 1973.
UK Chemicals Industry Facts, June 1995, Chemical Industries Association Ltd.
Chemicals Information Handbook, 1988, Shell International Chemical Company Ltd.
'World Chemical Outlook', *Chemical and Engineering News*, 1982, (Dec. 20), 45–67.

Note: The data for Tables 4.4, 4.5, 4.7, 4.8 and 4.9, and Figures 4.2 to 4.7 are taken from *U.K. Chemical Industry Facts*. June 1995, Chemical Industries Association Ltd., London.

Organizational Structures: A Story of Evolution

Jo McCloskey

5

5.1
Introduction

Organizational structure is designed to facilitate the implementation of the strategic objectives of the company. The structure that a company chooses will also determine how people behave in the workplace. A well designed organizational structure can act as the framework upon which economic success is based, while an ill-conceived one can impede or constrain corporate strategy. Hill and Jones[1] state that the purpose of organizational structure is twofold: to coordinate employees' activities so that they work together to most effectively implement a strategy that increases competitive advantage; and to motivate employees and to provide them with incentives to achieve superior efficiency, quality, innovation or customer responsiveness.

The structure of an organization is essentially a vehicle through which strategy is implemented and, as the corporate strategy of the company changes to respond to markets and environments, so too must the organizational design of the company alter.

This chapter will demonstrate the importance of evolving organizational structures to enable companies to meet their medium- and long-term objectives. It will also demonstrate how the chemical industry has recognized the importance of a dynamic and flexible organizational structures in sustaining a competitive edge in an increasingly demanding environment.

**5.2
The chemical industry in
the 1990s**

During the 1980s, the chemical industry generally pursued a strategy which was one of differentiation and diversification. In contrast, the strategy of the early 1990s has been characterized by streamlining: getting rid of unwanted businesses and cutting overhead costs.

As a result, most chemical corporations have, or are undergoing, restructuring or 're-engineering'. This latter term refers to a process which merges job functions, automates office procedures and makes more effective use of information. Chemical companies are also focusing on core businesses in order to consolidate and sustain competitive advantage. In general, they are now looking for innovative ways to cope within an environment of fluctuating economic cycles. In order to achieve these measures and reflect the changing strategic emphasis, the majority of chemical companies have had to undergo a re-structuring programme which, in the short term, has cut costs and reduced manpower.

The chemical industry in the 1990s has therefore been characterized by new organizational structures, not only to maximize opportunities within changing environments, but also to overcome fundamental structural problems that have developed during the last decade.

5.3
Why change organizational structures?

The role of organizational structure enables managers to coordinate the activities of the various corporate functions (or divisions) to fully exploit their skills and capabilities. For example, most chemical companies will wish to pursue a cost leadership strategy. To be successful it will need to operate within a structure that facilitates close coordination between the activities of research and development, manufacturing, and marketing to ensure that innovative products can be produced both reliably and cost-effectively. Managers must also ensure that mechanisms for sharing and transferring information, technical knowledge and expertise are in place to maximize synergistic opportunities. As chemical companies also pursue a global strategy, managers must create the right kind of organizational structure for managing the flow of resources and capabilities between domestic and foreign divisions.

The development of a structure that enables the company to achieve and sustain competitive advantage requires co-ordination, communications and cohesion if it is to allow the company to successfully pursue its objectives. However, the structure must also be flexible and organic; it must continually evolve as the organization responds to changing economic and market conditions. The differing strategies pursued by ICI during the 1980s and the 1990s necessitated a new structure of management which by the mid-1990s has enabled ICI to reinforce the competitive strengths of the group.

In the 1980s, ICI embarked on a programme in which it expanded its activities in bulk chemicals, explosives, fertilizers, paints, plastics and pharmaceuticals. By the end of the 1980s, ICI was the U.K.'s largest manufacturing enterprise and the world's fourth largest chemical company. However in 1990, bulk chemicals were still performing well while the speciality products were yielding small profit margins. ICI realized that it had become too diversified and any value added to the company by the speciality products was being swallowed up by the bureaucratic costs of sustaining many different businesses. In 1993 the company was split into two. ICI remained intact with industrial chemicals, materials, paints and explosives as its core businesses. The other part was called Zeneca, and it took control of ICI's bioscience businesses which included the drug, pesticide, seed and speciality chemical businesses. ICI then also developed a new management style and structure to facilitate the growth potential and other market opportunities available from the new strategy of focusing on core businesses.

5.4
Pre-structure decisions

There are a number of decisions that management must make in order to choose the structure that best fits the organization and will enable it to successfully implement and fulfil its medium and long-term objectives. Wilson and Rosenfield[2] argue that these decisions concern distinguishing between differentiation and integration. They describe two permutations of differentiation. The first is *vertical differentiation*, which addresses the issues of deciding the extent to which the organization is divided into specific levels of decision-making authority. *Horizontal differentiation* examines the extent to which overall tasks are performed in specialized units across the organization.

Differentiation is the way that a company allocates people and resources to organizational tasks in order to maximize market and economic opportunities. The greater the level of skill and specialization, the higher the level of differentiation.

Integration is the means by which a company seeks to coordinate people and functions to accomplish organizational tasks. An organizational structure must be such that it allows different functions and divisions to coordinate their activities to pursue a strategy effectively. Again, Wilson and Rosenfield[2] describe *vertical integration* as the extent to which there is co-ordination and control in the organizational hierarchy. *Horizontal integration* decides the extent to which there is coordination and control procedures across the different functions of the organization.

Both differentiation and integration are regarded from the vertical and horizontal perspectives, whereby 'vertical simply means viewing how much variation occurs at different levels hierarchically in the organization. Horizontal refers to variation across the organization at the same hierarchical level' (Wilson and Rosenfield[2]).

Differentiation concerns the division of authority; the aim is to specify the reporting relationships that link people, tasks and functions at all levels of a company. The organizational hierarchy establishes the authority structure from top to bottom of the organization. The *span of control* is defined as the number of subordinates a manager directly manages.

Integration is required at the operational and divisional levels of the organization. At divisional level, integration is crucial during planning, whereas at the operational level, the major thrust for integration is on implementation. Problems of communication and co-ordination can result in conflict, and where organizations are divided into several divisions, as the majority of chemical companies are, the problems of integration can be compounded. The more highly differentiated the organization is, the more integration mechanisms are required. Hill and Jones[1] state that the aim is to keep the structure as flat as possible, as problems occur when companies become too tall; communication is impeded and information becomes distorted.

5.5
Which type of structure?

There are several types of structure that companies can adopt, but this chapter will outline those that relate most closely to chemical industries today. The basic function of structure is to provide a means to process the division of labour, because all organizations must spread the amount of work that needs to be done. Some chain of authority then becomes necessary in order to ensure that each division of work is completed. This chain of authority not only allows certain individuals to delegate workloads, but also provides a network for communication. Greenley[3] states that organizations operate through the performance of people and the structure is the organization of human resources. An organization chart expresses the distribution of power because each function performed within the company needs a structure designed to allow it to develop its skills and become more specialized and productive.

5.5.1 *Functional structure*

This is the simplest form of structure and is normally used by small, entrepreneurial companies involved in producing one or a few related products for a specific market segment. Functional structures group people on the basis of their common expertise and experience, or because they use the same resources. The approach is to structure the company, or areas of responsibility that relate to traditional business functions such as marketing, manufacturing, engineering and finance.

Functional structures have several advantages. When people who perform similar tasks are grouped together, they can learn from one another. They can also monitor each other and ensure that each one is performing their tasks effectively and efficiently. As a result the work process becomes more efficient, reducing manufacturing costs and increasing operational flexibility. Managers have greater control of organizational activities because managing the business is much easier when different groups specialize in different organizational tasks and are managed separately.

There are also problems with functional structures because, in adopting such a structure, a company increases its level of personnel and horizontal differentiation to handle more complex tasks. If the company becomes geographically diverse, or it begins to produce a wide range of products, then control and coordination problems arise. Control becomes looser, lowering the company's ability to coordinate its activities and increasing bureaucratic costs.

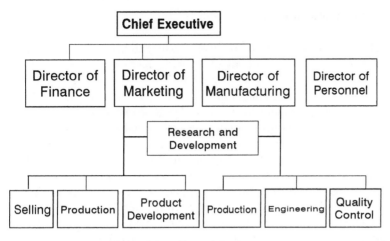

Figure 5.1 A functional structure.

As separate functional hierarchies evolve, functions become more remote from each other. It becomes increasingly difficult to communicate across functions and to coordinate activities. The various functions develop different orientations to the problems and issues facing the organization; they also have different time or goal orientations. For example, manufacturing see things in a short time frame and concentrate on achieving short-run goals such as reducing manufacturing costs. Others, like R&D, see things from a long-term perspective and their goals (innovation) and product development) may have a time horizon of several years. These factors may cause each function to develop a different view of the strategic issues facing the organization. Manufacturing may see a problem as the need to reduce costs; sales may see it as a need to increase customer responsiveness, and R&D may see it as a need to create new products!

Another problem that can occur is that, as the number of products proliferate, a company will find it difficult to gauge the contribution of each product, or group of products, to the overall profitability of the company. The company may be producing products that are unprofitable without realizing it. The situation is exacerbated if the company is producing or selling in different locations because the centralized system of control provided by the functional structure may not be flexible enough to respond to the needs of different regions or countries.

Experiencing this level of problem is a sign that the company needs to change its mix of horizontal and vertical differentiation in order to allow it to perform the organizational tasks that will enhance its competitive advantage. Essentially, the company has outgrown its organizational structure and it will need to develop a more complex structure which can meet the needs of its competitive strategy. Many companies chose to move to a matrix structure.

5.5.2 *Matrix structure*

Mintzberg[4] claims that innovative organizations cannot survive on standardized, functional bureaucracies with their clear divisions of labour and emphasis on planning and control. They must remain flexible with informal lines of communication if they are to be truly innovative and creative. Despite organizing around market-based projects, the organization must support and encourage some types of specialized expertise.

Matrix structures offer a combination of a functional and a divisional structure. In a matrix, experts are grouped together in functional units, such as production or research, and communications, to establish and facilitate a basic operational infrastructure. The experts are also deployed into project teams within which they can carry on with the development of innovation. To encourage coordination, the organization makes use of liaison devices to integrate the various teams and task forces. Managers tend to spend a great deal of their time liaising and coordinating the work laterally among the project teams. Matrix structures are flat, with few hierarchical levels and employees have two bosses; a functional boss who is head of the function and the project boss who is responsible for managing individual projects. Employees work on a project team with specialists from other functions.

Matrix structures can be successfully utilized by those divisions within chemical corporations who are developing radically new products such as CFC replacements. They are also effective where the economic competitive environment fluctuates making speed of product development and launch crucial. This type of structure allows team members to control their own behaviour and encourages synergy across the team. As the project goes through different stages, different specialists can be brought in as required.

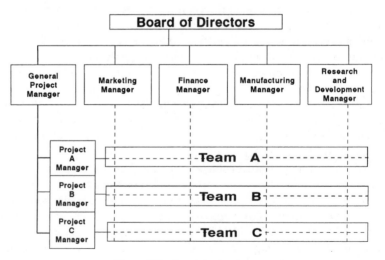

Figure 5.2 A matrix structure.

Mintzberg[4] outlines the major disadvantages to this structure which carries high bureaucratic costs in comparison to operating a functional structure. Employees tend to be highly skilled and salaries and overheads are high. The constant movement around the matrix means that the costs of building new team relationships and initiating a new project are also high. Sometimes conflict can arise due to overlapping authority and responsibility. Over time the project manager will assume the lead in planning and setting objectives, and the structure begins to operate like a divisional structure. The larger the organization, the more difficult it is to operate a matrix structure because the task and role relationships become very complex. When this occurs, the only option open to the organizations is to move on to a divisional structure.

5.5.3 *Multi-divisional structure*

Most multi-national corporations, including British Petroleum and ICI are now organized into a variety of divisional structures. In such a structure, the company as a whole may be split into several distinct divisions. Each distinct product line or business unit is placed in its own self-contained unit or division, with all support functions. The result is a higher level of horizontal differentiation.

This is an innovative structure which has similar advantages to a matrix structure but it is much easier and less costly to operate. Task activities are divided along product or project lines to reduce bureaucratic costs and to increase management's ability to monitor and control the manufacturing process. Functional specialists are organized into cross-functional teams on

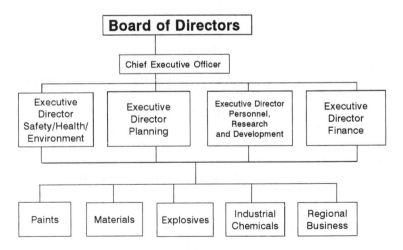

Figure 5.3   A product divisional structure.

a permanent basis and, as all functions have a direct input from the beginning, the process of innovation can be speeded up. A corporate headquarters is created to monitor divisional activities and to exercise financial control over each of the divisions. HQ staff contain corporate managers who oversee all divisional and functional activities, and it constitutes an additional level within the organizational hierarchy. Corporate headquarters staff, which includes members of the board of directors, as well as top executives, are responsible for overseeing the long-term plans and providing guidance for inter-divisional projects. This staff has strategic responsibility.

ICI adopted this kind of structure in the early 1990s when it embarked on a major re-shaping of its operations in order to increase profitability and achieve greater penetration of international markets. It now has five international groups of businesses: paints, materials, explosives, industrial chemicals and regional businesses.

In a divisional structure, each business operates as a self-contained business unit, each division possesses a full array of support services, although some services such as engineering and telecommunications are provided on a company-wide basis. For example, each has self-contained accounting, sales and personnel departments. Each division operates as a profit centre, making it much easier for corporate headquarters staff to monitor and evaluate each division's activities.

To illustrate divisional structure, ICI is headed by the Chief Executive who has four Executive Directors with overall strategic responsibility for operations, finance, personnel and planning. Each of the five businesses in ICI is headed by a chief executive (CE) who is accountable for its profits. The five CEs meet regularly with ICI's Chief Executive and his four Executive Directors to monitor performance and consider policies and objectives of the group.

The ICI board is supported in its international duties by headquarters based in London. In addition to its five international businesses, ICI has a number of regional businesses around the world, the strategy being to align them as closely as possible to the international businesses in order to reinforce the worldwide competitive strengths of the group. Where the regional businesses do not fit closely to the core businesses, ICI will harvest them. ICI, India disposed of its fertilizer and polyester fibres businesses, while ICI, Canada sold its Nitrogen Products and Agromart businesses, as these did not fit with the product portfolio that remained after ICI hived off its bioscience businesses to Zeneca.

However, during the same time period, ICI acquired a new ammonium nitrate plant in Australia, which will incorporate the latest technology and provide extra capacity in their explosives market. Similarly, they have recently completed a new tioxide plant in Malaysia which will become the hub of tioxide activities in Asia Pacific, thereby strengthening ICI's industrial chemicals business.

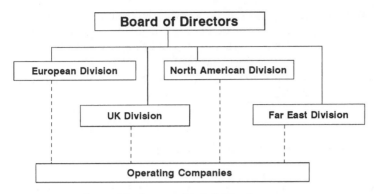

Figure 5.4 A geographical divisional structure.

Another variation of a divisional structure is one based on geographic divisions. Here, there is a recognition that different geographic areas will have different product and service requirements.

When a company operates as a geographical structure, organizational activities are grouped on a geographical basis. A company may divide its manufacturing operations and establish manufacturing plants in different regions of the country. This allows it to be responsive to the needs of regional customers and reduces transport costs. However, if a company diversifies into unrelated products or industries, the structure will not be able to cope and a move to a multi-business structure will be advised.

For historical reasons, Royal Dutch Shell's structure is organized on geographical divisions. Each country or region has its own companies and managers. These report back through layers of command to the corporate centre which is split between London and The Hague. At the top is the Committee of Managing Directors (CMD) which has four managing directors responsible for each of Shell's four regions and a number of businesses. These operations are supported by central companies providing legal, financial, information and other services.

The bureaucratic costs of operating a multi-divisional structure are high, the size of its corporate staff is a major expense and, as can be seen by the number of divestments and demergers currently operating within the top chemical corporations, are subject to continuing downsizing. Similarly, the use of product divisions, each with its own specialist support functions is a major expense. Constant performance review is essential to ensure that these costs are outweighed by increased profitability and enhanced reputation.

The benefits are that each division can adopt the structure that suits its needs best. For example an oil division could adopt a functional structure because its operations are standardized; the industrial chemicals could have a product team structure; and the plastics division could have a matrix structure.

Such a combination of self-contained divisions, with a centralized corporate management represents a higher level of vertical and horizontal differentiation. This can provide the extra control necessary to co-ordinate growth and diversification. When managed effectively at both the corporate and divisional levels, this organizational structure can increase profitability because it allows the organization to operate more complex kinds of corporate-level strategy.

Hill and Jones[1] have listed four major advantages in operating under a divisional structure:

(1) Enhanced corporate financial control

Because each division is its own profit centre, financial controls can be applied to each business on the basis of profit criteria. Typically, these controls involve establishing targets, monitoring performance on a regular basis, and selectively intervening when problems arise. Corporate headquarters is better placed to allocate funds among competing divisions. They can also identify the divisions in which investment would yield the greatest long-term returns.

(2) Enhanced strategic control

Corporate staff are released from operating responsibilities, giving them more time for contemplating wider strategic issues and for developing responses to environmental changes.

(3) Growth

It also enables the company to overcome an organizational limit to its growth. By reducing information overload at the centre, headquarters personnel can handle a greater number of businesses. They can consider opportunities for growth and diversification. Communications problems are reduced by applying accounting and financial control techniques as well as by implementing policies of 'management by exception', intervening only when problems arise.

(4) Stronger pursuit of internal efficiency

Within a functional structure, the interdependence of functional departments means that performance of functions cannot be measured by objective criteria. For example, the profitability of the finance function, marketing function or manufacturing function cannot be assessed in isolation, as they are only part of the whole. This means that a considerable amount of slack can go unnoticed and resources might be channelled into unproductive purposes. A divisional structure prescribes operating autonomy, therefore the division's efficiency can be directly observed and measured in terms of profit. Autonomy makes divisional managers accountable for their own performance and the general office is in a better position to identify inefficiencies.

Hill and Jones[1] counterbalance the advantages of the divisional structure by listing a number of disadvantages:

(1) Establishing a divisional–corporate authority relationship

There is a problem in deciding how much authority and control to assign to the operating divisions and how much to retain at corporate headquarters. This issue must be decided by each company in reference to the nature of the business and its corporate level strategies.

(2) Distortion of information

If corporate headquarters put too much emphasis on return on investment (ROI) (section 6.11.1) then the operating divisions will distort the information they send to top management. There could also be a temptation to maximize short-run profits (perhaps by cutting product development or marketing expenditures). On the other hand, if the divisional level exerts too much control, some of the more successful divisional managers may disguise performance in order to thwart attempts to use their profits to strengthen other divisions.

(3) Competition for resources

The rivalry between divisions competing for resources could prevent synergy gains from emerging and may lead to conflict and lack of co-ordination.

(4) Transfer pricing

Rivalry between divisions increases the problem of setting transfer prices between divisions. Each supplying division tries to set the highest price for its outputs in order to maximize its ROI. Such competition can undermine the corporate culture.

(5) Short-term research and development focus

If extremely high ROI targets are set by corporate HQ there is a danger that the divisions will cut back on research and development in order to improve the financial performance of the division. This will reduce the division's ability to innovate new products and lead to a decrease in long-term profits. Corporate HQ personnel need to carefully control their interactions with the divisions to ensure that both short- and long-term goals of the company are being achieved.

(6) Bureaucratic costs

Each division possesses its own specialized functions such as finance, research and development and multi-divisional structures which are expensive to run and manage. R&D is especially costly to run so some companies centralize such functions at the corporate level to serve all divisions. The duplication of specialist services is not a problem if the gains from keeping them separate outweigh the costs. Management must decide if duplication is financially justified. Activities are often centralized in times of economic downturn or recession, particularly advisory and planning functions. Divisions are, however, retained as profit centres.

The advantages of divisional structures must be balanced against their disadvantages, but the disadvantages can be managed by an observant, professional management team that is aware of the issues involved. The multi-divisional structure is the dominant one today (estimated to be the structure used by over 90% of all large U.S. corporations), which clearly demonstrates its usefulness as the means of managing the multi-business corporation.

5.6

Joint ventures and strategic alliances

Multi-divisional structures are often used to obtain synergies but, in line with the current global strategy of cost cutting and market expansion, there is a growing trend within the chemical industry to enter into joint ventures with other chemical companies. Usually each company takes a 50% joint stake in both ownership and operation of the business. DuPont and Dow Chemical have announced that they are planning to hive off their elastomer businesses into a joint venture in 1995. They have predicted that the new enterprise would produce annual sales of $1 bn which would rise to $5 bn within five years. The benefits to be obtained with a joint venture are shared research and development, technological knowledge and information. For example, Shell have signed agreements with Russia for the development of oilfields in the Salym region and with Kazakhstan for exploration activities. Shell will provide technological know-how and the local partners will provide knowledge of the host country's competitive and cultural environment and act as guides through the country's political and business systems.

Joint ventures are a useful strategy for entering new markets, particularly in countries where it would be difficult for companies to get permission to start up operations. For example, ICI opened a KLEA (HFA 134a) plant with Teijin in Japan where it was easier to set up operations with a Japanese partner, rather than if they had tried to set up on their own.

Chemical companies are seeking to expand their operations into emerging overseas markets, particularly in the Pacific Rim countries. While joint ventures are proving to be the most popular method of securing entry, there are a number of drawbacks. A company can lose control of its technology and it can be difficult to maintain quality processes if the host country demands a majority shareholding, usually based on a 51:49 split. Nonetheless, these types of business ventures can be cost effective and less risky, particularly in an unstable political and economic environment. Chemical companies will continue to make use of joint ventures and other forms of strategic alliances to enter new markets. How these enterprises will fit into the overall organizational structure in terms of monitoring and control will be a dominant question for the companies involved, as we proceed through the 1990s.

**5.7
Summary**

Companies can adopt a large number of structures to match the changes in their size and strategy over time. The structure that a company adopts will be the one whose logic of grouping activities best meets the needs of the business. It must choose a structure that encourages effective, flexible implementation of the company objectives and then make choices about levels in the hierarchy and degree of centralization and decentralization. It must select the appropriate level of integration to match its level of differentiation if it is to successfully coordinate its value creation activities. Since integration and differentiation are expensive, a company's goal is to economize on bureaucratic costs by adopting the simplest structure consistent with achieving its strategy.

Royal Dutch Shell demonstrates the need for corporations to change their structures as company strategies change to keep abreast with the increasingly competitive environments. In March 1995, it proposed a large re-structuring exercise that will divide the organization into five main businesses: exploration and production, oil products, chemicals, gas, and coal. These worldwide business units will be led by a committee of senior executives who will develop global and regional strategies as well as approve capital expenditure plans. The group's top four executives will not be affected by the changes. They will continue to review overall direction and policies and will be supported by corporate centres in The Hague and in London. The complex shareholding structure between the Dutch and British units of the company will not be affected, neither will Shell Oil, the group's U.S. arm.

As Shell has clearly indicated, an organization must choose the most appropriate form of differentiation to match its strategy; greater diversification requires that a company moves from a functional structure to a divisional structure. Differentiation is only the first organizational design decision to be made, the second concerns the level of integration necessary to make an organizational structure work effectively. Integration refers to the extent to which the organization seeks to coordinate its value–creation activities and make them interdependent.

Many chemical corporations worldwide have embarked on restructuring programmes during the 1990s. This policy has been implemented as a response to the changing strategic focus within the chemical industry. Throughout the 1980s many chemical companies acquired related businesses as a hedge against the cyclical commodity products of their core businesses. The trend was to buy into pharmaceuticals and other bioscience industries, with the common theme of spreading the risk by diversification.

However, as Peaff[5] pointed out, pharmaceuticals and biosciences are

R&D intensive and produce customer-specific products, whereas industrial chemicals are large-volume and capital-intensive businesses with a heavy emphasis on process technology. Given the two distinct cultures, it became impossible to establish sustainable competitive advantage and many chemical companies who could not realize the true potential have shed these businesses.

In the 1990s rising costs, increased competition and ever-changing environmental factors have caused the industry to focus on cutting costs and increasing productivity. The emphasis is now on growth through increased market share, low-cost production, and entering overseas markets, usually through joint ventures. Some chemical companies have focused on capacity expansion at existing plants in order to give themselves the flexibility to adjust their plans with changes in the economy. However, a number of companies, including Exxon Chemical, Phillips Petroleum and Shell Chemical, have committed to incremental expansions to go on-stream between now and 1997. On top of this, companies are continuing with their efforts to control costs and broaden the concept of competitiveness beyond cost.

These environmental and economic factors coupled with the chemical industry's efforts to re-focus and re-define its strategic direction has brought about a wholesale re-structuring process within the industry. Clarity of responsibility and accountability are the key issues when planning organizational design. Streamlined management structures, shorter lines of communication and measurable performance levels are paramount in facilitating quick responses to change brought about by fluctuating economic and market conditions.

A company structure changes as its strategy changes in a predictable way. However, as a company grows and diversifies, managers must try to keep the organization as flat as possible and follow the principle of the minimum chain of command, which states that an organization should choose a hierarchy with the minimum levels of authority necessary to achieve its strategy. The multi-divisional structure is proving to be the most flexible for implementing quality and reshaping management culture within the chemical industry today.

References

1. Hill, C. W. and Jones, G. R. (1995), *Strategic Management*, Houghton Mifflen.
2. Wilson, D. C. and Rosenfield, R. H. (1990), *Managing Organisations: Text, Readings and Cases*, McGraw-Hill.
3. Greenley, G. E. (1989), *Strategic Management*, Prentice Hall.
4. Mintzberg, H. (1990), *Mintzberg on Management*, Macmillan Publishers.
5. Peaff, G. (1994), Chemical Industry Debates Merits of Operating Pharmaceutical Units, *Chemical and Engineering News*, August 1st, pp. 19–23.
6. Chakravarthy, B. S. and Lorange, P., (1991), *Managing the Strategy Process: A Framework for a Multi-business Firm*, Prentice Hall.
7. Ainsworth, S. J. (1995), US Petrochemical Producers Anticipate Strong, More Stable 1995, *Chemical and Engineering News*, March 13th, pp. 15–23.

Technological Economics 6

Derek Bew

The basic function of any private company which aims to continue in business is to make a profit. Without this it cannot do other socially desirable things such as providing continuing employment, providing money for local social services through rates and local taxes and providing money for national activities through corporate taxes and through the taxes paid by the employees. In a planned economy, or a nationalized industry in the mixed economies currently operating in many countries, the profit motive is not the main imperative. Unprofitable activities may be continued for social or political reasons—to maintain employment in a depressed area or to maintain an industry to avoid dependence on imports in strategically important areas of manufacture. However, any such unprofitable operations represent a drain on the national resources and must be supported and subsidized by the taxes paid by profitable activities.

In the chemical industry, companies were founded, and continue to exist, to produce profits by making and selling chemicals. The products made are those for which a need has been identified and which can be produced for sale at a price which the consumer is willing to pay and which adequately rewards the producer. The processes which are chosen to manufacture the desired products are those which are believed to offer the best profit to the producer. They may not be the most chemically elegant routes to the products or the most scientifically interesting processes but in the context these factors are unimportant. In case it may be thought that the chemical producer operates his process to maximize his profits without concern for the effects on people or the environment it must be emphasized that, today, production of chemicals is governed by numerous regulations. Construction standards and operating safety of plant, the protection of the employees from dangerous levels of chemicals and the effect of effluents and products on the environment and people outside the plant boundaries are all subject to control by local and national legislation. These factors, particularly the control of atmospheric chemical levels within the plant and the safe disposal of gaseous and liquid effluents, can have a significant influence on the choice of process. They can also significantly increase both the capital cost of the plant and the cost of operating the process, as more equipment is required to meet increasingly stringent controls. The cost of operating a plant safely with minimum impact on the environment must ultimately be carried by the consumer in the higher product price needed to keep the producer in business. These aspects of chemicals manufacture are also considered in Chapter 9.

**6.2
Cost of producing a chemical**

The profit which the manufacturer obtains by producing and selling his chemicals can be measured in various ways as discussed later in the chapter. However, in order to know whether or not a profit is being made from an existing plant or product—or, in the case of a project at the planning stage, whether the new product would show an acceptable profit—we need to know the actual or estimated costs involved in producing the chemical. So let us consider how the cost of producing a chemical is built up. A considerable number of factors are involved in the production of a chemical, ranging from the supply and storage of the raw materials through to the storage and selling of the finished product. In between these steps is firstly all the complex and expensive equipment for carrying out the chemical processes, separating and purifying the product, and secondly the people who operate the plant and carry out maintenance work to keep the processes in operation. A simple way of combining information on these various cost factors into a useful economic model is the cost table. This presents information on an annual or 'per unit of product' basis and is useful for indicating the relative importance of the various factors which make up the cost. Table 6.1 shows an example for the production of cumene (isopropyl benzene) by reaction of propylene with benzene using excess benzene as solvent and phosphoric acid on a support as catalyst. Published information (*Hydrocarbon Processing* **55**, 3, March 1976, pp. 91–6) has been used in building up the table.

The reaction involves passage of propylene plus some propane diluent and an excess of benzene upwards over the supported acid catalyst at 230°C and 35 atm. Propylene conversion is high and the reactor product is flashed to recover propane plus a little propylene for recycle. Liquid product is then distilled to recover unreacted benzene, which is also recycled, and the crude product then distilled to give pure cumene and a small residue of fuel-value heavy ends.

For a process plant which is currently operating, the table can be drawn up using best data from plant records of operation under steady conditions and then used as a standard cost table to monitor subsequent operation of the unit. In the case of a new project being considered to manufacture a chemical by a new process under development, a comparable table can be drawn up using available information to indicate targets to be achieved to provide an economically attractive operation.

There are several ways of sub-dividing the components which make up the full cost of a chemical; the method used depends on the purpose for which the information is required. Traditionally accountants like to divide costs into variable and fixed elements and this split is useful when considering the costs of an integrated works, an individual production unit or a single product.

Table 6.1 Cost build-up for cumene production

Scale of Production 100 000 tonnes/year—operation at full capacity		
Operating costs	£000/*year*	£/*tonne Cumene*
Benzene 0·67 tonnes/tonne of product		
@ £310/tonne	20 770	208
Propylene 0·38 tonnes/tonne of product		
@ £305/tonne (propene)	11 590	116
Phosphoric acid catalyst + chemicals	140	1
Gross materials	32 500	325
Heavy end fuel credit 0·04 tonnes/tonne of		
product @ £55/tonne	(220)	(2)
Net materials	32 280	323
Energy inputs	710	7
*Plant fixed costs	1 150	12
Overhead charges	780	8
Depreciation (15 year life of fixed plant)	1 533	15
Works cost	36 453	365
Target return on total capital 10%	2 800	28
Required sales income	39 253	393
(net of packages and transport)		
Capital involved (early 1988 basis)		
Fixed plant	£23 million	
Working	£ 5 million	
Total	£28 million	

*Plant fixed costs are: operating and maintenance labour and supervision, maintenance materials, rates and insurance, and works overhead charges.

Variable costs comprise those factors which are only consumed (and therefore only charged to the operation) as product is being manufactured. As a result the total variable cost during an operating period—day, quarter, year—will vary directly as the plant output during that period, although the cost per unit of production will remain constant. Fixed costs are those charges which have to be paid at the same annual rate whatever the rate of production, in fact even if the plant is shut down for a short period for repairs, or any other cause, the fixed charges are still incurred. Since the total fixed costs are charged whatever the annual production from the plant, the cost per unit of output will increase significantly at output rates much below 100%. This problem will be discussed later in the chapter.

6.3
Variable costs

Let us now examine these cost components in more detail and look first at the variable costs. There are different views as to which items should be regarded as variable and which fixed, but the split described here is widely accepted.

Raw material costs
Energy input costs   } Variable cost elements
Royalty and licence payments   (total sum £000/year varies
                                 with plant output)

6.3.1 *Raw material costs*

Where a process plant is operating, the usages of feedstocks can be obtained by measurements of process flows during periods of steady operation. In the case of a new process at the research stage, or a project under development, raw material usages can be estimated from the process stoichiometry using assumed yields or yields obtained during process research experiments. In the latter case, if yields are based on analysis of the reactor product, allowance must be made for losses which occur during the product recovery and purification stages. Information on major chemical raw material prices is available in the techno-commercial literature (e.g. *European Chemical News, Manufacturing Chemist*) on a U.K. and European basis. Data on U.S. prices for a much wider range of materials is available (e.g. *Chemical Marketing Reporter*). In an industrial situation internal company data will usually be available for important feedstocks. Company technology strengths and strategic considerations can also influence the choice of feedstock and process route which will be selected to maximize commercial advantage. Thus a process which uses an internally available feedstock or a minor modification to an existing process could be preferred over a possibly better alternative using a purchased feedstock or requiring a major investment in process development work. The cost of catalysts and materials not directly involved in the process stoichiometry (solvents, acids or alkalis for pH adjustment, etc.) must also be included in the materials cost. Catalyst costs are based on loss of catalyst per cycle for homogeneous catalysts or cost of catalyst charge divided by output during charge life for a heterogeneous catalyst. Catalyst materials recovered from purge streams or by reprocessing a spent catalyst reduce the net catalyst cost. Solvent losses can be either physical (loss in off-gas streams, pump leaks) or chemical due to decomposition or slow reaction under process conditions. Such additional material consumptions are difficult to estimate in the absence of plant operating data but are usually relatively small cost items.

6.3.2 *Energy input costs*

This item in the cost table covers the multiple energy inputs necessary to carry out the chemical reaction and to separate the desired product(s) at the level of purity demanded by the market. Steam, at various pressure levels depending on the temperature required, is used to provide heat input to reactors and distillation column reboilers. Fuel oil or gas is used to provide higher temperature heat inputs either directly by heat exchange between process streams and hot flue gases, or indirectly by heating a circulating heat transfer medium. Electricity is used for motor drives for reactor agitators, pumps and gas circulators and compressors. Lighting of plant structures, tank farms and plant control rooms is an important, but usually minor, use of electricity. Cooling water is required to remove reaction exotherms and control reactor temperature, condense and cool still overhead streams and cool process streams. Inert gas—usually nitrogen—is used for purging equipment to provide an inert atmosphere over oxidizable or flammable materials. In the case of an operating process the consumption of energy inputs can again be obtained from process measurements during steady operation. For a process under development, estimates of major energy requirements can be made from the process energy balance and information on process conditions in reactors and operation trains. Price information on energy inputs is less readily available than raw material prices, although indicative figures do appear in chemical engineering journals. The energy inputs available (e.g. steam pressure available, use of fuel oil or gas) and the prices charged for them will vary from company to company and are very much site-dependent. A large integrated site could co-generate steam and electricity at high thermal efficiency in a site power station (see section 8.1.2). Use of waste streams (liquid or gaseous) could provide at least part of the fuel input, with the balance being purchased oil or gas. Steam would then be available at a number of pressure levels after staged let-down from boiler pressure through turbines to generate electricity. In contrast, a small site or isolated production unit would have to raise steam in a package boiler, using purchased fuel oil or gas, and buy electricity from the grid. The prices charged for energy inputs on the two sites would differ appreciably.

Massive rises in crude oil prices in the 1970s meant that hydrocarbon feed-stock costs and energy input costs became more important factors in process costs than previously. 'Energy conservation' became important and consider-able attention was paid to the integration of process energy requirements by using hot streams from one part of a process to provide heat in another area. This can often (but not always—see Lindhoff[1]) lead to increased plant capital requirements, and a balance must be struck between increased capital charges and lower energy charges. In a highly competitive industry, and also in the interests of environmental protection and resource conservation, efficient use of energy resources is important. Chapter 8 emphasizes the importance of energy requirements and conservation.

6.3.3 *Royalty/licence payments*

If the process being used for production (or to be used in the case of a planned project) is based on purchased technology rather than a process developed within the company, then a royalty or process licence fee will be incurred. This may be either a variable or fixed charge depending on the nature of the licence agreement. Usually a charge is made per unit of production and would appear in a cost sheet as a variable-cost item. Alternatively an annual licence fee related to the plant size (e.g. £1 million per 100 000 tonne installed capacity) will be charged and this would then appear as a fixed charge. The size of the royalty payment or licence fee is a matter of negotiation between the licensor and licensee and will be influenced by the nature of the technology, the advantages offered over competing processes and the number of alternative processes available for licensing.

6.3.4 *Effect of production rate on variable cost*

As has been indicated earlier, the total variable cost of a product is directly proportional to plant output and the variable cost/unit of production is normally constant. In the case of a continuously-operating process—the type of process used widely in the chemical industry—operation at low output rates or, conversely, higher than design rates can lead to process inefficiencies and an increase in the variable cost/unit of production. At low rates, increased residence time in the reactor can lead to overconversion and hence to increased usage of raw materials. The increase in by-product concentration in the product system leaving the reactor increases the *relative* demands on the product separation and purification system and hence results in increased energy consumption/unit of production. Operation at very high rates can result in increased levels of partial conversion products in the stream leaving the reactor. This, combined with reduced efficiency of distillation columns and other separation equipment when overloaded, can again lead to increased consumption of raw materials and energy inputs.

6.3.5 *Packaging and transport*

The costs involved in packaging and transport of a chemical product to the consumer are largely variable costs. However, such factors are not regarded as forming part of the production cost/income comparison. When considering process economics and profitability they are usually deducted from the money paid by the consumer to leave a net sales income which forms the revenue inflow to the producer to set against the production costs.

6.4
Fixed costs

The second category in the cost table, the fixed costs, can be divided as follows:

Operating labour and supervision
Maintenance labour and supervision
Analytical and laboratory staff   Fixed-cost elements
Maintenance materials   (total sum £'000/year
Depreciation   is fixed irrespective
Rates and insurance   of plant output)
Overheads—works overhead charges
          —general company overheads

6.4.1 *Labour charges*

The first three items represent the manpower cost associated with the process and includes the team of process workers who operate the equipment. In the case of large, continuous process units the operators work in shift teams to cover the 24 hrs per day, 7 days per week running of the unit. Maintenance labour includes fitters, electricians, plumbers, instrument artificers and other engineering workers who keep the process equipment in good working order. Usually maintenance work is planned and is generally carried out during normal day hours with only a small shift team to cover emergency breakdowns. Works analytical manpower is also included—again a shift team of analytical staff is involved to carry out checks on plant operation and product quality on a continuing basis.

The item for maintenance materials covers the cost of parts, replacement tools and similar items ranging from a replacement valve or pump to a new spanner. Only items to be expected in normal maintenance are included, the replacement, for example, of a reactor as a result of accident or fire would require a special appropriation of capital.

6.4.2 *Depreciation*

Depreciation is a term used in a number of different ways in different contexts. In the context of production costs for a chemical, the depreciation charge is regarded as an operating cost in the same way as material or energy usages. It represents the fact that the capital value of the plant is 'consumed' over the operating life of the plant. Calculation of the annual depreciation charge requires an estimate of the expected life of the equipment—a figure of 10 to 15 years is generally used in the heavy chemical industry. The annual charge can then be calculated by a simple straight-line method in which the same sum of money (equal to fixed capital costs ÷ expected life) is charged each year. Another method of calculating the depreciation charge is the declining balance or fixed percentage method. In this case the depreciation charge in a given year is a fixed percentage of the remaining undepreciated capital—thus in year 1 the depreciation charge would be $20\%C$ (for a 20% rate with fixed capital C). The next year the depreciation charge would be $20\% \times 0.8C$ and so on. This method gives higher depreciation during the early years of plant life but does not give depreciation to zero value at the end of expected service life. A high depreciation charge is then required in the final year of operating life to strike the balance.

A second interpretation of depreciation is as an allowance against tax. The annual income is reduced by an annual depreciation allowance before tax is charged thus reducing the tax payable. Calculation of the allowance is governed by the appropriate tax legislation and the depreciation shown for tax purposes may be a very different figure from that charged as an operating cost.

The use of depreciation as an allowance against tax forms part of net present value and discounted cash flow measures of profitability to be considered later.

6.4.3 *Rates and insurance*

This item covers the local rates or local authority tax levied in the area in which the plant is situated, together with the premium required to provide insurance cover for the facility. Actual charges will vary with the plant site and nature of the process being carried out (e.g. a plant with a high fire risk will carry a high insurance premium). Typical values for the rates and insurance item of cost lie in the range $0.5–2.0\%$ of plant capital.

6.4.4 *Overhead charges*

Overhead or general charges cover those items not associated with any particular product or process but which are an essential part of the functioning of an individual site or a whole company. These charges are usually divided into two broad classifications—local works overhead and general company overheads. The former category covers items such as the general management of a works (works manager, secretarial services, plant records, planning, security, safety organization, medical services, provision of offices and canteen facilities and so on). Company overheads include head-office charges, central research and development activities, legal, patent, supply and purchasing, and other company-wide activities. The allocation of company overheads to individual works and the further allocation of this charge and the works overhead to individual plants or products is something of an arbitrary process. Methods of allocation vary from company to company but are generally based on the plant capital and/or operating manning level for a plant relative to other plants in the works, and of the works in overall relation to the company.

**6.5
Direct, indirect and
capital related costs**

An alternative way of sub-dividing the components of full cost for a chemical product is into direct, indirect and capital-related costs. Direct charges are those arising directly as a result of the production operation and cover materials, energy, process labour and supervision costs, maintenance costs (labour and supervision and materials) and royalty payments. Indirect costs cover charges associated with, but not directly resulting from the process operation. They are essentially those charges termed overheads in the previous classification. Capital-related costs are those charges which result from the fixed-plant capital associated with the process. They consist of the annual depreciation charge and the rates and insurance item previously described. Since modern chemical processes (particularly in the heavy chemical and petrochemical industries) are highly capital-intensive, the capital-related charges form a significant part of the overall process cost. The relationship

Table 6.2 Relationship between variable and fixed, and direct and indirect, costs

Variable or fixed		Direct or indirect
V	Materials cost	D
V	Energy inputs	D
V	Royalty payments	D
F	Process and maintenance labour	D
F	Process and maintenance supervision	D
F	Maintenance materials	D
F	Rates and insurance	Capital-related
F	Works overhead	I
F	Site and company overhead	I
F	Depreciation	Capital-related

between the items classified as variable and fixed and the direct-indirect cost split is shown in Table 6.2.

6.6 Profit

Up to this point on the cost table (Table 6.1)—the works cost or total production cost—the cost build-up for the chemical allows only recovery of monies spent in producing the product and in maintaining the operation of the producing works and company. We now need to look at what profit will come from the sale of the product to provide for company growth, to reward shareholders and to pay taxes.

Profit can be measured in a variety of ways but two measures which are commonly used by accountants are gross profit (or gross margin) and net profit (net margin). The gross profit is obtained by deducting the direct production costs of the chemical (or other product) from the net sales income or revenue. It can be quoted on an annual basis or on a 'per unit of product' basis and represents the total cash available to pay for activities not directly concerned with production (indirect and capital charges in Table 6.2) and for growth, taxes and payment to shareholders. The net profit is obtained by deducting the total of direct and indirect production costs (works cost in Table 6.1) from the net sales income. This net profit is what is generally thought of as 'profit' and is essentially the money on which tax is levied leaving a net profit after tax to provide for growth of the company and pay dividends to the shareholders. For the simple annualized cost table, profit is shown as a return on the total capital involved (fixed plant and working) either derived from the sales income for an existing product or set as a target return for conceptual costing of a proposed new product. Thus the final total cost for our chemical product which will provide an acceptable return for the producer, and which must be met by the price paid by the customer, is much higher than the directly attributable production cost. This is the case both in the petrochemical industry where production is highly capital-intensive and in the speciality- and fine-chemicals areas where overhead charges are high to cover the extensive product and application development work required.

The build-up of the cost of producing a chemical has been shown for a single scale of production and with the plant operating at full capacity. Great emphasis has been placed in the chemical industry on the effect of scale and more recently, in the depressed economic climate, on the effect of operating chemical plants at low rates. What are the effects on product cost of an increase in scale or operation at less than full output?

6.7
Effects of scale of operation

6.7.1 *Variable costs*

In considering the production cost build-up the classification of costs as variable or fixed was made. The total sum of money in £'000 for the variable-cost items varies directly with the quantity produced. However, in unit cost terms—£/tonne of product or cents/lb—the variable cost is practically independent of the scale of operation. The yield in the reactor and the efficiency of product separation are not significantly affected by differing scale of operation so raw material usage and cost per unit of product will not be changed. Similarly the reactor energy requirement and separation energy inputs per unit of product will be largely unaffected by scale of operation since the endothermic or exothermic nature of the reaction is unchanged and the thermal efficiency of the separation stages does not change significantly. Thus, in producing, say, cumene from propylene and benzene, the raw-material costs and the cost of services will be the same (in £/tonne or cents/lb of cumene) whether the scale of production is 50 000 tonnes/year or 150 000 tonnes/year. When operating on a larger scale it may be possible to obtain raw materials at somewhat lower prices or it may be economic to install additional energy recovery equipment which would not be economically worthwhile on the smaller scale. These changes could reduce raw-material costs and energy costs but the effects are likely to be small and, in general, variable cost items are not significantly scale-dependent.

6.7.2 *Fixed costs*

Fixed-cost items (per unit of product) do, however vary significantly with scale of production. In the case of the labour element of fixed costs the number of process operators and maintenance workers are by no means directly proportional to scale. The precise relationship will depend on the type of process involved. A typical petrochemical industry process which involves largely fluid handling and is highly automated will have a relatively small process manning requirement and will require only a small increase in manning as process scale is increased. The number of maintenance workers will be greater than the number of process workers, but here again an increase in scale does not require a proportional increase in manpower; for example, a pump handling 50 m^3/hr will not require twice the maintenance man-hours of a pump handling $25 \, m^3/hr$. At the other end of the process spectrum, a small-scale batch process involving solids handling—typical of the dyestuffs industry—will have a higher manning level per unit of production and show an increase in manning requirements more nearly proportional to scale.

Attempts have been made to derive relationships between scale and manpower requirements, e.g. Wessel[2] derived a relationship of the form manpower $\propto$ no. of process steps (= process complexity) and (capacity)$^{0.24}$. Other studies[3,4] suggest even less dependence on capacity, with process complexity and company management philosophy being more important.

6.7.3 *Plant capital*

The major effect of change of scale is, however, on the plant capital requirement and consequently on the capital dependent charges per unit of production. Fixed-plant capital can be related to the production scale by an equation of the form

$$\frac{C_1}{C_2} = \left(\frac{S_1}{S_2}\right)^n$$

where C_1 and C_2 are fixed plant capital

             S_1 and S_2 are plant production scales

and n is a fractional power.

Early studies[5] in the petroleum and petrochemical industries derived a median value of n of 0·63. Subsequent studies confirmed a median value between 0·6 and 0·7 but showed values of n between 0·38 and 0·88 depending on the nature of the process and the operating scale. This is because some elements which go to build up the capital cost of a plant, such as engineering and supervision, electrical and instrument installations, are relatively unaffected by scale, whilst costs of machinery and equipment are scale-affected. In costs for these items a relationship of the above form is followed with $C_{1/2}$ being the item costs and $S_{1/2}$ being the item sizes.

For small plants or relatively small changes in scale only the cost of the plant equipment items will change significantly. Other factors, such as civil engineering work, support structures, installation costs, electrical and instrument costs, etc., will not change greatly and the overall scale factor will be below 0·6. If however, the increase in scale involves more units of equipment rather than a single larger item then the scale factor will be nearer unity. An example of this is shown in the production of ethylene by thermal cracking of hydrocarbons. Consider the initial capacity of the cracking furnace to be 30–35 000 tonnes/year of ethylene output. As the ethylene production capacity is increased, more furnace units are added and the scale-up factor for the cracking furnace section of the ethylene plant is 0·8–0·9. For comparison scale-up factors in the gas compression/treatment and distillation sections of such a cracker are 0·55.

Table 6.3   Effects of scale of production on cost

Product:	2-Ethylhexanol by carbonylation of propylene			
Basis: early 1988.	Plants at full capacity			
Scale tonnes/year	50 000		100 000	
Capital	£ million		£ million	
Fixed plant	67		102	
Working	12		19	
Total	79		121	
	£'000	£/tonne	£'000	£/tonne
Operating costs				
Propylene @ £305/tonne	12 688	254	25 376	254
Carbon monoxide @ £110/tonne	2 860	57	5 720	57
Hydrogen @ £700/tonne	2 800	56	5 600	56
Catalysts and chemicals	575	12	1 150	12
	18 923	379	37 846	379
Credit:				
iso-Butyraldehyde @ £280/tonne	(1 680)	(34)	(3 360)	(34)
By-product fuels @ £55/tonne	(318)	(6)	(637)	(6)
Net materials	16 925	339	33 849	339
Service costs (= energy costs)	2 150	43	4 300	43
Variable cost	19 075	382	38 149	382
Direct fixed costs	3 220	64	4 903	49
Depreciation (15 year life)	4 467	89	6 800	68
Overhead + indirect fixed	1 340	27	2 040	20
Works cost	28 102	562	51 892	519
10% return on total capital	7 900	158	12 100	121
Production cost + return	36 002	720	63 992	640

Table 6.3 gives an example of the effect of change in scale on production costs for two plants running at full capacity. Clearly, most of the reduction in unit cost on moving to the higher scale of production derives from the capital-dependent charges. Figure 6.1 shows the cost of production per tonne of 2-ethyl-hexanol based on propylene carbonylation assuming satisfactory sales of by-products and shows the continuing reduction in cost of production as production scale is increased. Production costs for most processes in the chemical industry—particularly in the petrochemical industry where continuous processes are operated—show a similar scale-dependence. It would seem that the only factor preventing the building of ever larger plants producing lower-cost product is the capacity of the market to absorb the products. However, by the late 1960s it was becoming apparent that the construction and operation of very large plants introduced new problems or intensified known problems. The larger plants involved more complex,

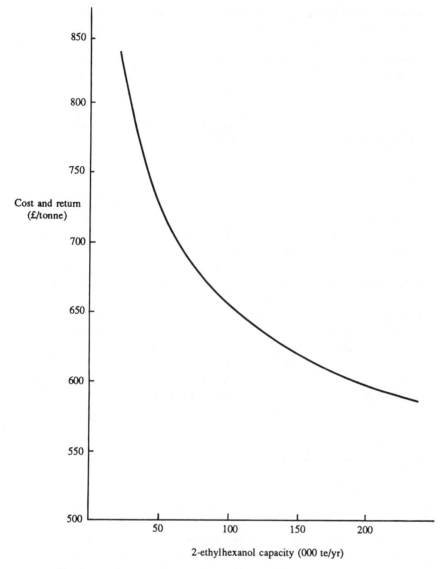

Figure 6.1 Cost versus production scale for 2-ethylhexanol production.

sophisticated process designs and greater process integration. Also large reactors and distillation columns had to be built on site, rather than built in a manufacture's workshop and then transported to the site. Partly as a result of this, and because of the sheer size of the projects, the large plants were taking longer to build and the very large amounts of fixed capital were committed for a longer period before producing any positive cash flow. Late start-up of very large plants can mean a project never achieves the expected cash flow and profit over its operating life. In order to maximize the benefit of increased scale on the plant capital, large plants are built on a 'single stream' basis on which the capacity of a process unit (reactor, compressor, distillation column) is increased by building a single larger unit rather than duplication of equipment. As a result of this, a breakdown of a single process unit could shut down the whole plant with very serious economic consequences. Largely because of these difficulties, the size of individual plants has not increased greatly since the late 1970s and it will probably require significant breakthroughs in process technology and project construction and management techniques to initiate further increases in plant scale.

6.8
Effect of low-rate
operation

Table 6.3 shows the production cost advantage of a larger plant when operated at full capacity. The larger plant provides a lower product cost whilst still meeting the desired return on total capital. However, if the unit suffers a breakdown or there is a market limitation such that the operator of the larger plant cannot sell all his potential production and has to operate at less than full output, then his cost per tonne of product will increase since his fixed costs have to be carried by the lower tonnage produced. The effect on product cost for the above 2-ethylhexanol plants operating with the larger unit achieving only 60% availability or an available market limitation of 60 000 tonnes/year is shown in Table 6.4. A sales income of £690/tonne of 2-ethylhexanol is assumed to simplify the picture.

The operator of the smaller plant could not meet all the potential market but could operate at full output showing a profit flow of £6.37 million at the going market price for 2-ethylhexanol and a reasonable 8·1% return on his

Table 6.4 2-Ethylhexanol from propylene effect of operation at below full capacity

	50 000		100 000	
Plant capacity tonnes/year	50 000		100 000	
Available market tonnes		60 000		
Sales possible tonnes	50 000		60 000	
Sales income @ £690/tonne (£'000)	34 500		41 400	
Operating costs	£'000	£/tonne	£'000	£/tonne
Materials	16 295	339	20 310	339
Service costs	2 150	43	2 580	43
Variable costs	19 075	382	22 890	382
Direct fixed costs	3 220	64	4 903	82
Depreciation	4 467	89	6 800	113
Overhead + indirect fixed	1 340	27	2 040	34
Works cost	28 102	562	36 633	611
Profit margin	6 398	128	4 767	79
Return on capital	8·1%		3·9%	

Cash flow = Sales income − (Variable cost + Direct fixed + Indirect fixed)
  = Profit margin + depreciation

£10 865 000   11 567 000

Table 6.5 Break-even production rate

Sales (kilotonnes/year)	Income (£m/year)	50 kilotonnes/year plant				100 kilotonnes/year plant			
		Variable cost (£m/year)	Fixed cost (£m/year)	Total cost (£m/year)	Profit (£m/year)	Variable cost (£m/year)	Fixed cost (£m/year)	Total cost (£m/year)	Profit (£m/year)
10	6·90	3·82	9·03	12·85	−5·95	3·82	13·74	17·56	−10·66
20	13·80	7·64		16·67	−2·87	7·64		21·38	−7·58
30	20·70	11·46		20·49	+0·21	11·46		25·20	−4·50
40	27·60	15·28		24·31	+3·29	15·28		29·02	−1·42
50	34·50	19·10		28·13	+6·37	19·10		32·84	+1·66
60	41·40					22·92		36·66	+4·74
70	48·30					26·74		40·48	+7·82
80	55·20					30·56		44·30	+10·90
90	62·10					34·38		48·12	+13·98
100	69·00					38·20		51·94	+17·06

Breakeven 29·3 kilotonnes/year
Return on capital at full capacity 8·1%

Breakeven 44·6 kilotonnes/year
Return on capital at full capacity 14·1%

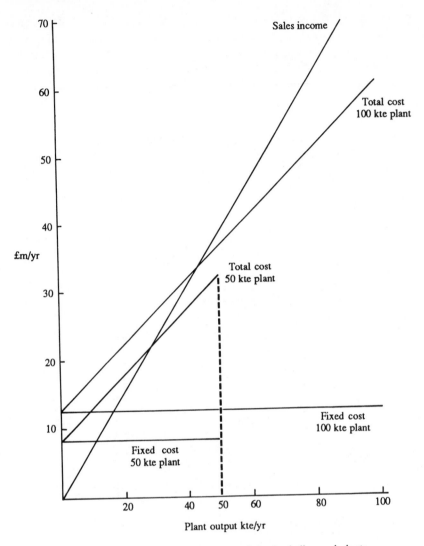

Figure 6.2 Break-even production rates for 2-ethylhexanol plants.

capital. The operator with the larger plant could satisfy all the available market demand but his production cost per tonne of product would be greater than that on the small plant and the profit flow and return on capital would be lower.

However, once the plant has been built the capital is regarded as 'sunk capital' and the important factor in the continuing viability of the business is the cash flow—defined as the sales income less total out-of-pocket expenses (Table 6.4). On this basis the larger plant provides more cash for the continuation of the business. Nevertheless, if the reduced market situation had been foreseen when the plants were being planned, the operator of the large plant would undoubtedly have built a unit of smaller capacity to match the lower market expectations and used the differential capital which would then be available to invest in more rewarding activities.

6.8.1 *Break-even production rate*

If we consider the situation, as above, where sales of 2-ethylhexanol show a net sales income of £690/tonne and the variable cost is £382/tonne, then there will be some level of production in the plants at which the sales income will cover all the costs (variable plus fixed including depreciation) but show no profit margin. This is the break-even production rate. Table 6.5 shows the build up of sales income and costs for the two plants and the figures are plotted in Fig. 6.2.

From the table and Fig. 6.2 it can be seen that the break-even production level, under the assumed conditions, for the 50 kilotonnes/year plant is 29 kilotonnes/year (58% of plant capacity) whereas the 100 kilotonnes/year plant must have sales of 45 kilotonnes/year (45% capacity) to break-even as a result of the higher fixed costs incurred by the larger plant.

6.9 Diminishing return

Although Figure 6.1 shows that the product cost and return at full output reduces as the production scale increases, the effect diminishes with increase in scale. Table 6.6 shows some of the data used to plot Figure 6.1, and also shows the incremental capital and reduction cost and return for each capacity step.

Examination of the figures shows that, as capacity is increased, the reduction in cost plus return for the next increment of capacity is diminished. Looking at the incremental reduction in cost plus return against the incremental fixed capital used to produce it, it can be seen that the additional capital has less effect as the scale is increased. This is to be expected since, with increase in production scale and reduction in fixed costs per tonne of product, the scale-independent variable cost will dominate the cost plus return. Thus from Table 6.3 for the 50 kilotonnes/year plant the variable cost represents 53% of the cost plus return, and fixed costs (including depreciation and return) the balance of 47%. For the 100 kilotonnes/year plant the relationship of

Table 6.6 Effect of increased production scale on cost and return

Scale kilotonnes/year	25		50		75		100		125
Fixed capital £ million	44		67		85		102		117
ΔFC £ million		23		18		17		15	
Cost + return £/tonne	826		720		668		640		618
ΔC + R £/tonne		106		52		28		22	
$\dfrac{\Delta C + R}{\Delta FC}$		4·6		2·9		1·6		1·5	

Table 6.7   Increased capacity effect on cost plus return of a single stream versus a twin-stream plant

	100 kilotonnes/year plant	125 kilotonnes/year plant	
		(a) single stream	(b) twinned reactors and furnaces
Fixed capital £million	102	117	122
Variable cost £/tonne	382	382	382
Fixed capital dependent charges £/tonne	216	198	207
Other fixed costs £/tonne	42	37	37
Cost + return	640	617	626
Δ C + R £/tonne of 125-kilotonne v. 100-kilotonne unit.		23	14

variable costs to fixed costs is 60%:40% and the reduction in fixed capital-dependent charges as scale is increased has a diminishing effect on the overall cost plus return. The effect is enhanced if a point is reached at which the next increase in capacity requires two reactors (or furnaces or stills) instead of a single larger unit. The increase in capital will then be greater than for a comparable single-stream plant with even less reduction in cost plus return. An illustration of this is given in Table 6.7.

This diminishing effect on cost plus return resulting from a uniform incremental increase in capacity as the total capacity level is increased is usually referred to in the chemical industry as the law of diminishing returns. In classical economics the same 'law' refers to the situation in which an increase in one of the factors of production (land, labour or capital) results in a reduction in output rather than an increase.

6.10
Absorption costing and marginality

When all the costs—direct and indirect charges—together with a profit margin are recovered by the sales income, the selling price is said to have been set on a full-cost or absorption-cost basis. Recovery of all costs (at least) from the revenue provided by sales of product is essential if the producing plant and the company as a whole are to continue to function in the long term. A profit margin over and above this level is needed to persuade investors to buy shares in the company or subscribe to loans which will enable the company to continue developing.

Although the total costs must be recovered by the product sales there may be possibilities to make additional sales at a price which is less than the full product cost. Such sales can be justified for strategic reasons such as to deter a competitor from building additional capacity or to develop a foothold in a new market. In order to avoid making such sales at a disastrously low price the concept of marginality and marginal cost must be developed.

Consider a manufacturer producing a chemical with a design capacity of 50 000 tonnes/year. Fixed costs (total labour plus supervision, maintenance, materials, depreciation, rates plus insurance and overhead allocation) amount to £3·0 million per year. Variable cost based on known usages and prices for raw materials and energy inputs is £415/tonne for outputs up to plant rated capacity. Suppose experience has shown that by operating at slightly higher rates throughout the year and using a little more catalyst, production could be increased to 52 000 tonnes/year with a variable cost of £418/tonne. Further increases in rates could raise capacity to 55 000 tonnes/year but at the expense of some loss of yield and reduced efficiency of operation on product separation and recovery which raises the variable cost to £428/tonne. Table 6.8 shows how the total production cost increases with increased output and how the average cost of product (or unit cost) falls as production is increased up to nominal capacity.

The figures of Table 6.8 are plotted graphically in Figure 6.3.

The marginal cost of production is the increase in total cost per unit increase in production: i.e. at any production level it is the cost of an additional tonne (or unit) of product. Table 6.8 and Figure 6.3 show that up to the normal plant capacity of 50 000 tonnes/year the marginal cost is constant and is equal to the variable cost.

This can easily be confirmed:

$$\text{Total cost at production rate } X \text{ tonnes/year} = \text{£3·0 million fixed}$$
$$+ (X \times \text{£415}) \text{ variable}$$
$$\text{Total cost at production rate } X + 1 = \text{£3·0 million fixed}$$
$$+ ((X + 1) \times \text{£415}) \text{ variable}$$
$$\text{Marginal cost} \quad \text{£415/tonne}$$

If, however, we want to push the operation of the plant above normal capacity, to produce 52 000 tonnes this year instead of 50 000 tonnes, the average cost/tonne increases slightly but the marginal cost of the extra output increases dramatically since to achieve it all the plant output will have been

Table 6.8 Effect of increased output on total production costs and unit cost of product

Production (tonnes/year)	Fixed costs (excludes profit) (£'000/year)	Variable cost (VC) (£/tonne)	Total VC (£'000/year)	Total cost (£'000/year)	Average cost (£/tonne)	Marginal cost (£/tonne)
5 000	3 000	415	2 075	5 075	1 015	
10 000		415	4 150	7 150	715	415
15 000		415	6 225	9 225	615	415
20 000		415	8 300	11 300	565	415
30 000		415	12 450	15 450	515	415
40 000		415	16 600	19 600	490	415
50 000		415	20 750	23 750	475	415
52 000		418	21 736	24 736	476	493
55 000		428	23 540	26 540	482	601

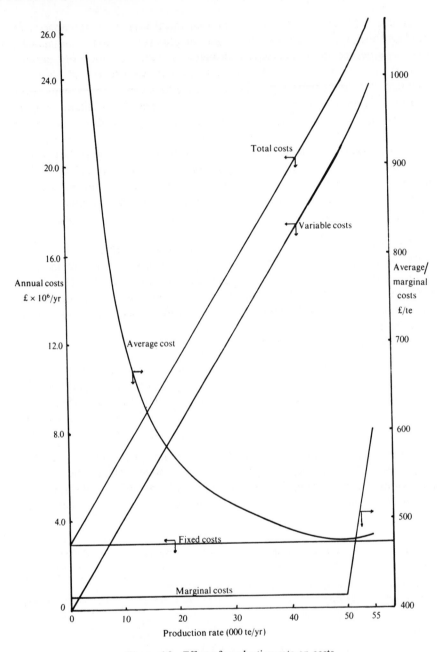

Figure 6.3   Effect of production rate on costs.

produced at slightly higher variable cost. So what does this tell us about operating our existing plant to maximize profit? Consider the situation where the marketing department expect to sell 40 000 tonnes/year of product in the coming year. A look at Table 6.8 or Figure 6.3 will tell us that if the average sales income is £490/tonne then the total of variable and fixed costs will be covered but there will be no profit. The output of 40 000 tonnes/year at an average sales income of £490/tonne represents a break-even situation. If the average sales income falls below £490/tonne at this level of output then the plant will not be making a full contribution to overhead charges. An average sales income greater than £490/tonne at our output level of 40 000 tonnes/year will generate a profit. Since the marginal cost of extra product at this level of output is equal to the variable cost (£415/tonne) then any additional sales giving a higher net sales income than £415/tonne will be beneficial. If the major sales are at a sufficiently high price to generate a profit then additional marginal sales at a sales income above marginal cost will increase profit. In the case where major sales are not producing a profit then marginal sales, made at above marginal costs, will reduce losses. To illustrate this, assume the major 40 000 tonnes/year sales are made at an average sales income of £520/tonne then total sales income is £20·80 million and total cost (Table 6.8) is £19·60 million and a profit of £1·20 million is produced. If an additional 1000 tonne could be sold (perhaps in an export market involving higher distribution costs) at a lower net sales income of £450/tonne (i.e. above marginal cost) the total income would be £21·25 million and total costs (£19·60 m + 1000 tonnes @ £415/tonne) would be £20·015 million giving a slightly larger profit of £1·235 million. Conversely, if the major 40 000 tonnes/year sales show an average sales income of £480/tonne, income would be £19·20 million against costs of £19·60 million—a loss of £0·4 million—and the manufacturing operation would not be making its full contribution to overall company costs. In this situation an addition of 1000 tonnes of sales at £450/tonne would raise income to £19·65 million against costs of £20·015 million, giving a slightly reduced loss of £0·365 million. These cases are summarized in Table 6.9.

Case (iii) in Table 6.9 shows the situation with the plant scheduled to operate at full design capacity to meet major sales of 50 000 tonnes/year at £520/tonne. Under these circumstances the marginal cost of additional sales rises rapidly (Figure 6.3) and the addition of 1000 tonnes of sales at £450/tonne—below the marginal cost at this level—is seen to reduce the total profit.

Table 6.9   Effect of additional sales at above marginal cost

Sales	Total sales income (£million)	Total variable cost (£million)	Total cost (£million)	Profit (loss) (£million)
(i) 40 000 tonnes @ £520/tonne	20·800	16·600	19·600	1·200
As (i) + 1 000 tonnes @ £450/tonne	21·250	17·015	20·015	1·235
(ii) 40 000 tonnes @ £480/tonne	19·200	16·600	19·600	(0·400)
As (ii) + 1 000 tonnes @ £450/tonne	19·650	17·015	20·015	(0·365)
(iii) 50 000 tonnes @ £ 520/tonne	26·000	20·750	23·750	2·250
As (iii) + 1 000 tonnes @ £450/tonne	26·450	21·318	24·318	2·132

To summarize this section on marginality and marginal costs: at any level of production and selling price, if marginal sales can be made at a sales income greater than marginal production costs, the marginal sales will make a positive contribution to income. However, for the overall operation of the production unit to generate a profit we must not lose sight of the fact that average sales income must be greater than average production cost at the given level of production.

6.11
Measuring profitability

In earlier sections of the chapter the term 'profit' has been used and in the present section a more detailed look will be taken at the ways in which profit is measured and the ways in which we can compare the profitability of investments.

6.11.1 *Return on investment*

Historically the yardstick used to measure profitability has been the percent return on capital or percent return on investment. This is defined as

$$\text{percent return on investment (ROI)} = \frac{\text{annual profit}}{\text{capital invested}} \times 100$$

The method and derived ratios are still widely used in basic accountancy, although they are being displaced by more up-to-date methods. This traditional method has the advantage of being simple and readily understood but is not very informative and can be misleading.

As used in Table 6.1 the target return on investment represents the cash income on an annual basis divided by the total capital invested. The cash income is obtained by deducting the total of direct and indirect annual production costs from the annual sales income, and the capital invested includes both fixed and working capital. Frequently the net cash income after tax is used as the numerator in the ROI ratio and this will, naturally, result in a lower return figure. One weakness of the return on capital as a measure of profitability is that the cash income generated by sales depends on the depreciation method used in the company. The annual charge for depreciation varies with the method and changes from year to year for methods other than simple straight-line depreciation. However, possibilities for confusion are multiplied by the choices available for the denominator in the ROI ratio. The capital value used can be total capital (i.e. fixed plus working) or fixed capital alone, and the fixed capital value used can be the original capital value, the current depreciated value, the average value over the life of the plant or the index-inflated current replacement value. The working capital element also depends on the methods used to value the feed and product stocks. Clearly, before using the ROI method to compare processes or companies using available published data the basis of the ratio and the definition of the factors involved must be carefully studied . Whilst all the above variations of the ROI ratio are used, the one most commonly encountered is (annual profit assuming straight-line depreciation) ÷ (total capital invested).

The traditional ROI method of expressing profitability presents a single year 'snapshot' of a process. No allowance is made for the often lengthy period when capital is being invested and no positive cash flow is generated. Distortions resulting from inflation and the effect of the time value of money are also ignored.

6.11.2 Use of inflated capital—current cost accounting

In periods of high inflation the use of the conventional ROI method to analyse profitability can seriously affect the financial strength of a company. Inflated prices for the product bring in apparent high income whilst capital charges, based on the original capital, will be low and ultimately insufficient for replacement of the plant. Apparent profits and ROI will then be high resulting in high taxes and expectations of good dividends by the shareholders. If capital charges are based on an updated valuation of assets a very different picture emerges. The example in Table 6.10 shows a plant for production of

Table 6.10 Comparison of the effect of using historic and inflated values for plant capital on rate of return

Scale: 100 kilotonnes/year	1982	1988	1988
Plant capital £100 million		Historic plant capital £100 million	Inflated plant capital £152 million
Working capital £20 million		£26 million[3]	£26 million[3]
Operation 1982: £/tonne at full output		1988: £/tonne at full output	1988: £/tonne at full output
Raw material	100	114[1]	114[1]
Services (energy)	25	29[1]	29[1]
Labour	10	16[2]	16[2]
Overheads	15	24[2]	24[2]
Plant capital dependent	160	160	243[4]
Works cost	310	343	426
Sales income	430	563[3]	563[3]
Profit margin	120	220	137
ROI before tax	10%	17·5%	7·7%

[1]Inflated, using Index for Materials and Fuels purchased, of the chemical and allied industries.
[2]Inflated, using Index for Earnings in the chemical and allied industries.
[3]Inflated, using Index for output prices of the chemical and allied industries.
[4]Inflated, using Plant Capital Index from *Process Engineering*.

100 kilotonnes of product with a fixed plant capital of £100 million in 1982 showing a 10% ROI at 1982 prices. For late 1988 with product prices and costs for materials, fuels and labour inflated using the appropriate indices but with capital on an historic basis the profit margin looks good and the ROI very attractive. However if the fixed capital is also inflated to a current value the profit is reduced and the ROI not particularly attractive. Furthermore, in the 'historic capital' case, tax would be levied on a 'profit' of which approximately 40% is needed to maintain the capital base of the company.

In the situation shown in the 'historic capital' column of Table 6.10 the tax paid and, on the whole company basis, the dividend paid resulting from the apparent profit shown by the above operation would in fact be taken from cash required to maintain the capital base of the company. To correct this highly undesirable situation the 'current cost accounting' approach used in column 3 of Table 6.10, in which all factors in the cost build-up are expressed in terms of current values, is being increasingly adopted. The Institute of Chartered Accounts supports a system of this type and many major companies have worked together to develop a mutually acceptable system.

6.11.3 *Payback time*

A second long-standing method of assessing profitability—usually used at the planning stage for a new project rather than valuing an existing asset—is the payback period required. This is not so much a direct measure of profitability but rather a measure of the time taken for the positive cash flows to recover the original fixed-plant capital investment. Payback time is usually based only on the fixed plant capital and the payback time is the time taken for the cumulative positive cash flows following plant start-up to balance the earlier negative cash flows of the fixed-plant investment. Cash flows are usually taken on an 'after-tax' basis and allowance must be made for grants or allowances which offset capital expenditure.

During the plant construction period all cash flows are negative, i.e. cash is flowing out of the company 'treasury' to pay for the plant construction. When the plant begins to make and sell product, cash inflow begins. The positive cash flow is: cash flow = (sales income) − (total direct and indirect production costs) + (depreciation). Depreciation is part of the positive cash flow since it is cash which has gone into the general company 'treasury' although it carries a label nominally reserving it for one specific function.

Figure 6.4 shows the cash outflow during the plant construction period and start-up. Cash inflow builds up after start-up as the plant output is increased until in the third year of operation full-rate operation is reached and a steady cash inflow is provided. Payback time is shown as just under five and a half years from plant start-up or seven and a half years from start-up of the project construction.

Care must be taken when using the concept of payback time to compare

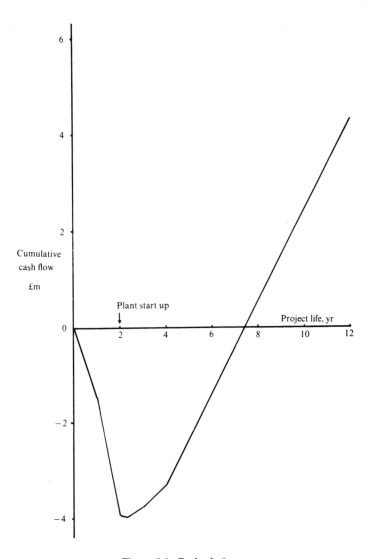

Figure 6.4 Payback time.

projects since the method does not allow comparison of projects over the total project life. In periods of high uncertainty the method has the advantage that it is dependent on the period immediately ahead during which the situation can be assessed with relatively more confidence than the distant future. If the payback time is short it offers a reasonable guarantee, that, at the worst, the fixed capital outlay will be recovered. The method makes no estimate of the profit to be earned once the payback point has been passed but to some extent this can be inferred as a continuation of the positive cash flow established in achieving payback. The method can be misleading if projects with different expected lives are compared since a project with (say) a five-year payback and continuing profit flows for ten years could overall be more attractive than an alternative with a three-year payback but only six-year profit flow. Because the method considers only short-term cash flows the payback time method of assessing profitability is usually used as a supporting technique with other methods of assessing overall profitability.

6.11.4 *Equivalent maximum investment period*

This method of assessing the relative attractiveness of projects by a time measurement was developed at Nottingham University[6]. It requires the same cumulative after-tax cash flow data as required by the payback time method but takes into account the pattern of cash flows in the early years. These are the most important ones in the life of a project and the years for which predictions are likely to be most reliable. The equivalent maximum investment period (EMIP) is defined from the cash-flow diagram (Figure 6.4) as the area under the cash-flow curve from the start of the project to the break-even point divided by the cumulative maximum expenditure. Since the area is measured in £ million × years and the expenditure is £ million, then the EMIP is expressed in years. EMIP represents the time period for which the whole maximum expenditure would be at risk if it were all incurred instantly and repaid by a single instant payment. In Figure 6.4 the area involved is $\sim$ 16·5 £million years and the maximum cash outflow £4 million, giving an EMIP of 4.1 years. The pattern of cash flows to the break-even point (the payback time) is taken into account since the area under the curve is measured. As a result of this two projects with the same payback time can have different EMIP values. As with the payback time, the value of EMIP which distinguishes an attractive project from an unattractive one is somewhat arbitrary. It depends on the nature of the industry, the degree of risk and type of project. Analysis of data from earlier projects within a company can set guidelines which can then be used to examine new projects. Although the method is simple and easy to apply it does not seem to have been widely used.

6.12
Time value of money

The profitability measures considered so far have not considered possible charges on the capital involved nor the effect of time on money. Since a sum of money available now can be invested to earn interest it is of greater value than if the same sum were received some time in the future. Modern methods of assessing profitability take account of this time value of money and the pattern of cash flow during the life of a production unit. The most widely used methods are the Net Present Value (NPV) and Discounted Cash Flow Return (DCF) methods. These methods are generally used to aid investment decisions and in comparing the relative merits of different projects before investing money and building plant. Once capital has been invested and the chemical plant has been built and is in operation then different criteria are involved. As previously indicated, the method of operation which maximizes cash inflow and profit is to run at as high a rate as possible provided the least profitable marginal sales are made at a price which gives a sales income above the marginal production cost.

6.12.1 *Net present value and discounted cash flow*

When considering a new project the capital for building the plant must obviously be invested before any cash inflow or profit appears. If the money is borrowed from a bank then interest will be charged until the positive cash flow from the project repays the loan. This interest must then be included as a cash outflow in assessing the overall project. If the project capital is available from internal company funds there will be no direct interest to be paid. However, the company has lost the opportunity to invest the money and earn interest on it, possibly by lending to a bank. This loss of possible investment income represents an opportunity cost which must be charged against the project. On the other hand, when the break-even line has been crossed and the project is providing a cash inflow this money could be invested and earn interest at the rate previously charged on the borrowed capital. The rate of interest to be charged on capital outflow or credited to cash inflows will depend on the source of the finance. A company can provide capital for a project in different ways—by raising a market loan, using retained profit and depreciation, borrowing from a bank or finance house. In general a company will raise money in several ways and have a 'pool' of available funds at an agreed average company interest rate to fund possible projects.

The NPV and DCF profitability estimates are derived from cash flow forecasts covering the total life of the project. The cash flow in any year is: (net sales income) − (total production cost excluding depreciation) − (capital invested) after tax allowances and deductions. Depreciation is excluded since recovery of capital is implicit in the calculation and the methods thus have the advantage of being independent of depreciation methods. Since there are considerable problems in forecasting costs and sales for more than two or three years ahead, the longer the time period, the less likely the forecasts are to be right. As future sums of money have different values year by year due to the interest charged or earned they must be adjusted—or discounted—to values at a single moment in time, usually the present. For simplicity NPV and DCF calculations are normally based on the assumptions that cash flows in a given year occur at the end of that year. This is not often the case and in practice cash flows—particularly payment for items of equipment during the plant construction and income from product sales—take place throughout the year. However, the difference between annual discounting and continuous discounting will not usually be very great and it is normally much smaller than errors in the cash flow data.

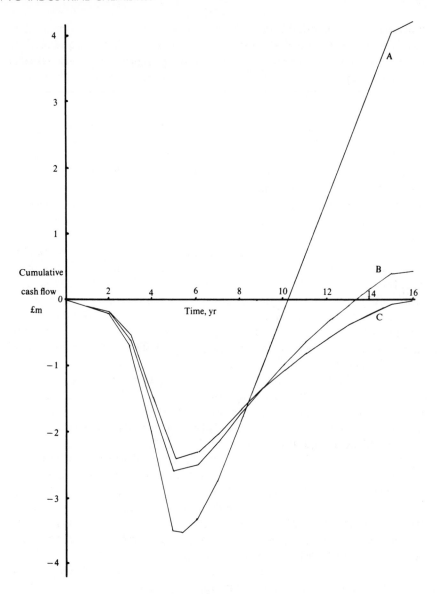

Figure 6.5 Discounted net cash flow pattern for a proposed project.

If we have a present sum of money P, its value n years in the future (F) will be $F = P(1 + i)^n$ where i is the accepted compound interest rate fraction. Conversely, the present-day value (P) of a sum of money (F) to be received n years in the future is given by the expression

$$P = \frac{F}{(1 + i)^n}.$$

Thus the individual cash flows in future years can be discounted to a common base (the present) and directly compared. From the form of the expression for present value it will be seen that a sum of money received in two years time is worth more than the same sum received ten years hence. This reduces the effect of the less certain cash flows in the long-term future and emphasizes the importance of early positive cash flows.

The net present value of a project or investment is the sum of the present values of each individual cash flow. To reduce the arithmetic involved the net cash flow in each year is determined and the present value of the annual cash flows is summed. For a project with net annual cash flow C_t in year t, the net present value;

$$\text{NPV} = \frac{C_0}{(1+i)^0} + \frac{C_1}{(1+i)^1} + \frac{C_2}{(1+i)^2} + \cdots \frac{C_t}{(1+i)^t} + \cdots \frac{C_n}{(1+i)^n}$$

or, in shorter form

$$\text{NPV} = \sum_{t=0}^{t=n} \frac{C_t}{(1+i)^t}$$

where n is the total project life.

The 'present' is usually the time of the evaluation which may be the start of the project or may be after a period of research and development but before significant capital outlay. Figure 6.5 shows the forecast net cash flow pattern for a proposed project. The forecast shows two years of research work at a net cost of £100 000 per year followed by one year of process and market development including process design and costing £500 000. Plant construction is forecast to require two years giving a total negative cash flow of £3·5 million by the end of year 4. Some £50 000 further expenditure is incurred during start-up but then the positive cash flow from sales begins and by the end of year 5 a positive flow of £150 000 is shown. Production and sales build up in the following year, reaching full capacity in year 7. After eight years of steady operation and positive cash flow, production ceases, the plant is scrapped and working capital recovered. The undiscounted net cash flows are shown in Table 6.11 and plotted as curve A in Figure 6.5. If the cost of capital—that is the interest rate to be charged on negative cash flows and credited to positive cash flows— is set at 10% as the average cost to the company, then the discounted cash flows are shown in Table 6.11 and as curve B in Figure 6.5.

Tables of discount factors which provide values of $1/(1+i)^n$ for different values of i and n are available and simplify manual calculations. Now, of course, computer programs are available to carry out discount calculations on any small computer. The project is seen to have a net present value of + £422 000 having recovered the capital and paid interest charges at the company rate of 10%. If, however, the company interest rate for raising money is 15%, then as shown in Table 6.11 the project shows a NPV of − £276 000. At this higher rate for borrowing money to carry through the project, a loss would be made and the project would be unattractive. The greater the

Table 6.11   Projected discounted cash flows

End of year	Net cash flow after tax (£000)	Discounted at 10% (£000)	15% (£000)
0	− 100	− 100	− 100
1	− 100	− 91	− 87
2	− 500	− 413	− 378
3	− 1 300	− 977	− 855
4	− 1 500	− 1 025	− 858
5	+ 150	+ 93	+ 75
6	+ 600	+ 339	+ 259
7	+ 850	+ 436	+ 320
8	+ 850	+ 397	+ 278
9	+ 850	+ 361	+ 242
10	+ 850	+ 328	+ 210
11	+ 850	+ 298	+ 183
12	+ 850	+ 271	+ 159
13	+ 850	+ 246	+ 138
14	+ 850	+ 223	+ 120
15	+ 150	+ 36	+ 18
Total	+ 4 200	+ 422	− 276

positive NPV of a project at the set discount rate the more economically attractive it is. A project showing a negative NPV is unprofitable at the set discount rate and would not be pursued.

6.12.2 *Discounted cash flow return*

As Figure 6.5 and Table 6.11 show, increasing the discount rate from 0 to 10% reduces the NPV from £4·2 million to £422 000. A further increase in discount rate to 15% produces a negative NPV showing a loss-making situation. At a discount rate somewhere between 10% and 15% the NPV will be zero, and this is the DCF value. DCF, which is also referred to as 'internal rate of return' is defined as that value of discount rate which results in a zero NPV for a project. The higher the value of the DCF the more economically attractive the project. A minimum acceptable level of DCF is set by the cost of capital to the company considering an investment. If the project shows a DCF greater than the cost of capital then the project will show a profit, conversely a DCF below the cost of capital indicates a loss-making project.

   Finding the DCF for a project is a matter of trial and error calculation. When carrying out manual calculations a value of i (the discount rate) is selected and the project NPV calculated. If this is positive the value of i chosen is lower than the project DCF, so a higher value of i is selected and a new NPV calculated. If a negative value of NPV is then obtained then the higher value of i is above the DCF for the project. Graphical interpolation between NPV values will enable the DCF value to be estimated. From the figures in Table

6.11 the project DCF rate lies between 10% and 15%. Interpolation gives a DCF value of 13% and curve C in Figure 6.5 shows the cash flows discounted at this rate. In fact, discounting the cash flows of Table 6.11 at 13% shows a small negative NPV, discounting at 12% shows a small positive NPV and interpolation gives a DCF rate of 12·7%. Although the arithmetic operations can be carried out to give an even more precise value of DCF—particularly if a computer is used—it must be remembered that the cash flows are based on forecasts of sales, prices, raw material costs and production costs for many years into the future. The cash flows of Table 6.11 represent an attempt to project 15 years into the future and the uncertainties in the data do not justify calculation of DCF rates to decimal places. This problem of uncertainty will be examined later when considering the assessment and evaluation of a new project.

The DCF method is not suitable for projects where a net negative cash flow takes place late in the project taking the cumulative cash flow back below the zero line. In most cases such a large cash outflow is a result of additional capital investment in a planned expansion of capacity. It is unlikely that a decision to commit funds to the expansion will be made at the same time as the decision on the original project so the two can be treated separately and individual cash flows evaluated.

6.12.3 *Use of NPV and DCF as profitability measures*

Since the same data and method of calculation are used to determine NPV and DCF it might be considered that one or other would suffice. There are, however, differences in the two measures and in their interpretation. Calculation of the project NPV by discounting the net cash flows at the company cost of capital gives a measure, in current money values, of the profit which the project will earn. This profit is produced after recovering the initial investment and paying all costs including the cost of 'borrowing' the capital from the company pool. It is not a ratio or relative value but a direct measure of the total profit expected from the project. The DCF for the project is a calculated rate of return on the invested capital at which the project breaks even, i.e. pays all expenses but shows no residual profit. It is a measure of the earning power of the project and also provides an indication of the efficiency with which the invested capital is used.

When a single project is being considered without reference to other possible projects the use of NPV or DCF will lead to the same conclusion on the economic acceptability of the project. This results from the fact that if a project shows a positive NPV, it must also show a DCF return higher than the discount rate used to calculate the NPV, that is a DCF rate above the cost of capital to the company. However, a more usual situation is one in which a choice must be made between alternative forms of a project which are mutually exclusive or in which a portfolio of projects must be selected. In this case the need for optimization arises and the company objectives and constraints must be taken into account.

In the former case, where a choice must be made between alternative forms of a project these will usually have different investments and operating costs. Thus although the income from sales will be the same the annual cash flows will differ. In order to maximize the profit from the project the proposal with the greater NPV at the cost of capital should be selected. This may not be the same as the proposal with the highest DCF since the return is measured on different levels of investment. A lower DCF on a larger capital investment could provide a higher NPV (i.e. a greater cash profit) than a higher DCF on a smaller investment.

When selecting a portfolio of projects the company aims must be considered. If the company objective in economic terms is to maximize the profit flow from the proposed list of projects, then this is an objective in money terms and can be attained by maximizing the total NPV of the projects chosen. This requires the acceptance of all independent projects which show a positive NPV at the company cost of capital. Such a solution is only possible if the resources available (supply of capital, project management, engineering facilities, etc.) are sufficient to carry out all projects with a positive NPV. If the company has more potentially suitable projects than can be financed with the available capital then the objective will be modified to maximizing the total NPV within the available capital restraint. Since the DCF is a measure of efficiency of use of capital, ranking projects in decreasing order of DCF and selecting projects in DCF order until the available capital budget has been taken up will give the desired maximum NPV. Of course, selection of projects is rarely as straightforward as implied above. Other resources besides capital could be inadequate to deal with all potentially attractive projects—more than one project might require the same technical manpower or engineering facilities. To resolve these problems, detailed network analysis and mathematical project programming techniques are required.

The NPV is generally more widely used in project evaluation studies than DCF measures. The NPV provides a direct measure in money terms of the attractiveness of a project and when dealing with more than one project NPVs are additive since they are measured at the same cost of capital. A DCF is, on the other hand, a rate of return and values cannot be combined for multiple projects since they are measured on differing capital investments.

6.13

Project evaluation

In the context of the chemical industry, a project can range from initial research studies (on a new product or new route to an existing product) to a capital project costing many millions of pounds or dollars for the installation of a new production unit. The common ground is the ultimate objective of generating profit at some time in the future by the use of present resources (skilled manpower, money). Economic evaluation of a project involves the consideration of those factors which can be measured and compared in money terms to provide information which will help decision making. Project evaluations may be carried out several times during the life of a project to assist in making decisions at the various stages. The initial decisions may involve continuation of a research programme, then to commit more resources to development of the process. Later the need to build a pilot plant to study scale-up problems may have to be studied, followed ultimately by a decision whether or not to invest the capital and build a full-scale unit to use the new process or manufacture the new product. The stages involved in project evaluation are summarized on p. 149.

6.13.1 *Comparison of process variable costs*

When considering a new route to an existing product the initial studies at an early stage must compare the proposed new route to the existing process or processes. Projects of this type are typical of the petrochemical or heavy chemical industries involving the production of bulk chemical intermediates, fertilizers and commodity polymers. Since the product is established and marketed then material from the new process must be able to compete, in quality and manufacturing cost, with that from existing processes. An initial evaluation of the proposed new route can then be made by comparing the cost plus return for the new process with that for existing technologies. It is essential that the comparison studies are made on the same basis and using the same methods to make the comparison as realistic as possible.

From the earlier part of the chapter it can be seen that, for large continuously operating units, the major factors in the build up of production cost are the variable costs (net raw materials and services) and capital-dependent charges. For smaller-scale batch production units the labour element can also become a significant factor. Hence, in order to compare our potential new process with existing methods we need to estimate the variable costs and the capital required for a plant to operate the process. As an example, consider maleic (cis-butenedioic) anhydride, the bulk of world production of which is currently manufactured by the oxidation of benzene. New catalysts are being developed by many companies to use *n*-butane as a feedstock. Let us assume that our research group is studying such a catalyst and the estimated production cost by the proposed butane oxidation needs to be compared with that of the benzene oxidation. Information on raw material and energy usages for the existing benzene oxidation process can be found in the literature[7], and numerous patents describe reaction conditions, yields, conversion of benzene and methods of product work-up, enabling the variable cost for maleic anhydride by benzene oxidation to be determined. Results from the research work will give indications of yield of maleic anhydride and *n*-butane coversion for the reaction step although data on recovery and recycle of unreacted feedstock are unlikely to be available. Purification by distillation or crystallization is normally very efficient, and in the absence of evidence for azeotropes or eutectics and making assumptions on separation efficiency (e.g. as good as in the benzene route) the overall raw material usage and cost for the new route can be estimated. If there is information on the expected life for the new catalyst this can be used to give an estimate of catalyst cost on an annual or per unit of product basis. In the absence of catalyst life data an acceptable charge for catalyst can be assumed and used to indicate a target catalyst life which must be achieved.

Estimation of the cost of energy inputs for the new process is difficult at the early research stage when process information is limited. One method which is useful at this stage is described by Marsden[8] and requires only data on the number of major process operations, process throughputs and chemical heats of reaction. It enables an estimate of the total energy requirement of the process to be made and inspection of the assumed reactor conditions and expected distillation temperatures will usually enable a split into steam, fuel and electricity consumption to be made.

6.13.2 *Estimation of plant capital*

Having made an estimate of the variable cost for the new process the next requirement is to produce an estimate of the plant capital requirement. Traditional methods of estimating capital, involving development of a process flowsheet and chemical engineering design, are time-consuming (expensive) and require a level of detailed information which is not available at the early research stage. Several workers have developed methods for the rapid estimation of plant capital at the predesign stage (see Bridgewater[9]) which require varying levels of input data. One method—'Process Step Scoring'[10] developed within ICI—requires only a simple flow diagram indicating the main conceptual process steps (react, filter, distill, separate etc.) with throughputs and estimated reaction times and conditions. The capital cost for a desired scale of operation can then be derived quickly at a level of accuracy sufficient for most preliminary evaluations. In order to compare the new process and the existing process on the same basis an estimate of capital requirement for the existing process should be prepared by the same method as that used for the new route. If capital estimates are available in the literature for the existing process they can serve as a cross-check on the estimate made. However, because of inflation, care must be taken to adjust earlier capital estimates to the same time base as the new estimate by use of the appropriate index. Suitable indexes showing change of plant capital cost with time are the CE index published in *Chemical Engineering* for American conditions, and the PE Index published in *Process Engineering* for UK conditions. The latter journal also publishes indexes for several other key countries, e.g. West Germany, Japan, France, Canada, and Australia.

6.13.3 *Process cost comparison*

With estimates of variable cost and capital requirement available the process cost table could be built up if an estimate of the labour requirement (process and maintenance and analytical) could be made. These items form part of the plant fixed costs and studies of a number of petrochemical processes show that an annual charge equivalent to 3–6% of the plant capital would be a suitable figure to include in a preliminary estimate for total plant fixed costs. For smaller batch operations a range of 8–12% of plant capital would be a reasonable figure to apply. A comprehensive cost table (Table 6.12) can then be built up.

This level of evaluation is sufficient at an early stage and would justify continuing research on catalyst and process developments. As more data become available the stage will be reached where a preliminary process flowsheet can be prepared and initial chemical engineering studies carried out to enable a more detailed estimate of capital and energy inputs to be made. This would involve sizing of the main items of the flowsheet using short-cut methods and building up to a full capital estimate using a factor method such

Table 6.12   Process cost comparison for the production of maleic (cis-butenedioic) anhydride by benzene and butane oxidation

Scale	25 000 tonnes/year	
Basis	Costs at January 1989 prices. Capital at a *Process Engineering* index level of 211 (1980 = 100)	
Process	*Benzene oxidation*	*n-Butane oxidation*
Plant capital	£30 million	£34 million
Working capital	£ 3 million	£ 2 million
	£33 million	£36 million
Operating costs per tonne of maleic anhydride	£	£
Benzene 1·13 tonnes/tonne of product @ £310/tonne	350	—
Butane 1·17 tonnes/tonne of product @ £185/tonne	—	216
Catalysts and chemicals	10	20
Materials	360	236
Net services (large steam credits/allowed)	6	14
Variable cost	366	250
Plant fixed costs 6% plant cap.	72	82
Depreciation 6.67% plant cap.	80	91
Overhead fixed costs 3% plant cap.	36	41
Worked cost	554	464
Target ROC 10%	132	148
	686	612

as that of Miller[11], Cran[12] or Sinha[12a]. At this point, process optimization studies can be valuable in determining the energy or capital-intensive steps in the process and indicating where efforts to make process improvements would be most rewarding.

However, as previously stated, the cost table provides only an instant picture of the process cost and profitability. As the stage is approached when capital will have to be allocated for pilot-plant studies there is a need for a view of the prospects for a project over its expected life. This requires the use of NPV and DCF methods of evaluation which in turn necessitate estimates of product sales, raw material prices and product selling prices over the life of the project.

In order to obtain the necessary information an estimate is required of the amount of the product which the market can take up and also an estimate of the share of the market which the Company carrying out the survey can expect to obtain. This will enable the size of the proposed new product unit to be fixed or set at a range of sizes to be evaluated in the NPV/DCF calculations.

6.13.4 *Estimating markets/prices*

The estimate of the future market size can be made in various ways— projection of historical data, comparison with other factors such as GNP where a relationship can be established (e.g. per capita usage of plastics and per capital GNP), and detailed user surveys.

Examination of consumption of many chemicals show that growth of consumption has three chief phases:

(a) Increasing rate of growth—an exponential curve.
(b) Constant rate of growth—a straight-line relation.
(c) Declining rate of growth—a curve with smaller slope than (b) and which may become negative.

A hypothetical product growth curve is illustrated in Figure 6.6. Clearly if the data available show exponential growth care must be taken in projecting over a long period. Techniques have been developed to modify the simple extra-polation[13] but require extensive data. The projection of past data can be improved by combining with detailed user surveys of major product outlets to provide a more rational basis for estimated demand. All methods of assessing future markets assume the absence of major perturbations, and events such as the OPEC actions on oil prices in 1973 and 1979 can make nonsense of any forecasts.

When considering a new product the product properties must first be used to determine the market areas which will be open to penetration and the markets for competitive products used to make initial estimates for the new product. Whether the product is new or already established, forecasts of markets and market share are at best uncertain. It is impossible to forecast the changing circumstances which will affect growth of consumption and even the most refined mathematical model cannot allow for the effect of unforeseen step

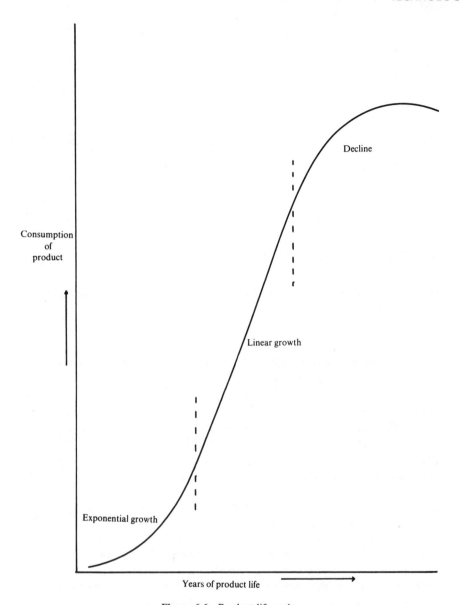

Figure 6.6   Product life cycle.

changes. Nevertheless, market evaluation is essential and will show trends and show the potential manufacturer the general direction if not the detailed path.

In forecasting price trends for established products the available data on prices and tonnages sold again forms the basis for extrapolation. Published information on many chemicals (e.g. US Tariff Commission reports, reports of Office Central des Statistiques de la CEE) enable a plot of log (price—adjusted to constant money values) against log (cumulative production) to be drawn.

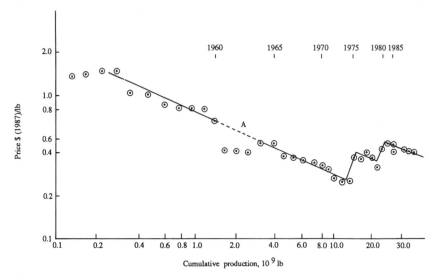

Figure 6.7 Acrylonitrile: U.S. production *v.* price. *A*: competition between established acetylene-based route and Sohio propylene-based route.

This is the so called 'Boston experience curve' (Figure 6.7, Table 6.13) which was introduced in 1968 (Boston Consulting Group)[14]. The concept of experience curves postulates that costs (or price) in constant money value declines by a characteristic amount for each doubling of experience (production or sales). From studies of rapidly growing sectors of industry the Boston group indicated a decline in price of 20–30% for each cumulative doubling of production. A further study[15] confirmed the characteristic decline for products showing rapid growth. However, for products with low growth rates the relationship between price and cumulative production is much more erratic and shows a much lower rate of decline than the expected 20–30%. In view of the low growth shown by many major chemicals in recent years the projection of future prices using historic data is open to question.

Having obtained forecasts of potential market share and price, the annual cash flow position for a selected plant size can be calculated using the capital estimate and operating cost estimates prepared from the literature or research data. The cash flows can then be discounted at the cost of capital to determine the NPV and the DCF rate can be estimated. The NPV and DCF for different plant scales can be calculated to determine the optimum size of plant and the competitive processes compared over expected project life.

6.13.5 *Effects of uncertainty*

As already stated, all the data used to calculate the NPV or DCF (capital estimate, operating costs, annual sales, selling price, etc.) are subject to

Table 6.13   Acrylonitrile: U.S. production (see Figure 6.7)

year	production × 10⁶ lb.	cumulative prod. × 10⁶ lb.	price/cents/lb.	const. 1987 price/cents/lb.
1950	25	129	36	146
1951	35	164	40	151
1952	50	214	43	159
1953	60	274	43	159
1954	70	344	31	112
1955	118	462	31	110
1956	141	603	27	93
1957	174	777	27	90
1958	180	957	27	88
1959	232	1 189	27	86
1960	229	1 418	23	72
1961	249	1 667	14·5	45
1962	360	2 027	14·5	45
1963	455	2 482	14·5	44
1964	594	3 076	17	50
1965	772	3 848	17	50
1966	716	4 564	14·5	42
1967	670	5 234	14·5	40
1968	1021	6 255	14·5	38
1969	1156	7 411	14·5	36
1970	1039	8 450	14.5	35
1971	979	9 429	14·5	33
1972	1115	10 544	13	29
1973	1354	11 898	13	28
1974	1411	13 309	14·5	28
1975	1214	14 523	23	40
1976	1518	16 041	24	39
1977	1642	17 683	27	43
1978	1752	19 435	27	39
1979	2018	21 453	25	34
1980	1830	23 283	37·5	46·3
1981	1996	25 279	44	50
1982	2041	27 320	46	49
1983	2146	29 466	43	45·7
1984	2201	31 667	45·5	47
1985	2346	34 013	45·5	46·6
1986	2314	36 327	45·5	47·2
1987	2550	38 877	34	34

uncertainty—particularly the commercial data of sales and prices. The value of a project evaluation using single point data as an aid to decision taking can be extended by carrying out a sensitivity analysis. This studies the effect on the economics of a project of changes in the major items contributing to the cash flows. It highlights areas which are most important and where uncertainty has the greatest effect on the project. The effect of a 10% change in fixed capital, change in sales or sales price, change in raw material cost, delay in start-up and other variations on the basic project estimates can be explored and the resulting changes in NPV and DCF calculated. A sensitivity analysis does not attempt to quantify the uncertainties in the different factors but explores the effects on the project of changes in these factors.

In order to quantify the uncertainties in a project it is necessary to indicate the relative chances that a variable (e.g. selling price) will have different values. This can be done subjectively on the basis of, say, a 10% chance that the price will be as low as x, a 10% chance that the price will be as high as z against an expected mid-value of y. If it is assumed that the variables lie in a normal distribution in the range considered then the subjective probability estimates can be used to define the total distribution. Having made estimates of subjective probability distribution for each of the major variables we need a method of combining the various inputs to the project cash flows to obtain the resulting distribution of NPV or DCF. One such method is the Monte Carlo simulation. If there are a number of independent inputs to the project (capital, materials costs, etc) each represented by a probability distribution of values, then there is an infinite number of possible outcomes. Representatives of these

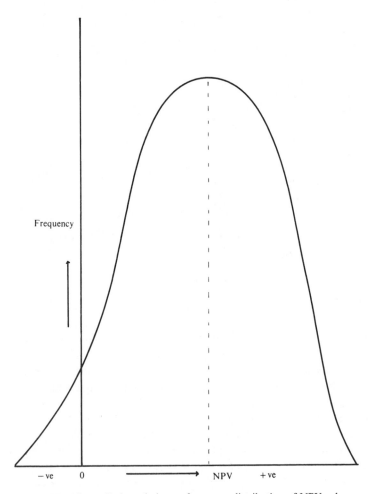

Figure 6.8   Monte Carlo technique—frequency distribution of NPV values.

can be calculated by selecting a value of each input from the range and calculating a value of NPV or DCF. Choosing a different value for any one input will lead to a different outcome.

The basis of the Monte Carlo technique is to carry out a large number of project evaluations with different input values selected from the individual distributions in a random way. The random selection is done in such a way that the number of times a value of a variable is selected is proportional to its probability. Clearly, the large number of calculations involved requires the use of a computer to make the method feasible. The result of the many calculations is a range of NPV or DCF values and, since the same value of NPV or DCF can result from different combinations of input data, a frequency distribution of the results can be plotted (Figure 6.8).

The example in Figure 6.8 shows the range of possible outcomes for the

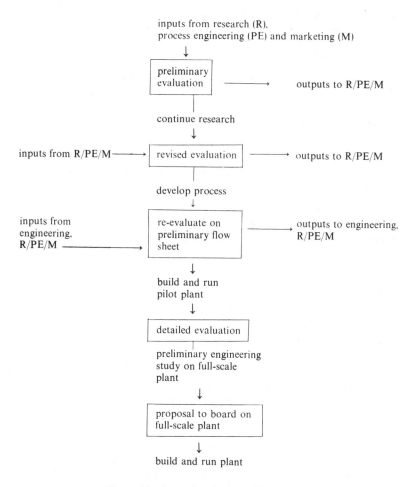

Figure 6.9 Stages involved in project evaluation.

project with a most likely value indicated by the dotted line. It also shows a small chance ($\sim 4\%$ based on ratio of negative area to total area) of a negative NPV at the company cost of capital. Having the distribution of possible outcomes enables the project manager to analyse the possible benefits and risks in the project in detail. Decisions can then be made by applying company policy with respect to risk, based on the indicated chances of gain or loss and the possible size of gain or loss.

The aim of project evaluation is to use the available data to provide information which will assist decision making on the future of the project. Use of sensitivity analysis and risk analysis techniques point up areas where uncertainty in the input data has greatest effect and indicates the effects of these uncertainties on the project outcome. Such evaluation does not eliminate the need for skilled judgement in the management team nor does it necessarily make the decision process any easier. However, it does ensure that a complete view of the project is available and makes clear the need for definitive company policy on risk and profitability criteria.

6.14
Conclusion

The objective of all the technical economic methods discussed in this chapter is to provide information to assist management decision-making. Use of standard cost sheets helps plant managers to check the short-term performance of the units under their control and prompts corrective action if raw material usages or utilities consumptions increase. Information on absorption cost and marginal costs—particularly when marginal cost increases as plant capacity is reached—enables the marketing manager to quote for additional business at a price which will make a positive cash-flow contribution. Finally the techniques of project evaluation including sensitivity analysis and risk analysis enable decisions on new projects to be taken based on the fullest information on the likely consequences interpreted in the light of company policy on risk taking and reward seeking.

Appendix

Examples of discounted cash flow (D.C.F.) calculations

1. *Cash flow comparison of benzene and n-butane routes to maleic anhydride*

Consider a 25 kilotonnes/year plant.
Sales forecasts: 10 kilotonnes in first year of operation, 20 kilotonnes in second
          year
          25 kilotonnes/year in subsequent years.
Assuming plants take 2 years to build, with sales income estimates shown below the project revenue inflow is shown in Table A for a ten year production life.

(1) *Benzene oxidation* (based on the data in Table 6.12).

Plant capital of £30 million is spent in two years with £1 million of second year expenditure on plant buildings. Plant equipment is allowable against tax on other company profits at 52% in the year following expenditure. Allowance

Table A

Year of Project Life	0	1	2	3	4	5	6	7	8	9	10	11
Sales Forecast kilotonnes/year	—	—	10	20	25	25	25	25	25	25	25	25
Sales Income £/tonne	—	—	690	690	685	685	680	680	675	670	670	665
Total Revenue Inflow £ thousands	—	—	6900	13 800	17 125	17 125	17 100	17 100	16 875	16 750	16 750	16 625

(Money values are all in £ of year 0)

Table B Benzene oxidation

Year	0	1	2	3	4	5	6	7	8	9	10	11	12
Fixed capital £000	-14 000	-16 000											
Depreciation allowance £000		+7 280	+8 080	+21	+21	+21	+21	+21	+21	+21	+21	+21	+51
Working capital £000			-1 200	-1 200	-600								+3 000
x Net capital flow £000	-14 000	-8 720	+6 880	-1 179	-579	+21	+21	+21	+21	+21	+21	+21	+3 051
Annual variable cost £000			-3 660	-7 320	-9 150	-9 100	-9 100	-9 050	-9 050	-9 050	-9 050	-9 050	
Fixed costs £000			-2 700	-2 700	-2 700	-2 700	-2 700	-2 700	-2 700	-2 725	-2 725	-2 725	
Revenue outflow £000			-6 360	-10 020	-11 850	-11 800	-11 800	-11 750	-11 750	-11 775	-11 775	-11 775	
Revenue inflow Table A £000			+6 900	+13 800	+17 125	+17 125	+17 000	+17 000	+16 875	+16 750	+16 750	+16 625	
y Net revenue inflow £000			+540	+3 780	+5 275	+5 325	+5 200	+5 250	+5 125	+4 975	+4 975	+4 850	
Tax at 52% NRI £000				-281	-1 966	-2 743	-2 769	-2 704	-2 730	-2 665	-2 587	-2 587	-2 522
NRI after tax £000			+540	+3 499	+3 309	+2 582	+2 431	+2 546	+2 395	+2 310	+2 388	+2 263	-2 522
z = x + y Net cash flow after tax £000	-14 000	-8 720	+7 420	+2 320	+2 730	+2 603	+2 452	+2 567	+2 416	+2 331	+2 409	+2 284	+529
Discount *z* at 10% £000	-14 000	-7 927	+6 132	+1 743	+1 865	+1 616	+1 384	+1 317	+1 127	+988	+929	+800	+168
Discount *z* at 5% £000	-14 000	-8 301	+6 730	+2 004	+2 247	+2 041	+1 829	+1 825	+1 635	+1 503	+1 479	+1 336	+295

Total +7341 ⎫ DCF
NPV −3858 ⎬ 6%
NPV +623 ⎭

Table C *n*-Butane oxidation

Year	0	1	2	3	4	5	6	7	8	9	10	11	12
Fixed capital £000	-16 000	-18 000											
Depreciation allowance £000		+8 320	+9 120	+21	+21	+21	+21	+21	+21	+21	+21	+21	+51
Working capital £000			-1 200	-1 200	-600								+3 000
x′ Net capital flow £000	-16 000	-9 680	+7 920	-1 179	-579	+21	+21	+21	+21	+21	+21	+21	+3 051
Annual variable cost £000			-2 500	-5 000	-6 250	-6 300	-6 300	-6 375	-6 375	-6 375	-6 425	-6 425	
Fixed costs £000			-3 075	-3 075	-3 075	-3 075	-3 075	-3 075	-3 075	-3 100	-3 120	-3 120	
Revenue outflow £000			-5 575	-8 075	-9 325	-9 375	-9 375	-9 450	-9 450	-9 475	-9 545	-9 545	
Revenue inflow Table A £000			+6 900	+13 800	+17 125	+17 125	+17 000	+17 000	+16 875	+16 750	+16 750	+16 625	
y′ Net revenue inflow £000			+1 325	+5 725	+7 800	+7 750	+7 625	+7 550	+7 425	+7 275	+7 205	+7 080	
Tax at 52% NRI £000				-689	-2 977	-4 056	-4 030	-3 965	-3 926	-3 861	-3 783	-3 747	-3 682
NRI after tax £000			+1 325	+5 036	+4 823	+3 694	+3 595	+3 585	+3 499	+3 414	+3 422	+3 333	-3 682
z′ = x′ + y′ Net cash flow after tax £000	-16 000	-9 680	+9 245	+3 857	+4 244	+3 715	+3 616	+3 606	+3 520	+3 435	+3 443	+3 354	-631
Discount *z′* at 10% £000	-16 000	-8 799	+7 636	+2 897	+2 899	+2 307	+2 039	+1 850	+1 644	+1 456	+1 329	+1 174	-201
Discount *z′* at 12% £000	-16 000	-8 644	+7 368	+2 746	+2 699	+2 106	+1 833	+1 630	+1 422	+1 240	+1 109	+963	-162

Total +15724 ⎫ DCF
NPV +231 ⎬ 10%
NPV −1690 ⎭

for buildings is spread over the project life with balancing adjustment in the final year.

Working capital is spent as sales develop and is recovered by clearance of stocks the year after plant shutdown.

Variable cost is £366/tonne in the initial years of operation, declining slightly over the project life due to minor savings in service usages at constant benzene and service prices.

Fixed cost excluding depreciation totals £2·70 million/year during most of the project life. No reduction is shown for this old, well developed process. Charges increase during the final three years life as increased maintenance is required to maintain full rate operation.

The build up of annual cash flows and NPV and DCF is shown in Table B. Discounting the net cash flows after tax at the assumed cost of capital (10%) the project shows a NPV of − £3·86 million. At the lower discount rate of 5% a NPV of + £0·62 million is shown and by interpolation the DCF rate is 6%. Since the project shows a negative cash flow at the cost of capital it is unattractive.

(2) *n-Butane oxidation* (based on the data in Table 6.12).

Plant capital of £34 million is spent in two years with £1 million of second year expenditure on buildings. Depreciation allowance as above.

Working capital is spent as sales develop and recovered after plant closure.

Variable cost is £250/tonne in the initial years of operation, increasing subsequently as improvements in the efficiency of the new process are offset by a rise (in constant money) in n-butane price.

Fixed cost excluding depreciation totals £3·075 million during most of the project life. No reduction is shown as a result of assumed operation problems with a new process. An increase is shown in the final three years as increased maintenance is needed.

The build-up of annual cash flows and NPV and DCF is shown in Table C. Discounting the net cash flows after tax at the cost of capital (10%) shows a NPV of + £0·23 million. At the higher rate of 12% the NPV is − £1·69 million and by interpolation the DCF rate is 10%. This represents a marginally attractive project i.e. DCF only just above the cost of capital. Further work to carry out sensitivity and risk analysis would be needed before a decision to go ahead could be made.

Undiscounted Cash Flows for each year for each process are plotted in Fig. A.1.

2. *Cumene production*

A company is planning to build a cumene plant with capacity 50 000 tonnes/year. Plant constructions will take two years with 40% of the fixed capital spent in year 0. The plant starts up in year 2 and operates at 50% capacity, building up to 75%, 90% and 100% in subsequent years.

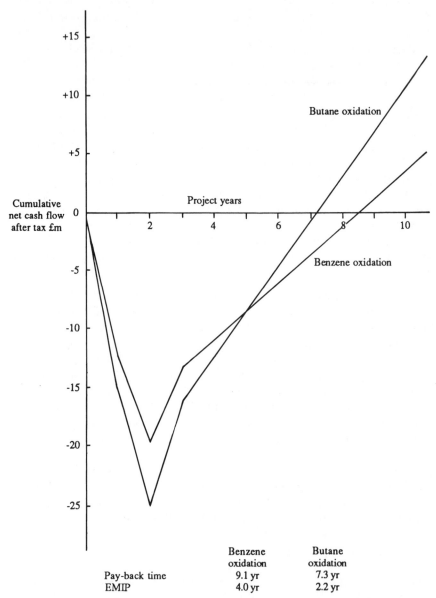

Figure A.1   Benzene and butane oxidation to maleic anhydride—cash flows.

The following information is available (all monies are in £ of year 0)

Fixed capital	£15 200 000
Working capital	£2 500 000 at full output
Sales income	£425/tonne cumene
Variable cost	£330/tonne cumene
Plant fixed costs	£760 000/yr
Overhead fixed costs	£514 000/yr

Table D Cumene-project—cash flow build up

	Year	0	1	2	3	4	5	6	7	8	9	10	11	12
	Fixed capital £000	−6080	−9120											
	Depreciation allowance £000		+3162	+4742										
	Working capital £000		—	−1250	−625	−375	−250							+2500
A	Net capital flow £000	−6080	−5958	+3492	−625	−375	−250							+2500
B	Production/sales tonnes			25000	37500	45000	50000	50000	50000	50000	50000	50000	50000	
	Sales income £000			+10625	+15938	+19125	+21250	+21250	+21250	+21250	+21250	+21250	+21250	
	Variable cost £000			8250	12375	14850	16500	16500	16500	16500	16500	16500	16500	
C	Fixed costs £000			1274	1274	1274	1274	1274	1274	1274	1274	1274	1274	
	Revenue outflow £000			9524	13649	16124	17774	17774	17774	17774	17774	17774	17774	
B−C	Net revenue inflow £000			+1101	+2289	+3001	+3476	+3476	+3476	+3476	+3476	+3476	+3476	
	Tax @ 52% NRI				−573	−1190	−1561	−1808	−1808	−1808	−1808	−1808	−1808	−1808
D	NRI after tax			+1101	+1716	+1811	+1915	+1668	+1668	+1668	+1668	+1668	+1668	−1808
A+D	Net cash flow after tax £000	−6080	−5958	+4593	+1091	+1436	+1665	+1668	+1668	+1668	+1668	+1668	+1668	+692
	Discount at 15% £000	−6080	−5181	+3472	+718	+821	+828	+721	+627	+545	+474	+412	+359	+129

NPW 7447 } DCF
−2155 } ~12%

The plant is expected to have a ten-year production life and scrap value will just cover the cost of demolition and site clearance. Working capital expenditure builds up to the maximum in proportion to production rate and is recoverable after plant ceases production.

If depreciation allowance and tax on net revenue are 52% payable a year in arrears, what is the payback period and EMIP for the project? If the company cost of capital is 15% does the proposed cumene project represent an attractive investment?

Table D shows the development of the cash flows. Fixed capital expenditure takes place in years 0 and 1 with depreciation allowance received in years 1 and 2. Working capital expenditure starts in year 2 and additional sums are required in years 3, 4 and 5 as production and sales build up. Sales income builds up as sales increase at the fixed sales income of £425/tonne cumene. Variable cost of production also increases as production builds up, and together with the fixed costs, gives the revenue outflow. The difference between sales income and revenue outflow provides the net revenue inflow. Tax at 52%

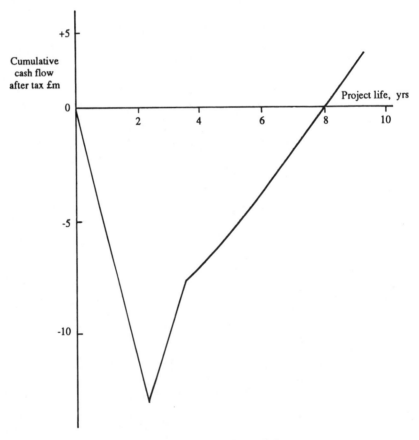

Figure A.2   Cumene project cash flows.

is deducted from the net revenue inflow one year in arrears to give the net revenue inflow after tax. Summations of this and the net capital flow gives the net cash flow after tax.

The cumulative net cash flow after tax is plotted against project life in Fig. A.2. This shows a payback time of almost 7 years from the start of the project. Measurement of the area under the curve to break-even shows just under 42 £ million years and with a maximum cumulative expenditure of £12·04 million the EMIP is almost 3·5 years. These values would have to be compared with target values set by known successes within the company.

Discounting the net cash flow after tax at the cost of capital shows a negative NPV. The proposed project would therefore be unattractive and ways must be sought to reduce cost or increase sales income.

3. Practice examples

Example 1. A company proposes to invest £500 000 this year in a plant to make a speciality chemical on a scale of 500 tonnes/year. Working capital is estimated to be £100 000 at full output and would be spent proportionally to output. Sales are forecast as 300 tonnes in year 1 rising to 400 tonnes and 500 tonnes in subsequent years. Product market life is expected to be five years before being replaced. A scrap value of £100 000 is expected for the plant after shutdown. The variable cost of the product is £350/tonne and fixed costs are £155 000/year. Depreciation allowance and tax rate are 52% payable one year in arrears and sales income is £1100/tonne. The company needs a 15% return; is the project viable? What is the DCF rate?

Yes; 16·5%.

Example 2. A study is being carried out on a proposal to produce a hydrocarbon solvent on a scale of 100 000 tonnes/year. The plant will take two years to construct with 45% expenditure in the initial year and the remaining 55% in the next year. Plant capital is estimated at £15·0 million in £ of the initial year. Production starts in the third year and build up is shown below; working capital is estimated at £3·0 million at full output, spent proportionally to production.

Variable cost of the solvent is £275/tonne and fixed costs £3·0 million/year. Sales income of £350/tonne solvent is expected to remain constant (in constant money values) over the plant life. A production life of ten years is forecast and there will be no residual scrap value. Depreciation allowance and tax rate are 52% paid one year in arrears.

The project is seen as a long-term strategic investment and the company is prepared to accept a rate of return as low as 12%. Should the project go ahead? What is the payback time and DCF?

Year of production	1	2	3	4
Production (tonnes)	60 000	75 000	90 000	100 000

Yes; 6.9 *years from start of construction*, 13.8%

Example 3. Hycar Resins plc is considering the introduction of a new resin. Fixed plant capital of £1·0 million will be spent this year for a plant to produce 3000 tonnes/year of product. Working capital at full output would be £300 000 and would be spent as sales build up. Variable cost of production is £300/tonne and fixed costs of £100 000/year. A ten-year production life is expected and net scrap value will be nil. Depreciation allowance and tax on net revenue inflow are 52% paid one year in arrears.

Market studies have shown that, at the target sales income of £470/tonne build-up of sales will be slow as shown:

Year	1	2	3	4	5	10
Sales/production (tonnes)	500	1000	1500	2300	3000	3000

The company requires a 15% return and wants to see a payback time of under six years from the start of construction. Is the project attractive? What DCF and payback time would be achieved?

No; DCF just under 15% (14·3%) payback 6·7 years

Example 4. In the above study market research shows that selling price is critical. If a slightly lower sales income of £450/tonne is accepted sales build-up would be much quicker as shown:

Year	1	2	3	10
Sales/production (tonnes)	1000	2000	3000	3000

How does the project look in terms of DCF and payback time in this situation?

Attractive; DCF 16·4%, payback 5·9 years.

Example 5. A project is under consideration for production of a detergent product on a scale of 100 000 tonnes/year. Market studies show sales build-up as follows:

Year of production	1	2	3	4
Sales (kilotonnes)	60	80	90	100

A sales income of £530/tonne is expected. Plant construction will take two years with fixed capital spent equally in each of the two years. Depreciation allowance and tax are 52% payable one year in arrears. Working capital outlay is proportional to production. Residual plant scrap value after ten years' production will be nil.

Two schemes are being studied:

(i) An available intermediate is purchased and reacted in a single step to give the desired product. Fixed capital in this case is £60 million and working capital at full output £8·0 million. Variable costs are £325/tonne and fixed costs £4·77 million/year.

(ii) A basic chemical is purchased and reacted in two stages to give the desired detergent product. Fixed capital requirement is greater at £85 million and working capital at full output increased to £11·0 million due to need for addition stocks and first stage catalyst. Variable cost is, however, only £250/tonne and fixed costs £5·00 million/year.

Capital is available internally at 13% interest. How do the schemes compare?

(i) *NPV @* 13% + £619 000, *DCF* 13·4%
(ii) *NPV @* 13% + £3 188 000 *DCF* 14·3%

If internal capital is limited to £60 million and the additional capital for scheme (ii) has to be borrowed at 15% interest rate, would the extra expenditure on scheme (ii) be justified?

Yes; *differential DCF* 16·2%.

References

1. B. Linnhoff, and J. A. Turner, *The Chemical Engineer*, 1980, 742.
2. H. E. Wessel, *Chemical Engineering*, 1952, **59** (July), 209.
3. W. L. Nelson, *Oil and Gas Journal*, 1977 (8 August), 61.
4. Stanford Research Institute, Process Economics Program report, 140.
5. C. H. Chilton, *Chemical Engineering*, 1949, **56** (June), 49.
6. D. H. Allen, *Chemical Engineering*, 1961, **74** (3 July), 75.
7. *Hydrocarbon Processing*, 1981, **60** (11), 179.
8. R. S. Marsden, P. J. Craven, and J. H. Taylor, *Transactions of the 7th International Cost Engineering Congress*, London, 1982, B10-1–B10-8.
9. A. V. Bridgewater, *Cost Engineering*, 1981, **23** (5), 293.
10. J. H. Taylor, *Engineering and Process Economics*, 1977, **2**, 259.
11. C. A. Miller, *Chemical Engineering*, 1965, **72** (13 September), 226.
12. J. Cran, *Chemical Engineering*, 1981, **88** (6 April), 65.
12a. V. T. Sinha, *Engineering Costs and Production Economics*, 1988, **14**, 259.
13. W. W. Twaddle and J. B. Malloy, *Chemical Engineering Progr.*, 1966, **62**, 90.
14. *Perspectives on Experience*, The Boston Consulting Group Inc. U.S.A., 1968.
15. J. H. Taylor and P. J. Craven, *Process Economics Internat.*, 1979, **1**, 13.

Bibliography

Introduction to Process Economics, F. A. Holland, F. R. Watson and J. N. Wilkinson, Wiley, 1974.
The Chemical Economy, B. G. Reuben and M. L. Burstall, Longmans, 1973.
Institute Francais du Pétrole, *Manual of Economic Evaluation of Chemical Processes*, McGraw-Hill, 1981.
A Guide to the Economic Evaluation of Projects, 2nd Edition, D. H. Allen, Institute of Chemical Engineers, London, 1980.
Managing Technological Innovation, B. Twiss, Longmans, 1974.
Assessing Projects: A Programme for Learning, ICI Methuen, 1972.
Preliminary Chemical Engineering Plant Design, W. D. Baasel, Elsevier, 1976.
Modern Decision Analysis, G. M. Kaufmann and H. Thomas (eds.), Penguin Modern Management Readings, Penguin, 1977.

7 Chemical Engineering

Richard Szczepanski

**7.1
Introduction**

The aim of this chapter is to provide an introduction to the basic principles of chemical engineering and some of the techniques used by engineers in the analysis and design of chemical processes. It is impossible to be comprehensive within the bounds of a single chapter, and some important topics such as safety, materials and chemical reaction engineering have been omitted. The reader is referred to the general bibliography at the end of this chapter which lists books covering the whole of the subject.

The role of the chemical engineer is not just to carry out chemistry on a large scale. Typically, the engineer is involved in the design of a chemical plant, its construction, commissioning and its subsequent operation. All this requires a wide range of skills and knowledge and hence chemical engineering is a very broadly-based discipline. The practising engineer must be able to apply basic principles of mathematics, physics and chemistry to solve complex problems on a large scale. Information available may be unreliable or incomplete and there are usually many constraints including those of time, money and legal requirements. The ability to analyse such problems and render them tractable by appropriate simplifying assumptions is one of the most important skills of the engineer.

The analysis of chemical engineering systems is based on the principles of conservation of mass, energy and momentum. The following sections are mostly concerned with the quantitative application of these simple principles. Many small-scale problems are used as illustrative examples and these provide an idea of the routine calculations carried out by chemical engineers. It must be remembered, however, that basic technical skills are only one component in tackling engineering problems.

**7.2
Material balances**

7.2.1 The flowsheet

A process flowsheet is a schematic representation of a process which shows the equipment used and its interconnections. The flowsheet of a prototype chemical process is shown in Figure 7.1. The stages involving preparation of feed and separation of products usually occupy most of the equipment in a

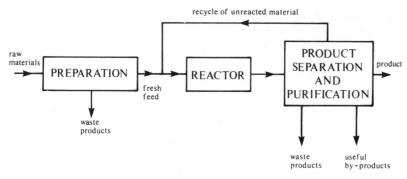

Figure 7.1   Prototype chemical process.

plant. The reaction stage is of course crucial but is generally confined to a single vessel. An example of an actual process flowsheet is shown in Figure 7.2. The process for production of vinyl acetate (ethenyl ethanoate) is typical in having lengthy and complex separation stages.

The type of information shown on a flowsheet depends on its purpose. Figure 7.2 is a useful aid to understanding a process description. In order to analyse process performance or specify equipment more information is necessary. It is convenient to have the flowrate, composition, temperature and

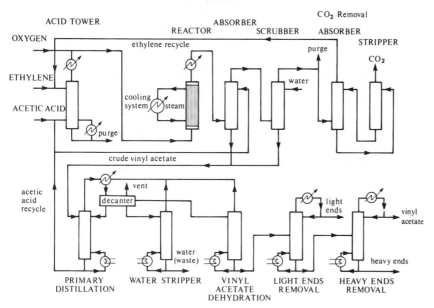

Figure 7.2   Vinyl acetate process.

pressure of each stream displayed on the flowsheet. For a process of any complexity this information would be in tabular form next to the drawing—for an example see Coulson and Richardson[1]. These authors also list the flowsheet symbols recommended by B.S.I.

7.2.2 *General balance equation*

The material balance establishes relationships between flows and compositions in different parts of a plant. It provides the principal tool for analysing a process.

Consider the system shown in Figure 7.3. Input and output of material are related by the equation.

$$\begin{bmatrix} \text{input of} \\ \text{material} \end{bmatrix} + \begin{bmatrix} \text{generation of material} \\ \text{within system} \end{bmatrix} - \begin{bmatrix} \text{output of} \\ \text{material} \end{bmatrix}$$
$$- \begin{bmatrix} \text{consumption of material} \\ \text{within system} \end{bmatrix} = \begin{bmatrix} \text{accumulation of material} \\ \text{within system} \end{bmatrix} \qquad (7.2.1)$$

Equation 7.2.1 may be applied in different ways according to the precise definition of 'material' and the way in which the process is operated.

If the balance is applied to the total mass or to the mass of each element entering and leaving the system, the generation and consumption terms are zero (excluding nuclear reactions) and equation 7.2.1 becomes

$$\begin{bmatrix} \text{total mass} \\ \text{in} \end{bmatrix} - \begin{bmatrix} \text{total mass} \\ \text{out} \end{bmatrix} = \begin{bmatrix} \text{accumulation} \\ \text{of mass} \end{bmatrix} \qquad (7.2.2)$$

or

$$\begin{bmatrix} \text{mass of element} \\ i \text{ in} \end{bmatrix} - \begin{bmatrix} \text{mass of element} \\ i \text{ out} \end{bmatrix} = \begin{bmatrix} \text{accumulation of mass} \\ \text{of element } i \end{bmatrix} \qquad (7.2.3)$$

When considering quantities (mass or number of moles) of individual molecular species, material may be produced or consumed by chemical

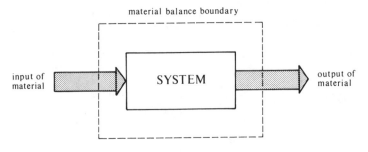

Figure 7.3

reaction (section 7.2.5). In the absence of chemical reactions equations like 7.2.2 apply also to the total number of moles of each molecular species.

Processes may be classified as *continuous* (open) or *batch* (closed). Most modern high volume processes operate with a continuous feed and form product continuously. In a batch process, materials are charged to a vessel and products withdrawn when the reaction is complete. Sometimes, as in batch distillation, products may be withdrawn continuously. Batch operation is usually used for low volume products, e.g. manufacture of pharmaceuticals.

The material balance equation for a batch process must necessarily include an accumulation term. Continuous processes are often assumed to operate at *steady-state*, i.e. process variables such as flows do not change with time. There is therefore no accumulation and the general balance equation becomes

rate of input + rate of generation = rate of output + rate of consumption

$$(7.2.4)$$

or, for the total mass,

$$\text{mass flow in} = \text{mass flow out} \qquad (7.2.5)$$

Alternatively, when considering the operation of a continuous steady-state process for a fixed period of time, each of the terms in equation 7.2.4 may be expressed simply as a mass or number of moles. Equation 7.2.2 thus becomes

$$\text{total mass in} = \text{total mass out}$$

The following is an example of a material balance calculation.

Example 7.1

$10 \, \text{kg s}^{-1}$ of a 10% (by mass) NaCl solution is concentrated to 50% in a continuous evaporator. Calculate the production rate of concentrated solution (C) and the rate of water removal (W) from the evaporator.

Solution

The flowsheet is shown below (Figure 7.4). The steady state mass balance is given by equation 7.2.5. Balances can be made for individual species or total

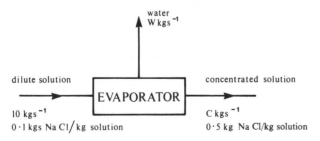

Figure 7.4

mass. Because no chemical reaction takes place we can write balances for molecular species rather than considering individual elements.

For each species, flowrate = mass fraction × total flowrate.

Balance on NaCl:

$$0.1 \times 10 = 0.5 \times C$$
$$\therefore C = 2 \, kg \, s^{-1}.$$

Total balance:

$$10 = W + C$$
$$\therefore W = 8 \, kg \, s^{-1}.$$

7.2.3 *Material balance techniques*

It is usually helpful to follow a systematic procedure when tackling material balance problems. One possibility is outlined below.

1. *Draw and label the process flowsheet*—organize information into an easy to understand form. If possible show problem specifications on the flowsheet. Label unknowns with algebraic symbols.
2. *Select a basis for the calculation*—the *basis* is an amount or flowrate of a particular stream or component in a stream. Other quantities are determined in terms of the basis. E.g. in Example 7.1 the flowrates of product and water were obtained on a basis of $10 \, kg \, s^{-1}$ of feed. It is usually most convenient to choose an amount of feed to the process as a basis. Molar units are preferable if chemical reactions occur, otherwise the units in the problem statement (mass or molar) are probably best.
3. *Convert units/amounts*—as necessary to be consistent with the basis.
4. *Write material balance equations*—for each unit in the process or for the overall process. In the absence of chemical reactions the number of independent equations for each balance is equal to the number of components.
5. *Solve equations*—for unknown quantities. This can be difficult, particularly if non-linear equations are involved. Overall balances usually give simpler equations. For complex flowsheets computer methods offer the only practical solution.
6. *Scale the results*—if the basis selected is not one of the flowrates in the problem specification the results must be scaled appropriately.

The following example illustrates the use of this procedure.

Example 7.2

An equimolar mixture of propane and butane is fed to a distillation column at the rate of $67 \, mol \, s^{-1}$. 90% of the propane is recovered in the top product which has a propane mole fraction of 0.95. Calculate the flowrates of top and bottom products and the composition of the bottom product.

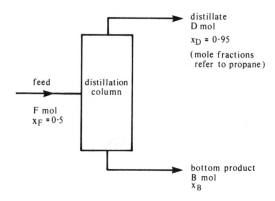

distillate
D mol

$x_D = 0.95$

(mole fractions
refer to propane)

feed

distillation
column

F mol
$x_F = 0.5$

bottom product
B mol
x_B

Figure 7.5

Solution

1. Flowsheet (Figure 7.5)
2. Basis: 10 mol of feed, i.e. $F = 10$ mol. This basis is somewhat easier to work with than the specified feed flowrate. By selecting a *quantity* of feed (rather than a flowrate) we are, implicitly, considering the operation of the process for a fixed period of time.
3. Conversion of units—not necessary, all units are molar.
4. Equations—there are no reactions and the steady-state balance equation for total or component flows is

$$\text{input} = \text{output}$$

(i) Total material balance

$$F = D + B$$

(ii) Component balances
    propane: $Fx_F = Dx_D + Bx_B$
    butane: $F(1 - x_F) = D(1 - x_D) + B(1 - x_B)$

It is evident that these equations are not independent. The total balance can be obtained by summing the component balances. Inserting specified values into the first two equations gives

$$10 = D + B \tag{1}$$

$$5 = 0.95D + Bx_B \tag{2}$$

A third independent equation is provided by the specification on propane recovery:

$$x_D D = 0.90 F x_F$$

$$\text{or} \quad 0.95D = 0.90 \times 10 \times 0.50$$

$$\therefore 0.95D = 4.5 \tag{3}$$

5. Solution
   From (3) $D = 4 \cdot 74 \, \text{mol}$

   Substituting in (1), $B = 5 \cdot 26 \, \text{mol}$

   and equation (2) is now solved for the bottom-stream composition, $x_B = 0 \cdot 095$.
6. Scaling—compositions are unchanged by the choice of basis. The quantities obtained above must be scaled by

$$\frac{67 \, \text{mol s}^{-1}}{10 \, \text{mol}} = 6 \cdot 7 \, \text{s}^{-1}$$

to solve the problem as stated, i.e.

$$\text{Bottom flowrate} = 5 \cdot 26 \times 6 \cdot 7 = 35 \cdot 2 \, \text{mol s}^{-1}$$
$$\text{Top flowrate} \quad = 4 \cdot 74 \times 6 \cdot 7 = 31 \cdot 8 \, \text{mol s}^{-1}$$

7.2.4 Multiple unit balances

Most processes consist of many interconnected units. In analysing such processes material balance equations may be written for each unit, for groups of units, or for the whole plant. To obtain a unique solution the number of equations describing a process must be equal to the number of unknown variables. If the analysis leads to fewer equations it is necessary to specify extra *design variables*. In the case of an actual plant where values of process variables are obtained by direct measurements the number of equations may exceed the number of unknowns. In such circumstances calculations should be based on the most reliable measurements.

Example 7.3

The flowsheet (Figure 7.6) shows part of a process for the production of pure ethanol from aqueous solution by azeotropic distillation. Known quantities are shown on the flowsheet. Determine the remaining flowrates and compositions.

Solution

Basis: 1 second, i.e. 10 kg of feed.
   Four possible material balance boundaries are shown in Figure 7.7. The equations and unknowns associated with each balance will now be considered.

A: Overall balance

The notation F_j will be used to represent the total (mass) flowrate of stream j and $x_{i,j}$ denotes the mass fraction of component i in stream j. Component 1 is ethanol, 2, water, and 3, benzene.

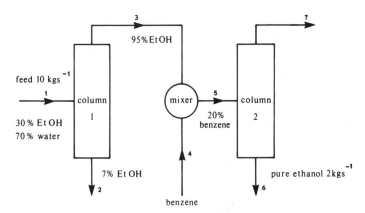

Figure 7.6

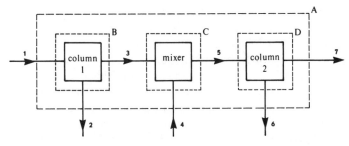

Figure 7.7

Equations:

$x_{1,2}F_2 + x_{1,7}F_7$
$\quad = x_{1,1}F_1 - x_{1,6}F_6 = 3 - 2 = 1 \text{ (ethanol)}$
$x_{2,2}F_2 + x_{2,7}F_7 = 7 \text{ (water)}$
$F_4 = x_{3,7}F_7 \text{ (benzene)}$
$x_{1,2} + x_{2,2} = 1$
$x_{1,7} + x_{2,7} + x_{2,7} = 1$

_____

total 5

Variables:

$F_2 \; x_{1,2} \; F_7 \; x_{1,7}$
$x_{2,2} \; x_{2,7}$
$F_4 \; x_{3,7}$

_____

total 8

The overall balance does not give enough equations to solve for all the variables. Three more unknowns would have to be specified before this balance could be used.

B: Balance on column 1

Equations:

$0.95F_3 + 0.07F_2 = 3 \text{ (ethanol)}$
$0.05F_3 + 0.93F_2 = 7 \text{ (water)}$

_____

total 2

Variables:

$F_3 \; F_2$

_____

total 2

Equations resulting from balance **B** may be solved immediately to give $F_2 = 7\cdot39$ kg, $F_3 = 2\cdot61$ kg.

As an exercise confirm that the numbers of equations and variables involved for the other unit are as follows:

C: Balance on mixer

4 equations   5 variables: F_3 F_4 F_5 $x_{1,5}$ $x_{2,5}$

D: Balance on column 2

5 equations   7 variables: F_5 $x_{1,5}$ $x_{2,5}$
                                    F_7 $x_{1,7}$ $x_{2,7}$ $x_{3,7}$

Only balance **B** yields enough equations to solve directly for the unknowns. However, solving **B** for F_2 and F_3 eliminates F_3 from the unknowns in balance **C** and leaves 4 equations in 4 unknowns. Solving these gives $F_4 = 0\cdot65$ kg, $F_5 = 3\cdot26$ kg, $x_{1,5} = 0\cdot75$ and $x_{2,5} = 0\cdot04$. Balance **D** equations can now be solved to give $F_7 = 1\cdot26$ kg, $x_{1,7} = 0\cdot48$, $x_{2,7} = 0\cdot10$, $x_{3,7} = 0\cdot52$.

Example 7.3 illustrated the technique of working forwards through a process, solving balance equations unit by unit. This is usually possible for processes without recycle streams provided that the feed is specified. If no individual balance yields enough equations it is necessary to solve simultaneously equations arising from balances on two or more units. Most multiple unit processes involve recycle streams but the treatment of these will be postponed until section 7.2.5.3.

When no reactions occur the number of independent equations contributed by each balance is equal to the number of components in the streams. Balances in terms of total flows and composition fractions (as in Example 7.3) must also satisfy a constraint equation on the sum of fractions in each stream j,

$$\sum_i x_{i,j} = 1 \qquad (7.2.6)$$

Overall balances are simply the sum of all unit balances and do not introduce extra independent equations. However, it is often advantageous to use overall balances and discard some of the unit balances. This is because most information is usually known about feed and product streams.

7.2.5 *Chemical reactions*

Many examples of industrially important reactions are described in other parts of this book, particularly in Chapters 2, 11 and 12.

The treatment of material balances for systems in which chemical reactions take place involves some new considerations. Generation and consumption terms must be included for molecular species and the stoichiometric constraints must be observed.

7.2.5.1 *Stoichiometry*. The *stoichiometric equation* of a reaction defines the ratios in which molecules of different species are consumed or formed in the reaction, e.g.

$$C_2H_6 + \tfrac{7}{2}O_2 \rightarrow 2CO_2 + 3H_2O$$

For the purposes of defining *stoichiometric coefficients* v_i, it is convenient to write the stoichiometric equation with all species, i, on the right-hand side, i.e.

$$0 = 2CO_2 + 3H_2O - C_2H_6 - \tfrac{7}{2}O_2$$

The stoichiometric coefficients are

$$v_{CO_2} = 2 \qquad v_{C_2H_6} = -1$$

$$v_{H_2O} = 3 \qquad v_{O_2} = -\tfrac{7}{2}$$

A general stoichiometric equation may be written as

$$\sum_i v_i A_i = 0 \qquad\qquad (7.2.7)$$

where A_i are the participating species and v_i is negative for reactants, positive for products, and zero for inerts (substances unchanged in the reaction).

7.2.5.2 *Extent of reaction*. It is useful to have a measure of the amount of material consumed or produced in a chemical reaction. The most convenient quantity is the *extent*. Consider a material balance for a chemical reactor (Figure 7.8). The extent ξ is defined by the equation

$$f_{i,out} = f_{i,in} + v_i \xi \qquad\qquad (7.2.8)$$

where $f_{i,out}$ is the number of moles of species i leaving the reactor
          $f_{i,in}$ is the number of moles of i entering
          v_i is the stoichiometric coefficient.

i.e.
$$\xi = \frac{f_{i,out} - f_{i,in}}{v_i} \qquad\qquad (7.2.9)$$

The extent has the same units as f_i, i.e. mol or mol/unit time. It is always a positive quantity because of the sign convention for v_i and has the same value for all species because $(f_{i,in} - f_{i,out})$ is proportional to v_i.

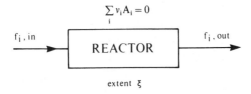

Figure 7.8   Material balance for a chemical reactor.

Most chemical reactions do not result in complete conversion of reactants to products. The thermodynamic relationships which govern the extent of a reaction are discussed in section 11.3. The maximum possible extent depends on the equilibrium constant but it is often more convenient to define a *fractional conversion* based on the quantities present in the feed:

$$\text{fractional conversion of species } i \text{ (a reactant)} = \frac{f_{i,in} - f_{i,out}}{f_{i,in}} \qquad (7.2.10)$$

Unless the feed composition is stoichiometric the fractional conversion will be different for each species. Even for a reaction which goes to completion the fractional conversion will not be unity except for the limiting reactant.

The following is a simple example of a material balance calculation for a reactive system.

Example 7.4

200 mol of ethane are burned in a furnace with 50% excess air. A conversion of 95% is achieved. Calculate the composition of the stack gases.

Solution

Flowsheet: (Figure 7.9)

The stated quantity of ethane will be used as a basis. The stoichiometric equation is

$$C_2H_6 + \tfrac{7}{2}O_2 \rightarrow 2CO_2 + 3H_2O$$

A stoichiometric feed would require $200 \times \tfrac{7}{2} = 700$ mol O_2, thus 50% excess air provides 1050 mol O_2. Taking air to be 21% (molar) O_2 and 79% N_2 gives 3.76 mol N_2/mol O_2, i.e. 3948 mol N_2 in the feed.

The conversion is 95%; since no reactant is specified we assume that it refers to the limiting reactant—ethane in this case.

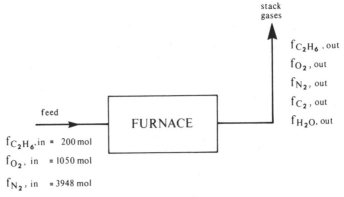

Figure 7.9

$$\text{conversion} = \frac{f_{C_2H_6,\text{in}} - f_{C_2H_6,\text{out}}}{f_{C_2H_6,\text{in}}} = 0.95$$

$$\therefore f_{C_2H_6,\text{out}} = 10\,\text{mol}.$$

$$\text{Extent } \xi = \frac{f_{C_2H_6,\text{out}} - f_{C_2H_6,\text{in}}}{\nu_{C_2H_6}} = \frac{10 - 200}{-1} = 190\,\text{mol}$$

The material balance is summarized in the table below

	IN		OUT	
	mol	mol		mole fraction
C_2H_6	200	$200 - \xi$	10	0·0019
O_2	1050	$1050 - \frac{7}{2}\xi$	385	0·0727
N_2	3948	3948	3948	0·7459
CO_2	0	2ξ	380	0·0718
H_2O	0	3ξ	570	0·1077
		$5198 + \frac{1}{2}\xi$	5293	1·0

7.2.5.3 *Recycles.* Processes in which part of a product stream is separated and recycled back to the feed are very common in the chemical industry. The protoype chemical process in Figure 7.1 shows a recycle stream. Material balance techniques are, in principle, the same as for non-recycle processes. However, because the recycle stream is usually unspecified, a large number of equations may have to be solved simultaneously.

Recycles are an essential part of most processes involving chemical reactions because it is usually difficult to achieve near-equilibrium conversion in a *single pass* of reactants through a reactor. It may also be the case that equilibrium conversion is very low, e.g. in ammonia or methanol synthesis (see section 12.4). The *overall* conversion for the reactor-separator-recycle system can be much closer to 100%. The following example illustrates this.

Example 7.5

A methanol synthesis loop with a stoichiometric feed of CO and H_2 is to be designed for a 95% overall conversion of CO. All the methanol formed leaves in the product stream. Not more than 2% of the CO and 0·5% of the H_2 emerging from the reactor is to leave in the product stream—the remainder is recycled. Calculate the single pass conversion, the recycle ratio and the composition of the product.

Solution

Flowsheet: (Figure 7.10)
A basis of 100 mol CO in the feed will be used.

$$\text{Reaction: } CO + 2H_2 \rightarrow CH_3OH, \qquad \text{extent } \xi\,\text{mol}.$$

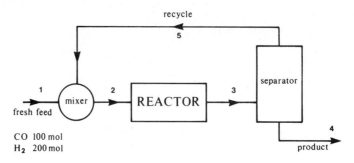

Figure 7.10

Let $f_{i,j}$ represent the number of moles of component i in stream j. Components are numbered as follows: $1, CO$; $2, H_2$; $3, CH_3OH$.

Overall balances
The overall balance equations have the same form as the reactor balance (equation 7.2.8) because the only consumption and generation terms are due to the chemical reaction.

$$CO: f_{1,4} = f_{1,1} - \xi$$

$$H_2: f_{2,4} = f_{2,1} - 2\xi$$

$$CH_3OH: f_{3,4} = \xi$$

Inserting basis quantities for the feed gives

$$f_{1,4} = 100 - \xi \tag{1}$$

$$f_{2,4} = 200 - 2\xi \tag{2}$$

$$f_{3,4} = \xi \tag{3}$$

Mixer

$$100 + f_{1,5} = f_{1,2} \tag{4}$$

$$200 + f_{2,5} = f_{2,2} \tag{5}$$

$$f_{3,5} = f_{3,2} = 0$$

Separator

$$f_{1,3} = f_{1,5} + f_{1,4} \tag{6}$$

$$f_{2,3} = f_{2,5} + f_{2,4} \tag{7}$$

$$f_{3,3} = f_{3,4} \tag{8}$$

Specifications

$$\text{Overall conversion of CO} = \frac{\begin{array}{c}\text{moles in} \\ \text{fresh feed}\end{array} - \begin{array}{c}\text{moles in} \\ \text{product}\end{array}}{\text{moles in fresh feed}} = 0.95$$

i.e.

$$\frac{100 - f_{1,4}}{100} = 0.95 \tag{9}$$

product purity:

$$f_{1,4} = 0.02 f_{1,3} \tag{10}$$

$$f_{2,4} = 0.005 f_{2,3} \tag{11}$$

Solution of equations

Note that the overall balances replace balances over the reactor.

Equation (9) gives $f_{1,4} = 5$ mol and thus from (1) $\xi = 95$ mol. The remaining equations are solved in the order (2), (3), (10), (11), (6), (7), (8), (4), (5). The following table summarizes the results.

Component	*Stream (quantities are in mol)*				
	1	2	3	4	5
1. CO	100	345	250	5	245
2. H_2	200	2190	2000	10	1990
3. CH_3OH	0	0	95	95	0
total	300	2535	2345	110	2235

$$\begin{array}{c}\text{Single pass} \\ \text{conversion of CO}\end{array} = \frac{\text{CO in reactor feed} - \text{CO in reactor outlet}}{\text{CO in reactor feed}}$$

$$= \frac{345 - 250}{345} = 0.275$$

The single pass conversion of 27% is increased to 95% overall by recycling unreacted CO.

$$\text{Recycle ratio} = \frac{\text{recycle flowrate}}{\text{fresh feed flowrate}}$$

$$= \frac{2235}{300} = 7.45 \text{ mol recycle/mol feed}$$

Composition of product stream is 4.5% CO, 9.1% H_2 and 86.4% CH_3OH.

7.2.5.4 *Multiple reactions.* If more than one reaction occurs, equation 7.2.8 becomes

$$f_{i,out} = f_{i,in} + \sum_k v_{i,k}\xi_k \tag{7.2.11}$$

where $v_{i,k}$ is the stoichiometric coefficient of component i in reaction k and ξ_k is the extent of reaction k.

The terms *selectivity* and *yield* are used to describe how far a particular (desired) reaction proceeds relative to other (undesired) reactions.

$$\text{Selectivity} = \frac{\text{moles desired product formed}}{\text{moles undesired product formed}}$$

$$\text{Yield} = \frac{\text{moles desired product formed}}{\text{moles of specified reactant fed or consumed}}$$

The following example includes two competing reactions and also introduces the concept of a *purge* stream. A fraction of the recycle leaves the process in the purge—it may be treated or simply vented. The purge is a means of removing substances which do not leave the system elsewhere. If a purge was not used in Example 7.6 methane would accumulate in the system and eventually prevent any useful reaction. It is desirable to keep the purge flowrate as small as possible to minimize loss of useful recycled materials.

Example 7.6

The catalytic dealkylation of toluene to benzene involves recycling of unreacted toluene after removal of by-product phenylbenzene. Using the information shown on the process flowsheet (Figure 7.11) determine

(a) quantities of recycle streams,
(b) quantity and composition of purge stream,
(c) quantity of hydrogen make up.

Solution

Basis: 100 mol toluene in reactor feed—this is the most convenient basis since most information is provided about stream 2.

Reactions:

$$\text{toluene} + H_2 \rightarrow \text{benzene} + CH_4, \text{extent } \xi_A$$

$$2 \text{ benzene} \rightarrow \text{phenylbenzene} + H_2, \text{ extent } \xi_B$$

Balance on reactor

Let $f_{i,j}$ denote the number of moles of component i in stream j. Components are numbered: 1, toluene; 2, hydrogen; 3, benzene; 4, methane; 5, phenylbenzene.

$$f_{1,3} = 100 - \xi_A$$

$$f_{2,3} = 500 - \xi_A + \xi_B$$

$$f_{3,3} = \xi_A - 2\xi_B$$

$$f_{4,3} = 500 + \xi_A \quad \text{(quantity of } CH_4 \text{ into reactor is}$$

$$f_{5,3} = \xi_B \quad\quad\quad \text{equal to quantity of } H_2\text{)}$$

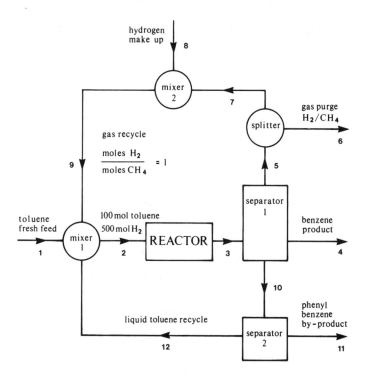

Conversion of toluene per pass = 25%

Yield of benzene (based on toluene consumed) per pass = 75%

Figure 7.11

$$\text{Conversion} = \frac{\xi_A}{100} = 0.25$$

$$\therefore \xi_A = 25 \text{ mol.}$$

$$\text{Yield} = \frac{f_{3,3}}{100 - f_{1,3}} = \frac{\xi_A - 2\xi_B}{\xi_A} = 0.75$$

$$\therefore \xi_B = 3.12 \text{ mol.}$$

Overall balances

$$0 = f_{1,1} - \xi_A$$

$$f_{2,6} = f_{2,8} - \xi_A + \xi_B$$

$$f_{3,4} = \xi_A - 2\xi_B$$

$$f_{4,6} = \xi_A \tag{1}$$

$$f_{5,11} = \xi_B$$

Equation (1) expresses the requirement that all methane formed in the reactor must leave in the purge. Substituting for the extents gives

$$f_{1,1} = 25 \, \text{mol}$$

$$f_{2,6} = f_{2,8} - 21 \cdot 88 \, \text{mol} \qquad (2)$$

$$f_{3,4} = 18 \cdot 75 \, \text{mol}$$

$$f_{4,6} = 25 \, \text{mol}$$

$$f_{5,11} = 3 \cdot 12 \, \text{mol}$$

Purge

All the hydrogen and methane leaving the reactor are in stream 5, thus

$$f_{2,5} = f_{2,3} = 478 \cdot 12 \, \text{mol}$$

$$f_{4,5} = f_{4,3} = 525 \, \text{mol}.$$

The fraction of this methane leaving in the purge (stream 6) is

$$\frac{f_{4,6}}{f_{4,5}} = \frac{25}{525} = 0 \cdot 0476$$

The same fraction of the H_2 in stream 5 leaves in the purge

$$\therefore \frac{f_{2,6}}{f_{2,5}} = \frac{f_{2,6}}{478 \cdot 12} = 0 \cdot 0476$$

$$\Rightarrow f_{2,6} = 22 \cdot 77 \, \text{mol}$$

Answers

(a) Gas recycle (stream 9) = 1000 mol
   (500 mol H_2, 500 mol CH_4)
   Liquid toluene recycle (stream 12):
   25 mol of fresh feed ($f_{1,1}$) make up the 100 mol of toluene fed to the reactor. Thus, recycle = 75 mol.

(b) Purge = 47·44 mol (52% methane, 48% hydrogen).

(c) Hydrogen make up: from the overall balance (equation (2))
   $f_{2,8} = f_{2,6} + 21.88$. i.e. make up = 44·65 mol H_2.

7.3

Energy balances

The energy balance is based on the principle of conservation of energy. It provides an important additional technique for analysing processes. An energy balance is used to determine the energy requirements of a process or unit in terms of heating, cooling or work (pumps, compressors, etc.). Although for many purposes it is possible to carry out material and energy balances independently this is not always the case, e.g. in the design of a chemical reactor the extent is strongly dependent on the temperature and the two balances must be solved simultaneously.

7.3.1 *Energy balance equations*

The general balance equation (7.2.1) may be applied to energy but the generation and consumption terms are always zero. Energy is neither generated nor consumed in chemical reactions—there is merely a difference in energy associated with chemical bonds in reactants and products.

$$\begin{bmatrix} \text{energy} \\ \text{in} \end{bmatrix} - \begin{bmatrix} \text{energy} \\ \text{out} \end{bmatrix} = \begin{bmatrix} \text{accumulation of} \\ \text{energy within system} \end{bmatrix} \qquad (7.3.1)$$

7.3.1.1 *Steady-state flow systems.* Energy flows for a steady-state system are represented in Figure 7.12. The forms of energy important in chemical processes and their relative magnitudes are discussed in more detail in section 8.2. The internal energy term includes both 'chemical' energy of bonding and 'thermal' energy due to molecular motion and intermolecular interactions. The 'flow work' term represents the work done by the surroundings or system in transferring material across system boundaries—see Himmelblau[2]. In addition to energy associated with streams, energy may also be transferred between the system and its surroundings in the form of heat or by doing work. We use the IUPAC convention that *heat transferred to the system and work done on the system is defined as positive.*

For a system at steady state

$$[\text{energy in}] = [\text{energy out}]$$

hence

$$\sum_{\substack{\text{inlet} \\ \text{streams}}} f(u + Pv + P' + K') + Q + W$$

$$= \sum_{\substack{\text{outlet} \\ \text{streams}}} f(u + Pv + P' + K') \qquad (7.3.2)$$

where f is the flowrate, P is the pressure and u, v, P' and K' are specific

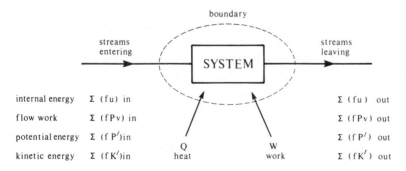

Figure 7.12 Energy flows for a continuous chemical process.

internal energy, volume, potential energy and kinetic energy. The difference between potential and kinetic energy terms in inlet and outlet streams is usually small compared with the remaining terms. In such cases

$$\sum_{\text{outlet}} f(u + Pv) - \sum_{\text{inlet}} f(u + Pv) = Q + W$$

or in terms of the enthalpy $h = u + Pv$,

or
$$\left.\begin{array}{c} \sum_{\text{outlet}} f_i h_i - \sum_{\text{inlet}} f_i h_i = Q + W \\[2ex] \Delta H = Q + W \end{array}\right\} \qquad (7.3.3)$$

This form of the energy balance equation is the one most frequently used.

Note that lower-case symbols are used for specific quantities and upper case for total quantities, i.e. h has units of J mol^{-1} or J kg^{-1} whilst H has units of J.

7.3.1.2 *Closed systems.* A closed system has no flow of matter in or out. Batch processes are usually operated in this way. Changes in the internal energy of the system must be due solely to heat flow and work done. Hence

$$U_{\text{final}} - U_{\text{initial}} = Q + W$$

or

$$\Delta U = Q + W \qquad (7.3.4)$$

where U is the total internal energy of the system.

7.3.2 *Estimation of enthalpy changes*

To apply equation 7.3.3 the enthalpy of inlet and outlet streams must be obtained. For a few common substances tables of thermodynamic properties are available (see reference 3 for a reasonably up-to-date list). In most cases a correlation or estimation method has to be used. The thermodynamic background is outside the scope of this chapter but is well covered in standard texts[3]. Reid *et al.*[4] have produced a compendium of correlation/estimation methods for a wide range of thermophysical properties. Simple correlations for use in energy balances have been compiled by Himmelblau[2]. Most of the data used in this section are taken from that source.

It is necessary to evaluate enthalpy changes for the following elementary processes:

    (i) change of temperature at constant pressure
    (ii) change of pressure at constant temperature
    (iii) change of phase
    (iv) mixing of pure substances
    (v) chemical reaction (section 7.3.3)

Only the first of these will be elaborated on in this section—for details of (ii) to (v) see references 2–5.

7.3.2.1 *Use of heat capacities*.

The change in enthalpy at constant pressure due to a change in temperature from T_0 to T_1 is given by

$$\Delta h = \int_{T_0}^{T_1} c_p \, dT \qquad (7.3.5)$$

where c_p is the isobaric heat capacity. Heat capacity data for many substances have been correlated as functions of temperature by simple empirical equations.

Example 7.7

Determine the heat load required to heat a stream of nitrogen, flowing at $50 \, \text{mol min}^{-1}$, from 20°C to 100°C.

Solution

This is a steady-state flow system, hence $\Delta H = Q + W$. Assuming that work done is negligible in comparison with the heating

$$\Delta H \approx Q$$

The heat capacity of nitrogen gas at low pressures is represented[2] by

$$c_p = 29 \cdot 0 + 2 \cdot 2 \times 10^{-3} \, T + 5 \cdot 7 \times 10^{-6} \, T^2$$
$$- 2 \cdot 87 \times 10^{-9} \, T^3 \, \text{J K}^{-1} \, \text{mol}^{-1} \qquad (7.3.6)$$

where T is in °C.

$$\Delta h = \int_{20}^{100} c_p(T) \, dT$$

Performing the integration gives $\Delta h = 2332 \, \text{J mol}^{-1}$. The flowrate is $50 \, \text{mol min}^{-1}$.

$$\therefore Q = 50 \times 2332 = 116 \cdot 6 \, \text{kJ min}^{-1}$$
$$= 1 \cdot 94 \, \text{kW}$$

The pressure was not specified in Example 7.7 but it has little effect on the answer. Experimental data for nitrogen are tabulated below.

P/MPa	$\Delta h(20°C \rightarrow 100°C)/J \, mol^{-1}$
10^{-6}	2330
10^{-1}	2333
1	2358
10	2594
100	2922

At low pressures the heat capacity changes little with pressure. Up to about 1 MPa the use of low pressure equations like 7.3.6 involves little error.

7.3.3 Reactive systems

Energy changes which accompany chemical reactions are usually large and therefore form an important component in the energy balance. Equation 7.3.3 is applicable to reactive systems but care must be taken to use appropriate and consistent expressions for the enthalpy.

7.3.3.1 *Enthalpy*. For the purposes of energy balances the enthalpy of a substance i at a pressure P and temperature T may be evaluated by expressing it in the form

$$h_i(P, T) = (h_i(P, T) - h_i^\circ) + \Delta h_{f,i}^\circ \qquad (7.3.7)$$

where $\Delta h_{f,i}^\circ$ is the standard enthalpy change of formation of i from its elements and $h_i^\circ = h_i$ (101325 Pa, 298 K) i.e. the enthalpy at standard conditions. Equation 7.3.7 is based on the convention that enthalpies of elements at 101325 Pa and 298 K are zero. The total enthalpy is thus defined as the sum of two enthalpy *changes* which are separately evaluated. Enthalpies of formation are tabulated[2,3]. The enthalpy change between standard conditions and process conditions may be evaluated by the methods referred to in section 7.3.2.

7.3.3.2 *Balance equations*. The energy balance equation for the continuous steady state process shown below (Figure 7.13) is

$$\Delta H = H_{out} - H_{in} = Q + W$$

Using equation 7.3.7 and assuming ideal mixing the total enthalpy of the input stream(s) is given by

$$H_{in} = \sum_{input} f_{i,in}[(h_i(P, T)_{in} - h_i^\circ) + \Delta h_{f,i}^\circ],$$

where f_i is the flowrate of component i and the sum is over all components in all input streams. A similar equation may be written for the output stream(s).

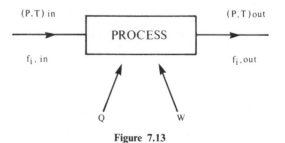

Figure 7.13

The energy balance equation thus becomes

$$\sum_{\text{output}} f_{i,\text{out}}[(h_i(P, T)_{\text{out}} - h_i^\circ) + \Delta h_{f,i}^\circ]$$

$$- \sum_{\text{input}} f_{i,\text{in}}[(h_i(P, T)_{\text{in}} - h_i^\circ) + \Delta h_{f,i}^\circ] = Q + W \qquad (7.3.8)$$

Equation 7.3.8 is applicable to both reactive and non-reactive systems. For the latter the standard state enthalpies and enthalpies of formation cancel because input and output flowrates are equal. A more convenient form of equation 7.3.8 for reactive systems is obtained by introducing the enthalpy change for the reaction. For a reaction described by the general stoichiometric equation (equation 7.2.7), the standard enthalpy change is defined as

$$\Delta h_r^\circ = \sum_i v_i \Delta h_{f,i}^\circ \qquad (7.3.9)$$

Using the material balance (equation 7.2.8) and equation 7.3.9 to eliminate $\Delta h_{f,i}^\circ$ terms from equation 7.3.8 gives

$$\sum f_{i,\text{out}}[h_i(P, T)_{\text{out}} - h_i^\circ] + \xi \Delta h_r^\circ$$

$$- \sum f_{i,\text{in}}[h_i(P, T)_{\text{in}} - h_i^\circ] = Q + W \qquad (7.3.10)$$

Note that the above equations apply to reactions taking place at *any* pressure and temperature—they are not restricted to standard conditions.

7.3.4 *Energy balance techniques*

The following steps are involved in carrying out an energy balance:

(i) Determine flowrates from a material balance.
(ii) Find values of $\Delta h_{f,i}^\circ$ and evaluate Δh_r°.
(iii) Evaluate enthalpy difference between process conditions and standard conditions for each component i.e. $[h_i(P, T) - h_i^\circ]$.
(iv) Use above information to solve energy balance equation for the desired quantity.

Example 7.8

$100 \, \text{mol min}^{-1}$ of methane are mixed with air in stoichiometric proportions and used to fuel a boiler. The methane is at 25°C and the air is preheated to 100°C. The reaction products leave at 500°C. What is the rate of heat generation in the boiler? A 90% conversion of methane is achieved.

Solution

Flowsheet: (Figure 7.14)
Basis: 100 mol methane in feed (i.e. 1 min of operation)
Reaction: $CH_4(g) + 2O_2(g) \rightarrow CO_2(g) + 2H_2O(g)$

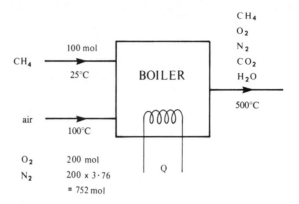

Figure 7.14

Material balance

There is a 90% conversion of CH_4, hence $\xi = 90$ mol. The quantity of each substance leaving is obtained from the equation

$$f_{i,out} = f_{i,in} + v_i \xi$$

It is convenient to set up a table containing all required flowrates and enthalpy differences. As quantities are evaluated they should be entered in the table; the completed table is shown below.

Substance	$\Delta h_f^{\ominus}$	IN		OUT	
		f_i	$h_i - h_i^{\ominus}$	f_i	$h_i - h_i^{\ominus}$
$CH_{4(g)}$	$-74\cdot85$	100	0	10	$23\cdot10$
$O_{2(g)}$	0	200	$2\cdot23$	20	$15\cdot03$
$N_{2(g)}$	0	752	$2\cdot19$	752	$14\cdot24$
$CO_{2(g)}$	$-393\cdot5$	—	—	90	$21\cdot34$
$H_2O_{(g)}$	$-241\cdot8$	—	—	180	$17\cdot00$

Enthalpies are in $kJ\,mol^{-1}$; values of f_i are in mol.

Enthalpy of reaction

$$\Delta h_r^{\ominus} = \Delta h_{f,CO_2}^{\ominus} + 2\Delta h_{f,H_2O}^{\ominus} - \Delta h_{f,CH_4}^{\ominus} - 2\Delta h_{f,O_2}^{\ominus}$$

From $\Delta h_f^{\ominus}$ values in the table[2], $\Delta h_r^{\ominus} = -802\cdot3\,kJ\,mol^{-1}$

Enthalpy changes

Using heat capacity equations[2] of the form of equation 7.3.6 the following values are obtained.

CH_4

$$h(500°C) - h(25°C) = 23\cdot10\,kJ\,mol^{-1}$$

O_2

$$h(100°C) - h(25°C) = 2.23 \, \text{kJ mol}^{-1}$$
$$h(500°C) - h(25°C) = 15.03 \, \text{kJ mol}^{-1}$$

N_2

$$h(100°C) - h(25°C) = 2.19 \, \text{kJ mol}^{-1}$$
$$h(500°C) - h(25°C) = 14.24 \, \text{lkJ mol}^{-1}$$

CO_2

$$h(500°C) - h(25°C) = 21.34 \, \text{kJ mol}^{-1}$$

$H_2O_{(g)}$

$$h(500°C) - h(25°C) = 17.00 \, \text{kJ mol}^{-1}$$

Energy balance
Assuming work done is negligible

$$Q = \sum f_{i,out}[h_{i,out} - h_i^°] + \xi \Delta h_r^° - \sum f_{i,in}[h_{i,in} - h_i^°]$$

$$= 16\,220.7 + 90(-803.3) - 2092.9 \, \text{kJ}$$

$$= -58\,080 \, \text{kJ}$$

This is for one minute's operation, hence the rate of heat production is 968 kW.

Note that a negative value for Q indicates heat is transferred from the system.

The final example of the section illustrates the calculation of an adiabatic reaction temperature. This type of calculation is often used to estimate the maximum temperature which can be attained in a reaction, e.g. maximum flame temperature.

Example 7.9

10 000 mol hr^{-1} of limestone (pure $CaCO_3$) are calcined continuously in a kiln by burning 12 000 mol hr^{-1} of a fuel gas, with 18 000 mol hr^{-1} of air, in direct contact with the limestone. The limestone enters the kiln at 25°C, the fuel gas at 400°C and the air at 25°C. The gaseous products leave the kiln at 200°C and consist of CO_2, N_2 and O_2 only. Assuming there is no heat loss, determine the temperature of the solid product (pure CaO) leaving the kiln.

The molar composition of the fuel gas is CO 60%, CO_2 13% and N_2 27%. The mean heat capacity of the CaO product is 47 J mol^{-1}K^{-1}.

Solution

Flowsheet: (Figure 7.15)
Basis: 10 mol limestone in feed.

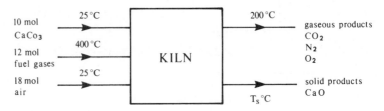

Figure 7.15

Material balance
The following reactions take place,

calcination: $\qquad CaCO_3 \rightarrow CaO + CO_2 \qquad$ (1)

combustion: $\qquad CO + \frac{1}{2}O_2 \rightarrow CO_2 \qquad$ (2)

The fuel gas contains $2 \cdot 7\,mol\,CO$, $1 \cdot 56\,mol\,CO_2$ and $3 \cdot 24\,mol\,N_2$. Air contains $3 \cdot 76\,mol\,N_2/mol\,O_2$ i.e. $3 \cdot 78\,mol\,O_2$ and $14 \cdot 22\,mol\,N_2$ are supplied in the air stream. Both reactions go to completion thus $\xi_1 = 10\,mol$ and $\xi_2 = 7 \cdot 2\,mol$. Quantities pertaining to inlet and outlet streams are entered in the table below. Temperature T in °C, flowquantities f in mol, Δh_f° in kJ mol^{-1}, $\Delta h = h(P, T) - h^\circ$ in kJ mol^{-1}.

Substance	$\Delta h_f^\ominus$	IN			OUT		
		T	f	Δh	T	f	Δh
$CaCO_3$	$-1206 \cdot 9$	25	10	0	—	—	—
CaO	$-635 \cdot 6$	—	—	—	T_s	10	see text
CO	$-110 \cdot 5$	400	7·2	11·25	—	—	—
CO_2	$-393 \cdot 5$	400	1·56	16·35	200	18·76	7·08
N_2 (fuel gas)	0	400	3·24	11·35 }	200	17·46	5·13
N_2 (air)	0	25	14·22	0			
O_2	0	25	3·78	0	200	0·18	5·13

Enthalpies of reaction
Reaction 1 (calcination)

$$\Delta h_{r.1}^\circ = -393 \cdot 5 - 635 \cdot 6 - (-1206 \cdot 9)$$
$$= 177 \cdot 7\,kJ\,mol^{-1}$$

Reaction 2 (combustion)

$$\Delta h_{r.2}^\circ = -393 \cdot 5 - (-110 \cdot 5)$$
$$= -283 \cdot 0\,kJ\,mol^{-1}$$

Enthalpy differences
Enthalpy differences between process conditions and standard conditions have been entered in the table. These are evaluated in the usual way from

heat capacity correlations. Note that nitrogen in the fuel gas (at 400°C) and nitrogen in air (at 25°C) must be treated separately when calculating the input enthalpy.

For CaO in the outlet stream the temperature is not known. Using the mean heat capacity we obtain the following expression

$$\Delta h_{CaO} = 0.047(T_s - 25°C)\,kJ\,mol^{-1}$$

where T_s is the outlet temperature of the CaO in °C.

Energy balance
The counterpart of equation 7.3.10 for the case of multiple reactions is

$$\sum f_{i,out}[h_{i,out} - h_i°] + \sum_{\substack{reactions \\ k}} \xi_k \Delta h_{r,k}°$$

$$- \sum f_{i,in}[h_{i,in} - h_i°] = Q + W$$

Assuming that the reactor operates adiabatically, $Q = 0$. The work term is taken to be negligible.

Using data from the energy balance table

$$\sum f_{i,in}[h_{i,in} - h_i°] = 7.2 \times 11.25 + 1.56 \times 16.35 + 3.42 \times 11.15$$
$$= 144.64\,kJ$$

$$\sum f_{i,out}[h_{i,out} - h_i°] = 18.76 \times 7.08 + 17.44 \times 5.13 + 0.18$$
$$\times 5.13 + 10 \times 0.047(T_s - 25)$$
$$= 211.46 + 0.47\,T_s\,kJ$$

$$\sum_k \xi_k \Delta h_{r,k}° = 10 \times 177.8 - 7.2 \times 283.0$$
$$= -259.6\,kJ$$

Inserting terms in the energy balance equation

$$211.46 + 0.47\,T_s - 259.6 - 144.64 = 0$$

∴ Temperature of solid product $\underline{T_s = 410°C}$

**7.4
Fluid flow**

Materials are usually transferred between different parts of a chemical plant in the fluid state. Large quantities of gases, liquids and fluidized solids may be easily and economically transported by pumping along a network of pipes. In process design and analysis it is necessary to determine pump duties, pipe sizes, pressure drops, flowrates etc. This section reviews the basic principles of fluid mechanics on which such calculations depend.

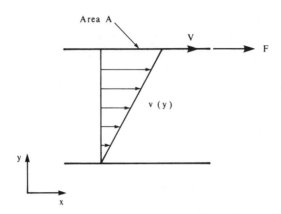

Figure 7.16   Velocity profile for a Newtonian fluid subject to shear.

7.4.1 *Types of fluid*

The term 'fluid' covers a wide range of materials—from gases and simple liquids to polymeric materials and semi-solid slurries. Fluids may be classified as either *compressible* or *incompressible*. The density of a compressible fluid depends on the pressure. Although this is true for all real fluids, the compressibility of liquids is very small under most conditions and they may be considered incompressible. The flow of gases must usually be treated as compressible unless pressure changes are small.

A second type of classification may be made according to the behaviour of a fluid subject to shear stress. Consider a fluid contained between parallel plates as shown in Figure 7.16. The upper plate is moved in the x direction at a constant velocity V by a constant force F. The shear stress exerted on the fluid by the moving plate of area A is $\tau = F/A$. For a *Newtonian fluid* in laminar flow* the velocity profile in the y direction is linear and the shear stress is proportional to the velocity gradient,

$$\tau = \mu \frac{dv}{dy} \tag{7.4.1}$$

The constant of proportionality μ is the *viscosity* of the fluid and equation 7.4.1 is known as *Newton's law of viscosity*. The viscosity of a fluid is a measure of how easily momentum is transferred through the fluid. A high viscosity indicates rapid momentum transfer from the moving plate to the stationary plate and hence a high resistance to shear. The units of shear stress are the same as those of pressure, i.e. Nm^{-2} or Pa, therefore viscosity has units of Pa s. The viscosity of water at room temperature is approximately 10^{-3} Pa s whereas that of air is about 2×10^{-5} Pa s.

---

*see section 7.4.2.1

Gases and most simple liquids (e.g. water) are Newtonian fluids. Non-Newtonian fluids do not obey equation 7.4.1; they include many materials of industrial importance. Examples are polymeric substances, slurries and suspensions. For non-Newtonian fluids the viscosity depends on the shear stress and may also depend on the previous deformation history of the fluid.

The treatment of fluid flow in this chapter is restricted to incompressible flow of Newtonian fluids. For a wider coverage see textbooks detailed in references 6, 7 and 8.

7.4.2 Flow regimes

7.4.2.1 *Laminar and turbulent flow.* The flow pattern which develops when a given fluid flows through a pipe or channel of fixed dimensions depends on the velocity. At low velocities the pattern tends to be *laminar* or *streamline*—adjacent layers of fluid move past each other with no mixing in the direction normal to the flow. At higher velocities the flow becomes *turbulent*—smooth laminar flow breaks up and eddies form. These eddies are of varying sizes and move rapidly in random directions. Turbulent flow results in a high degree of mixing between fluid elements and hence leads to high rates of convective heat transfer (see section 7.5.1.2). In laminar flow, by contrast, mixing only occurs by diffusion and heat transfer by conduction—both of which are relatively slow processes.

7.4.2.2 *Reynolds number.* Experiments show that the velocity at which the transition from laminar to turbulent flow occurs depends on the physical properties of the fluid and the geometry of the flow. The nature of the flow is indicated by a dimensionless group[9] known as the *Reynolds number Re*. The Reynolds number represents a ratio of inertial forces (rate of change of momentum of fluid elements) to viscous shear forces acting in a fluid. For flow in a pipe the Reynolds number is defined as

$$Re = \frac{\bar{v}\, d\rho}{\mu} \qquad (7.4.2)$$

where $\bar{v}$ is the mean velocity, d is the pipe diameter, ρ is the density of the fluid and μ its viscosity. When these quantities are expressed in a consistent set of units Re is a pure number, i.e. dimensionless.

The following criteria for flow regimes within smooth pipes have been established

$$Re < 2000 \qquad \text{laminar flow}$$
$$2000 < Re < 4000 \qquad \text{transition between periods of laminar and turbulent flow}$$
$$Re > 4000 \qquad \text{turbulent flow}$$

Large values of Re correspond to highly turbulent flow with inertial forces dominant: because of high velocity, high density, large diameter or low viscosity. At low Re viscous forces are dominant and hence the flow is laminar.

7.4.2.3 *Boundary layer*. For all flow regimes, whether laminar or turbulent, the effects of viscous shear forces are greatest close to solid boundaries. Fluid actually in contact with a surface usually has no relative motion; the so called *no-slip* condition. There is therefore a region extending from the surface to the bulk of the fluid within which the velocity changes from zero to the bulk value. This region is known as the *boundary layer*.

The velocity distribution for flow over a large, smooth, flat plate is shown schematically in Figure 7.17. The boundary layer thickness δ is conventionally (arbitrarily) defined as the distance from the surface at which the velocity is 0·99 of the bulk velocity.

In air δ would be of the order of several mm. The boundary layer is an important concept in fluid mechanics because it allows a fluid to be separated into two regions; (a) the boundary layer, which contains the whole of the velocity gradient and all viscous effects; (b) the bulk fluid in which viscous forces are small compared with other forces. Flow in the boundary layer may be laminar or turbulent; however, very close to the wall viscous forces always dominate in a thin region called the *viscous sub-layer*.

When a fluid flows through a pipe, boundary layers form at the entry point and grow in thickness along the length of the pipe until they meet in the centre. The boundary layer thus fills the entire pipe and the flow is termed *fully developed*. If the boundary layer is still laminar then laminar flow persists.

7.4.3 *Balance equations*

The important concepts of material and energy balances were introduced in previous sections. They are restated below in forms more appropriate for fluid

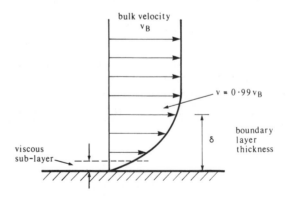

Figure 7.17   Velocity profile in flow over a flat plate.

flow. The momentum balance gives an additional conservation equation. Together these three balance equations provide the basis for studying fluid flow phenomena.

7.4.3.1 *Material balance—continuity.* In fluid mechanics the material balance is usually known as the continuity equation. Figure 7.18 shows fluid flowing into a region through plane 1, area A_1, with a mean velocity $\bar{v}_1$, and density ρ_1. Fluid leaves across plane 2. For steady flow there is no accumulation of mass in the region and hence

$$\dot{m} = \rho_1 \bar{v}_1 A_1 = \rho_2 \bar{v}_2 A_2 \tag{7.4.3}$$

where $\dot{m}$ is the mass flow rate. For a fluid of constant density equation 6.4.3 reduces to

$$\bar{v}_1 A_1 = \bar{v}_2 A_2 \tag{7.4.4}$$

7.4.3.2 *Energy balance.* The energy balance applied to fluid flow must include the kinetic and potential energy terms of equation 7.3.2. Changes in these quantities are no longer negligible. Again referring to Figure 7.18, the kinetic energy per unit mass entering the control region at plane 1 is $(\frac{1}{2}\bar{v}_1^2)\alpha_1$, where $\alpha = \overline{v^3}/\bar{v}^3$, the ratio of the mean cubed velocity to the mean velocity cubed. α is usually referred to as the *kinetic energy correction factor.* It is equal to unity if the velocity is uniform over the whole area. The gravitational potential energy per unit mass is gz_1, where g is the gravitational acceleration and z represents the height (relative to an arbitrary datum). The energy balance equation for steady flow becomes

$$(\bar{v}_1 A_1 \rho_1)\left[u_1 + \frac{P_1}{\rho_1} + gz_1 + \frac{\alpha_1}{2}\bar{v}_1^2 \right] + Q + W$$

$$= (\bar{v}_2 A_2 \rho_2)\left[u_2 + \frac{P_2}{\rho_2} + gz_2 + \frac{\alpha_2}{2}\bar{v}_2^2 \right] \tag{7.4.5}$$

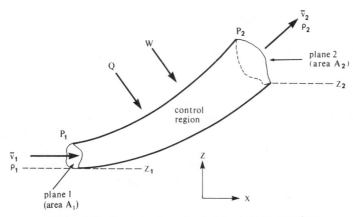

Figure 7.18   Control volume for deriving balance equations.

Note that Pv in equation 6.3.2 has been written as P/ρ. By the continuity equation $\bar{v}_1 A_1 \rho_1 = \bar{v}_2 A_2 \rho_2$, thus

$$u_1 + \frac{P_1}{\rho_1} + gz_1 + \frac{\alpha_1}{2}\bar{v}_1^2 + Q' + W'$$
$$= u_2 + \frac{P_2}{\rho_2} + gz_2 + \frac{\alpha_2}{2}\bar{v}_2^2 \tag{7.4.6}$$

where Q' and W' are heat transfer and work per unit mass.

Energy required to overcome viscous forces within the fluid and friction at the walls is dissipated as heat. Some of this energy goes out through the walls (Q') and the rest goes into raising the internal energy of the fluid ($u_2 - u_1$). The above energy changes may be grouped together in a 'loss' term

$$l = u_2 - u_1 - Q'$$

where l represents the loss of mechanical energy by the system (transformed into thermal energy). Equation 7.4.6 may therefore be written as

$$\frac{P_2}{\rho_2} + gz_2 + \frac{\alpha_2}{2}\bar{v}_2^2 = \frac{P_1}{\rho_1} + gz + \frac{\alpha_1}{2}\bar{v}_1^2 + W' - l \tag{7.4.7}$$

This is one of the many forms of the *Bernoulli equation*.

Energy associated with fluid flow is often discussed in terms of the *equivalent head* of liquid. Dividing equation 7.4.7 by g gives

$$\frac{P_2}{\rho_2 g} + z_2 + \frac{\alpha_2}{2g}\bar{v}_2^2 = \frac{P_1}{\rho_1 g} + z_1 + \frac{\alpha_1}{2g}\bar{v}_1^2 + \frac{W'}{g} - h_l \tag{7.4.8}$$

where each term now has units of length. The mechanical energy losses are expressed as a head loss h_l.

7.4.3.3 *Momentum equation.* The momentum equation introduces the force acting on a body or fluid element. Newton's second law states that the sum of forces acting on a body is equal to the rate of change of momentum of the body. Force and momentum are vector quantities—we shall treat each coordinate direction separately in the following derivation of the momentum equation.

Consider a steady flow through the control volume in Figure 7.18 with uniform velocities over each plane at positions 1 and 2. We shall first carry out a momentum balance in the x direction. Let v_{1x} denote the x component of the velocity at plane 1, etc. The mass flowrate into the control volume is $v_1 \rho_1 A_1$ and hence the rate at which x momentum enters is $v_{1x}(v_1 \rho_1 A_1)$. The rate at which x momentum leaves through plane 2 is $v_{2x}(v_2 \rho_2 A_2)$. The rate of change of x momentum is equal to the net force on the fluid in the x direction:

$$F_x = v_{2x}(v_2 \rho_2 A_2) - v_{1x}(v_1 \rho_1 A_1)$$

Using the continuity equation (equation 6.4.3) gives

$$F_x = \dot{m}(v_{2x} - v_{1x}) \qquad (7.4.9)$$

with corresponding equations for the y and z directions.

In general the velocity across any sizeable area is not uniform and a momentum correction factor β must be introduced to account for this; $\beta = \overline{v^2}/\bar{v}^2$. The general momentum equation is

$$F_x = \dot{m}(\beta_2 \bar{v}_{2x} - \beta_1 \bar{v}_{1x}) \qquad (7.4.10)$$

Values of the momentum and kinetic-energy correction factors depend on the details of the velocity distribution for a particular flow. For flow in circular pipes the following values are obtained[7]:

	α	β
laminar	2	4/3
turbulent	≈ 1	≈ 1

In the majority of practical cases flows are turbulent and the corrections may then be omitted.

7.4.4 Flow in pipes

In this section we shall apply the balance equations to a selection of problems concerned with flow of fluids in pipes.

7.4.4.1 *Laminar flow.* We shall derive equations for flow in a horizontal straight pipe of circular cross-section. It will be assumed that the flow is steady and fully developed, i.e. the velocity distribution does not change with time or distance along the pipe.

Figure 7.19 shows a cylinder of fluid, radius r and length L, within a pipe of radius R. Consider a momentum balance in the axial (x) direction. Because the flow is fully developed $\bar{v}_{1x} = \bar{v}_{2x}$ and $\beta_1 = \beta_2$; hence from equation 7.4.10 the total force on the cylinder in the x direction is zero. That part of the force due to

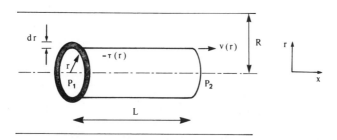

Figure 7.19   Notation used in pipe-flow equations.

the pressure difference acts on the ends of the cylinder and is given by

$$\pi r^2 P_1 - \pi r^2 P_2$$

acting in the $+x$ direction.

There is also a viscous shear force which acts on the fluid at the outer surface of the cylinder and is given by a product of the shear stress and surface area, i.e.

$$-\tau(r)2\pi rL$$

where $\tau(r)$ is the shear stress at radius r. The negative sign indicates that the force exerted on the cylinder by the fluid outside is in the $-x$ direction. The total force on the cylinder is, therefore,

$$\pi r^2(P_1 - P_2) - \tau(r)2\pi rL = 0$$

$$\therefore \tau(r) = \frac{r(P_1 - P_2)}{2L} \tag{7.4.11}$$

But for laminar flow the shear stress is given by equation 7.4.1 which, for flow in a pipe, may be written as

$$\tau(r) = -\mu\frac{dv(r)}{dr}$$

where $v(r)$ is the velocity in the x direction at radius r (the sign is changed because here r is measured in the direction *towards* the wall in contrast to Fig. 7.16).

Substituting in equation 7.4.11,

$$\frac{dv}{dr} = -\frac{(P_1 - P_2)r}{2L\mu}$$

Integrating the above equation gives

$$v = -\frac{(P_1 - P_2)r^2}{4L\mu} + c_1$$

where c_1 is a constant of integration. Applying the no-slip boundary condition, $v = 0$ at $r = R$

$$\therefore 0 = -\frac{(P_1 - P_2)R^2}{4L\mu} + c_1$$

hence

$$v(r) = \left(\frac{P_1 - P_2}{4\mu L}\right)R^2(1 - (r/R)^2) \tag{7.4.12}$$

The parabolic velocity profile represented by equation 7.4.12 is shown in Figure. 7.20. It will now be used to evaluate the flowrate through the pipe. The volumetric flowrate through the shaded annular area in Figure 7.19 is

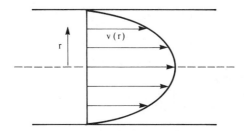

Figure 7.20   Velocity profile for laminar flow within a circular pipe.

given by the product of the velocity at radius r and the area,

$$dQ = v(r)2\pi r dr$$

$$= \left(\frac{P_1 - P_2}{4\mu L}\right)2\pi(R^2 r - r^3)dr$$

The total flowrate is obtained by integrating over the entire cross-section:

$$Q = \int_0^R dQ = \frac{\pi(P_1 - P_2)}{2\mu L}\int_0^R (R^2 r - r^3)dr$$

$$\therefore Q = \frac{\pi R^4(P_1 - P_2)}{8\mu L} \tag{7.4.13}$$

This result is known as the *Hagen-Poiseuille equation*.

7.4.4.2 Friction in pipes. For straight horizontal pipes resistance to flow arises because of viscous shear or friction at the wall. Dimensional analysis[8] leads to the conclusion that for smooth pipes the pressure drop (scaled to make it dimensionless) is a function only of the Reynolds number. The dimensionless group containing the pressure drop is known as the *friction factor*:

$$f = \frac{\Delta P d}{2L\rho\bar{v}^2} \tag{7.4.14}$$

where ΔP is the magnitude of the pressure drop, L is the pipe length and d is the diameter. The friction factor defined in equation 7.4.14 is the *Fanning friction factor*—friction factors defined in other ways are sometimes encountered, e.g. $4f$ or $f/2$.

For laminar flow the mean velocity may be obtained from equation 7.4.13:

$$\bar{v} = Q/\pi R^2$$

$$= \frac{d^2 \Delta P}{32\mu L}$$

$$\therefore f = \frac{16\mu}{\rho\bar{v}d}$$

or using equation 7.4.2

$$f = 16/Re \tag{7.4.15}$$

There is no simple analytic treatment for turbulent flow. Experimental measurements of pressure drops have been used to determine the relationship between f and Re as shown in Figure 7.21. For $Re \leq 10^5$ the data are well represented by the *Blasius equation*:

$$f = 0.079\,Re^{-1/4} \tag{7.4.16}$$

In practice pipes are not smooth and for turbulent flow it is found that f also depends on surface roughness. Friction factor charts for rough pipes are available[7].

A relationship between the friction factor and mechanical energy losses in a flowing fluid may be obtained from the energy balance in equation 7.4.7. For a horizontal pipe

$$\frac{P_2}{\rho} + \frac{\alpha_2 \bar{v}_2}{2} = \frac{P_1}{\rho} + \frac{\alpha_1 \bar{v}^2}{2} - l_f$$

If the flow is steady and fully established

$$\bar{v}_1 = \bar{v}_2 \quad \text{and} \quad \alpha_1 = \alpha_2$$

$$\frac{P_1 - P_2}{\rho} = \frac{\Delta P}{\rho} = l_f$$

From the definition of the friction factor (Equation 7.4.14)

$$l_f = \frac{f 2\bar{v}^2 L}{d} = \frac{\bar{v}^2}{2}\left(\frac{4fL}{d}\right) \tag{7.4.17}$$

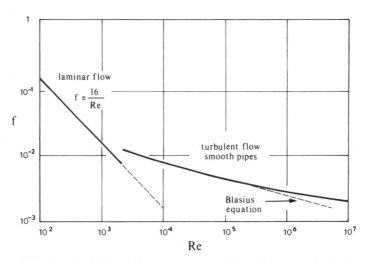

Figure 7.21 Fanning friction factor as a functioning Reynolds number for circular pipes.

The head loss due to friction is therefore

$$h_f = \frac{f 2 \bar{v}^2 L}{gd} = \frac{\bar{v}^2}{2g}\left(\frac{4fL}{d}\right) \qquad (7.4.18)$$

Pipe flow problems are typically of two types:

(i) Pipe size given—determine pressure drop for specified flow rate. Re can be evaluated immediately and hence f is found from the chart or correlating equations.

(ii) Pipe size given—estimate flow for specified pressure drop. An iterative solution is necessary because the flowrate is unknown; guess Re, estimate f, calculate $\bar{v}$ from energy equation and repeat until convergence.

The following examples illustrate these procedures.

Example 7.10

Crude oil (density $800\,\mathrm{kg\,m^{-3}}$, viscosity $4 \times 10^{-3}\,\mathrm{Pa\,s}$) is pumped through a smooth pipe of internal diameter $0.305\,\mathrm{m}$ at a flowrate of $0.152\,\mathrm{m^3\,s^{-1}}$. If the pipe is horizontal estimate the pressure drop along a length of $2100\,\mathrm{m}$.

Solution

The mean velocity is $\bar{v} = Q/\pi R^2 = 0.152/\pi(\cdot305/2)^2 = 2.08\,\mathrm{m\,s^{-1}}$

$$Re = \frac{\rho \bar{v} d}{\mu} = \frac{800 \times 2.08 \times 0.305}{4 \times 10^{-3}} = 1.27 \times 10^5$$

The flow is therefore turbulent. From the Blasius equation

$$f = 0.079\,Re^{-1/4} = 4.2 \times 10^{-3}$$

$$\Delta P = \frac{f 2L\rho \bar{v}^2}{d} = \frac{4.2 \times 10^{-3} \times 2 \times 2100 \times 800 \times (2.08)^2}{\cdot305} = 2 \times 10^5\,\mathrm{Pa}.$$

Although the Blasius equation has been used for $R_e > 10^5$ the extrapolation is small and it is clear from Figure 7.21 that no significant error is involved.

Example 7.11

Water at a temperature of $20°C$ is flowing through a smooth pipe of internal diameter $0.2\,\mathrm{m}$. The pressure drop over a horizontal test section of length $305\,\mathrm{m}$ is measured as $0.21 \times 10^5\,\mathrm{Pa}$. Estimate the flowrate. (Density of water at $20°C = 988\,\mathrm{kg\,m^{-3}}$, viscosity $= 10^{-3}\,\mathrm{Pa\,s}$.)

Solution

From equation 7.4.14,

$$\bar{v} = \left(\frac{\Delta P d}{2L\rho f}\right)^{1/2}$$

hence, inserting values of known quantities,

$$\bar{v} = \left(\frac{6\cdot96 \times 10^{-3}}{f}\right)^{1/2} \text{m s}^{-1} \quad \text{and} \quad Re = 1\cdot976 \times 10^{5}\bar{v}.$$

Iteration 1
Assume flow is turbulent, guess $Re = 10^{5}$. From the Blasius equation $f = 4\cdot44 \times 10^{3}$, thus $\bar{v} = 0\cdot8 \text{ m s}^{-1}$.

Iteration 2
$Re = 1\cdot976 \times 10^{5} \times 0\cdot8 = 1\cdot5 \times 10^{5}, \quad f = 0\cdot079 Re^{-1/4} = 3\cdot96 \times 10^{-3},$
$\bar{v} = 0\cdot75 \text{ m s}^{-1}$

Iteration 3
$Re = 1\cdot49 \times 10^{5}, f = 4\cdot02 \times 10^{-3}, \bar{v} = 0\cdot76 \text{ m s}^{-1}$

Iteration 4
$Re = 1\cdot50 \times 10^{5}, f = 4\cdot01 \times 10^{-3}, \bar{v} = 0\cdot76 \text{ m s}^{-1}$
   The calculation has converged. $Q = \pi R^{2}\bar{v} = 2\cdot3 \times 10^{-2} \text{ m}^{3}\text{ s}^{-1}$ or $\dot{m} = \rho Q = 23\cdot6 \text{ kg s}^{-1}$.

7.4.4.3 *Bends and fittings.* Piping systems seldom consist of straight pipes of constant diameter. It is therefore necessary to account for pressure drops and mechanical energy losses associated with bends, changes in diameter, valves and other fittings.
   For turbulent flow, losses are proportional to the square of the mean velocity. It is therefore customary to express pipeline losses in terms of *velocity heads* $\bar{v}^{2}/2g$. The loss associated with a particular fitting may be written as

$$h_{\mathrm{f}} = K_{\mathrm{f}}\frac{\bar{v}^{2}}{2g} \qquad (7.4.19)$$

Table 7.1 Loss coefficients for bends and fittings in turbulent flow

Type	K_{f}
45° elbow	0·3
90° elbow	0·8
90° square elbow	1·2
coupling, no diameters change	~0
side outlet of T piece	1·8
gate valve—fully open	~0·2
gate valve—$\frac{1}{2}$ open	~5
globe valve—fully open	~5
globe valve—$\frac{1}{2}$ open	~10
entry to pipe from large vessel	~0·5
sudden expansion, cross-section $A_{1} \rightarrow A_{2}$	$(1 - A_{1}/A_{2})^{2}$

where K_f is the *loss coefficient*. Some typical values of loss coefficients are given in Table 7.1. Although these are mostly determined by experiment an approximate analytical treatment is possible in some simple cases; this is illustrated in the next example.

Example 7.12

Use the conservation equations to determine the loss coefficient for a sudden expansion in a horizontal pipe.

Solution

Consider the flow arrangement shown in Figure 7.22. We shall first apply the momentum equation to find the pressure difference between planes 1 and 2, followed by the energy equation to obtain the losses. Taking components of force and velocity in the direction of flow

$$F = \dot{m}(\beta_2 \bar{v}_2 - \beta_1 \bar{v}_1)$$

where F is the total force on the fluid between planes 1 and 2. This force is made up of three contributions:

(i) **Pressure/area difference**, $F_p = P_1 A_1 - P_2 A_2$
(ii) **Shear stress at wall**, F_w
(iii) **Force exerted on fluid by fitting**, F_f

Because planes 1 and 2 are close together the stresses at the wall are small compared with other terms, thus $F_w \simeq 0$. If required, an approximate value for F_w may be evaluated using equations 7.4.11 and 7.4.14.

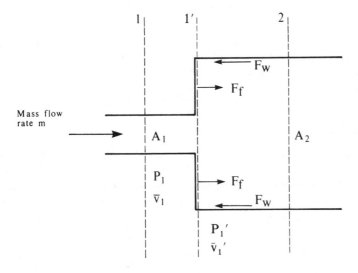

Figure 7.22

The force exerted by the fitting (at plane 1') is equal and opposite to the force exerted by the fluid, i.e. $F_f = P'_1(A_2 - A_1)$. At plane 1', just inside the expansion, the flow is still confined to an area A_1. Plane 2 is chosen to be at a point where the flow is fully expanded. Thus by continuity $\bar{v}'_1 = \bar{v}_1$, and from the energy equation 7.4.7 $P'_1 = P_1$ (assuming friction losses are negligible). It follows that $F_f = P_1(A_2 - A_1)$. Substituting in the momentum equation and setting $\beta_1 = \beta_2 = 1$ for turbulent flow gives

$$(P_1 A_1 - P_2 A_2) + P_1(A_2 - A_1) = \dot{m}(\bar{v}_2 - \bar{v}_1)$$

$$\Rightarrow (P_1 - P_2) = \frac{\dot{m}}{A_2}(\bar{v}_2 - \bar{v}_1) \tag{1}$$

For turbulent flow in a horizontal pipe the energy equation gives

$$h_f = \frac{(P_1 - P_2)}{\rho g} + \frac{(\bar{v}_1^2 - \bar{v}_2^2)}{2g} \tag{2}$$

From the continuity equation

$$\dot{m} = \rho \bar{v}_1 A_1 = \rho \bar{v}_2 A_2 \tag{3}$$

$$\therefore \bar{v}_2 = \bar{v}_1 A_1 / A_2 \tag{4}$$

Substituting (1), (3) and (4) in (2)

$$h_f = \frac{\bar{v}_1^2}{2g}\left(1 - \frac{A_1}{A_2}\right)^2$$

hence from the definition of the loss coefficient in equation 7.4.19

$$K_f = \left(1 - \frac{A_1}{A_2}\right)^2$$

Note that this value is based on the upstream velocity $\bar{v}_1$. A different expression is obtained if the coefficient is based on $\bar{v}_2$.

In the final example the methods of this section are applied to estimate a pump duty.

Example 7.13

Water is pumped at the rate of 35 kg s^{-1} from the collection basin of a cooling tower and delivered to the storage tank of a chemical plant at a higher level. The flow arrangement and process parameters are shown in Figure 7.23. Estimate the power input required for a pump which has an efficiency of 60%. ($\rho = 10^3$ kg m^{-3}, $\mu = 10^{-3}$ Pa s.)

Solution

The energy equation will be applied between the two control points shown in

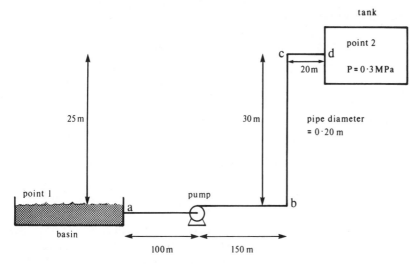

Figure 7.23

the diagram. From equation 7.4.7 the work done (per unit mass) is

$$W' = \frac{P_2 - P_1}{\rho} + \left(\frac{\alpha_2 \bar{v}_2^2}{2} - \frac{\alpha_1 \bar{v}_1^2}{2} \right) + g(z_2 - z_1) + l \qquad (1)$$

$P_1 \simeq 10^5$ Pa since the collection basin is open to the atmosphere. $\bar{v}_1 = \bar{v}_2 = 0$. The velocity in the pipeline is

$$\bar{v} = \frac{\dot{m}}{\rho \pi R^2} = \frac{35}{10^3 \pi (0 \cdot 1)^2} = 1 \cdot 1 \, \text{m s}^{-1}$$

$Re = \dfrac{\bar{v} \rho d}{\mu} = 2 \cdot 23 \times 10^5$, the flow is turbulent.

Losses are made up of frictional losses in the pipe and 'minor losses' associated with fittings.

$$l = \frac{\bar{v}^2}{2} \left(\frac{4fL}{d} + K_a + K_b + K_c + K_d \right)$$

Loss coefficients are taken from Table 7.1. Point a is a pipe entry, $K_a = 0 \cdot 5$; b and c are 90° elbows, $K_b = K_c = 0 \cdot 8$; d is a sudden expansion with $A_1/A_2 \simeq 0$, $K_d = 1$.

Assuming that the pipe is smooth $f = 0 \cdot 079 Re^{-1/4} = 3 \cdot 6 \times 10^{-3}$, hence

$$\frac{4fL}{d} = \frac{4 \times 3 \cdot 6 \times 10^{-3} \times 300}{0 \cdot 2} = 21 \cdot 6$$

$$\therefore l = \frac{\bar{v}^2}{2} (21 \cdot 6 + 0 \cdot 5 + 0 \cdot 8 + 0 \cdot 8 + 1)$$

$$= 24 \cdot 7 \bar{v}^2 / 2$$

Substituting values in equation (1):

$$W' = \frac{(3 \times 10^5 - 1 \times 10^5)}{10^3} + 0 + 9{\cdot}8(25) + \frac{1{\cdot}24}{2}(24{\cdot}7)$$

$$\begin{array}{cccc} = 200 & +0 & +245 & +15{\cdot}3 \\ \text{pressure} & \text{kinetic} & \text{height} & \text{losses} \\ \text{difference} & \text{energy} & \text{difference} & \end{array}$$

$$= 461 \, \text{J} \, \text{kg}^{-1}$$

The mass flow rate is $35 \, \text{kg} \, \text{s}^{-1}$ thus the power requirement is $16{\cdot}1 \, \text{kW}$. For a pump with 60% efficiency the power input is $26{\cdot}8 \, \text{kW}$.

7.5
Heat transfer

Energy transfer in the form of heat is part of almost all chemical processes. Estimating and controlling the rate of heat flows in chemical reactors, separation processes, furnaces and boilers represent important problems in plant design and operation. Efficient energy utilization is also of some importance and this too requires a basic understanding of heat transfer. The scope of this section is restricted to simple heat transfer without change of phase in the heating or cooling medium. More comprehensive treatments are available elsewhere[10-12].

7.5.1 Mechanisms

The driving force for heat transfer is temperature difference: heat will only flow from a hotter to a colder part of a system. The mechanisms of heat transfer may be conveniently classified as conduction, convection and radiation.

7.5.1.1 Conduction.
Heat flow by conduction is a result of transfer of kinetic and/or internal energy between molecules in a fluid or solid. The basic equation of conductive heat transfer is *Fourier's law*

$$q = -k\frac{dT}{dx} \tag{7.5.1}$$

where q is the heat flux (rate of heat transfer per unit area), dT/dx is the temperature gradient in the x direction and k is the thermal conductivity of the material through which heat is flowing. Like viscosity, thermal conductivity is an intrinsic thermophysical property: values at $300 \, \text{K}$ range from $400 \, \text{W} \, \text{m}^{-1} \, \text{K}^{-1}$ for copper, through $0{\cdot}6 \, \text{W} \, \text{m}^{-1} \, \text{K}^{-1}$ for water, to $0{\cdot}03 \, \text{W} \, \text{m}^{-1} \, \text{K}^{-1}$ for air.

Steady-state conduction through a slab of material is represented in Figure 7.24. For a material of constant thermal conductivity the

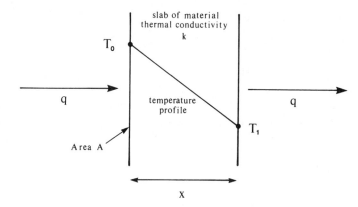

Figure 7.24 Conductive heat transfer through a slab.

temperature profile must be linear because the heat flux at all points is the same. Equation 7.5.1 may therefore be written as

$$q = k\frac{(T_0 - T_1)}{x} \qquad (7.5.2)$$

The rate of heat transfer is equal to the product of the heat flux and the area

$$Q = qA = \frac{kA}{x}(T_0 - T_1) \qquad (7.5.3)$$

Materials of different thermal conductivity are frequently used in adjacent layers to provide thermal insulation. Consider a flat composite slab made of layers as shown in Figure 7.25. The heat flux through each layer and the

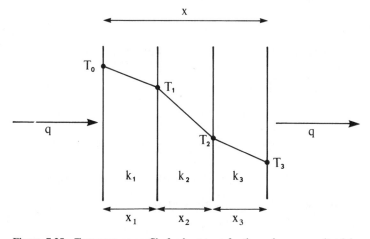

Figure 7.25 Temperature profile for heat transfer through a composite slab.

whole slab is equal, hence

$$q = h_1(T_0 - T_1) = h_2(T_1 - T_2) = h_3(T_2 - T_3)$$
$$= U(T_0 - T_3) \tag{7.5.4}$$

where $h_1 = k_1/x_1$ is the *heat transfer coefficient* of layer 1 etc. and U is the *overall heat transfer coefficient* of the slab. From equation 7.5.4 the temperature difference across layer i is given by

$$\Delta T_i = q/h_i$$

and the overall temperature difference is $\Delta T = q/U$
since $\Delta T = \Delta T_1 + \Delta T_2 + \Delta T_3$

$$\Rightarrow \frac{1}{U} = \frac{1}{h_1} + \frac{1}{h_2} + \frac{1}{h_3} \tag{7.5.5}$$

i.e. the reciprocals of heat transfer coefficients (thermal resistances) are additive.

Example 7.14

A composite furnace wall consists of 0.30 m hot face insulating brick and 0.15 m of building brick. The thermal conductivities are 0.12 and $1.2 \, \text{W m}^{-1} \text{K}^{-1}$ respectively. The inside wall of the furnace is at $950°C$ and the surrounding atmospheric temperature is $25°C$. The heat transfer coefficient from the brick surface to air is $10 \, \text{W m}^{-2} \text{K}^{-1}$.

Estimate

  (i) the surface temperature of the outside wall of the furnace
  (ii) the thickness of insulating brick required to reduce heat loss from the furnace by 10%.

Solution

The problem statement is summarized in Figure 7.26.

  (i) The heat flux is $q = U\Delta T = U(950-25)$
      From equation 7.5.5

$$\frac{1}{U} = \frac{0.30}{0.12} + \frac{0.15}{1.2} + \frac{1}{10} = 2.725$$

$$\therefore U = 0.37 \, \text{W m}^{-2} \text{K}^{-1}$$

$$q = 0.37 \times 925 = 339.4 \, \text{W m}^{-2}$$

But the flux through the wall is equal to that from the outer surface to the air, hence

$$q = h_3(T_2 - 25) = 339.4 \, \text{W m}^{-2}$$

Inserting the value of h_3 gives the outside wall temperature as 59°C.

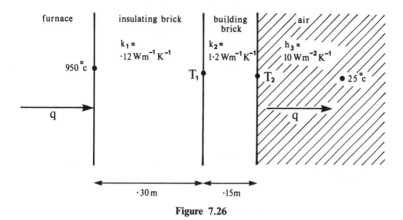

Figure 7.26

(ii) Let the insulation thickness required to reduce the heat flux by 10% be x_1.

$$\frac{1}{U'} = \frac{x_1}{0\cdot12} + \frac{0\cdot15}{1\cdot2} + \frac{1}{10} = \frac{x_1}{0\cdot12} + 0\cdot225$$

$$q' = 0\cdot9q = 305\cdot5 = U'925$$

$$\frac{1}{U'} = 3\cdot028 = \frac{x_1}{0\cdot12} + 0\cdot225$$

$$\text{hence } \underline{x_1 = 0\cdot336\,\text{m}}$$

7.5.1.2 *Convection.* Convective heat transfer is a result of fluid motion on a macroscopic scale. A fluid element moving across a boundary carries with it a quantity of energy and hence gives rise to a heat flux. The phenomenon is termed *natural convection* or *forced convection* depending on the nature of the forces which are responsible. In natural convection, currents are generated in the fluid by differences in density which themselves result from temperature differences. In forced convection, currents and eddies are produced by work done on the system, e.g. stirring a vessel or pumping a fluid in turbulent flow through a pipe. Rates of heat transfer by forced convection are generally much higher than in natural convection.

Heat transfer by convection is a complex process but the analysis is simplified by the boundary layer concept. All resistance to heat transfer on the fluid side of a hot surface is supposed to be concentrated in a thin film of fluid close to the solid surface. Transfer within the film is by conduction. The temperature profile for such a process is shown in Figure 7.27. The thickness of the thermal boundary layer is not generally equal to that of the hydrodynamic boundary layer. The heat flux could be expressed as

$$q = \frac{k_b}{x}(T_w - T_B)$$

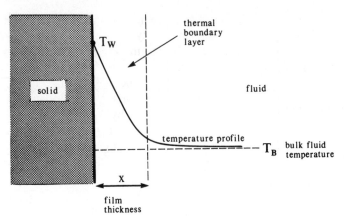

Figure 7.27   Convective heat transfer from a solid wall.

where k_b is the thermal conductivity in the boundary layer. However, because the film thickness is usually not known, it is more common to express the heat flux in terms of a *film heat transfer coefficient h*, thus

$$q = h(T_w - T_B).$$

Values of film transfer coefficients are usually obtained from correlations between dimensionless groups. For forced convection (turbulent flow) within circular tubes the relevant groups are:

$$\left.\begin{array}{l} \text{Nusselt number, } Nu = \dfrac{hd}{k} \\[2em] \text{Prandtl number, } Pr = \dfrac{c_p\mu}{k} \\[2em] \text{Reynolds number, } Re = \dfrac{\rho \bar{v} d}{\mu} \end{array}\right\} \qquad (7.5.6)$$

where h is the heat transfer coefficient, d is the tube diameter, k is the fluid thermal conductivity, c_p is the heat capacity (mass units) and μ is the viscosity. Experimental data have been correlated by the equation

$$Nu = 0.023\, Re^{0.8}\, Pr^{0.33} \qquad (7.5.7)$$

Equations relevant to other situations such as flow across tubes are given in textbooks.[10–12]

7.5.1.3 *Radiation.* Radiant heat transfer occurs in the form of electromagnetic radiation with wavelengths from about 10^{-7} to 10^{-4} m; this is known as *thermal radiation*. The visible spectrum is centred on wavelengths of 5×10^{-7} m. The total energy flux emitted by a body at an absolute temperature T is given by

$$E = \varepsilon \sigma T^4 \qquad (7.5.8)$$

where σ is the Stefan–Boltzmann constant ($5 \cdot 67 \times 10^{-8}\,\mathrm{W\,m^{-2}\,K^{-4}}$) and ε is the *emissivity* of the body. The emissivity lies in the range 0 to 1. For polished metals it is low ($\simeq 0 \cdot 05$) whereas most non-metals and oxidized metal surfaces have emissivities of about $0 \cdot 8$.

The net heat flux from a body to its surroundings is equal to the emitted energy (equation 7.5.8) minus the energy absorbed from the surroundings. Assuming (i) emitted energy is totally absorbed by the surroundings; and (ii) the fraction of incident radiant energy absorbed by a body is equal to ε (a reasonable assumption for metals), the net flux is given by

$$q = \varepsilon \sigma (T^4 - T_0^4) \qquad (7.5.9)$$

where T_0 is the temperature of the surroundings.

It is evident that radiative heat transfer is not an important contribution to the total heat flux for small temperature differences. When large temperature differences exist, e.g. in furnaces, radiation may be the dominant mechanism of heat transfer.

7.5.2 *Shell and tube heat exchangers*

Shell and tube heat exchangers are probably the most common type of heat transfer equipment used in the process industries. They provide a high heat transfer area per unit volume and can be designed to fulfil most duties. Thermal and mechanical design procedures are well established—a fairly comprehensive account is given in Coulson and Richardson[13]. The internal layout of a simple shell and tube exchanger is shown in Figure 7.28. One process

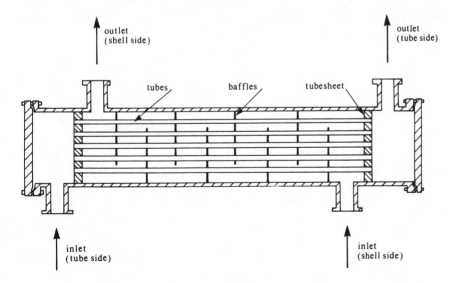

Figure 7.28

fluid flows through the tubes and the other inside the cylindrical shell. The shell side fluid is made to flow across the tube bundle by a series of baffles.

7.5.2.1 *Basic equations.* The rate of heat transfer in an exchanger with a heat transfer area A is given by

$$Q = qA = UA\Delta T_{\mathrm{m}} \qquad (7.5.10)$$

where U is the overall heat transfer coefficient and ΔT_{m} is a mean temperature difference between the two fluids. The definition of ΔT_{m} depends on the flow configuration within the exchanger.

Figure 7.29 (*a*) shows a *countercurrent* flow arrangement. Temperature profiles for the hot and cold streams are plotted against heat transferred in Figure 7.29 (*b*): it is assumed that the overall heat transfer coefficient and heat capacities do not vary significantly with temperature. It can be shown[10] that the heat transferred is given by

$$Q = UA\Delta T_{\mathrm{lm}} \qquad (7.5.11)$$

where ΔT_{lm} is the *log mean temperature difference*.

$$\Delta T_{\mathrm{lm}} = \frac{\Delta T_{\mathrm{in}} - \Delta T_{\mathrm{out}}}{\ln(\Delta T_{\mathrm{in}}/\Delta T_{\mathrm{out}})} \qquad (7.5.12)$$

If the two fluids flow in the same direction the arrangement is termed *cocurrent* (Figure 7.30). The temperatures of the two streams approach each other along the exchanger but the outlet temperature of the cold stream cannot exceed the inlet temperature of the hot stream. The appropriate mean temperature difference for cocurrent flow is also ΔT_{lm}. However, for a given set of inlet and outlet temperatures for hot and cold streams ΔT_{lm} is larger in countercurrent flow and hence the exchanger will be smaller.

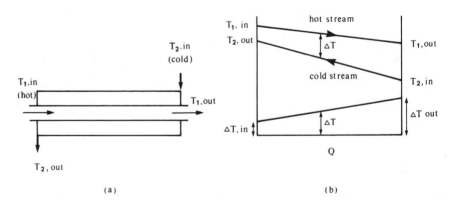

Figure 7.29 Flow arrangement and temperature profiles for countercurrent heat exchanger.

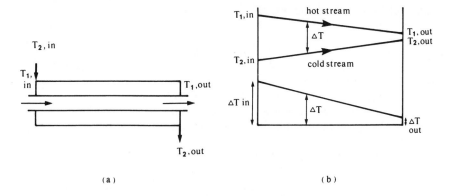

Figure 7.30   Flow arrangement and temperature profiles for cocurrent heat exchanger.

Flow arrangements in industrial exchangers are rarely of the simple 1 tube pass, 1 shell pass type shown in Figure 7.28. Fluids are usually passed through the shell and tubes more than once (in different directions) in order to increase the flow velocity and hence the heat transfer for a given area. In some sections of these *multipass* exchangers the flow is countercurrent and in others co-current. Correction factors to ΔT_{lm} for multipass exchangers are available[13].

7.5.2.2 *Overall heat transfer coefficient.* In an exchanger heat is transferred from one fluid to another through a tube wall. An idealized temperature profile is shown in Figure 7.31. The overall heat transfer coefficient is

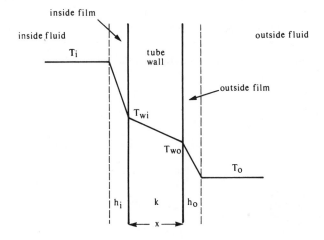

Figure 7.31   Temperature profile for heat transfer across a tube wall.

obtained by summing the resistances.

$$\frac{1}{U} = \frac{1}{h_i} + \frac{x}{k} + \frac{1}{h_0} \qquad (7.5.13)$$

where h_i is the inside film transfer coefficient,
k is the thermal conductivity of the tube wall,
x is the wall thickness,
h_0 is the outside film transfer coefficient.

In equation 7.5.13 it is assumed that inside and outside surface areas are approximately equal, i.e. the tube is thin-walled.

The inside transfer coefficient may be estimated from equation 7.5.7. The shell side coefficient is difficult to estimate because of the complex flow patterns within a baffled shell. For further details see Coulson and Richardson[13].

Some typical values of overall heat transfer coefficients are given in Table 7.2. These values are very approximate and should only be used for order-of-magnitude estimates.

7.5.2.3 *Fouling factors*. In use, heat transfer surfaces tend to become 'fouled' by deposits of scale, dirt or other solids. These deposits represent an additional resistance to heat transfer and allowance for them must be made in the design of exchangers. A modified expression for the overall heat transfer coefficient is generally used

$$\frac{1}{U} = \frac{1}{h_i} + \frac{1}{h_{d,i}} + \frac{x}{k} + \frac{1}{h_{d,0}} + \frac{1}{h_0} \qquad (7.5.14)$$

where $h_{d,i}$ and $h_{d,0}$ are inside and outside dirt coefficients or fouling factors.

Some typical values of fouling factors are given in Table 7.3. The values are subject to considerable uncertainty and accurate figures can only be obtained from operating experience.

The following example illustrates the calculation of heat transfer area for different flow arrangements.

Table 7.2 Typical values of overall heat transfer coefficients for shell and tube heat exchangers

Fluids	Process	$U/Wm^{-2}K^{-1}$
water/water	heating/cooling	600–1800
oil/oil	heating/cooling	100–400
gas/gas (0·1 MPa)	heating/cooling	25–50
gas/gas (5 MPa)	heating/cooling	250–500
steam/water	heating	1400–4000
steam/aqueous solution	evaporation	1000–3000
water/organic vapours	condensation	200–1000

Table 7.3   Typical values of fouling factors

Fluid	$h_d/W\,m^{-2}\,K^{-1}$
River water	2000–10 000
Cooling tower water	5000
Boiler feed water	6000
Condensing steam	6000
Clean gases	5000–10 000
Heavy oils	2000

Example 7.15

A shell and tube heat exchanger is to be used to cool $100\,kg\,s^{-1}$ of 98% sulphuric acid from 60°C to 40°C. Cooling water is available at 10°C and a flowrate of $50\,kg\,s^{-1}$. The overall heat transfer coefficient is $500\,W\,m^{-2}\,K^{-1}$. Determine the required surface area for (a) countercurrent and (b) cocurrent flow. Assume that the heat transfer coefficient is the same in both cases. (Heat capacities: water, $4200\,J\,kg^{-1}\,K^{-1}$; 98% sulphuric acid, $1500\,J\,kg^{-1}\,K^{-1}$.)

Solution

(a) The exit temperature of the water is obtained from an energy balance:

$$Q = F_1 c_{p_1}(T_{1,in} - T_{1,out}) = F_2 c_{p_2}(T_{2,out} - T_{2,in})$$

where Q is the heat transfered, F_1 the flowrate of sulphuric acid and F_2 the flowrate of water.

$$\Rightarrow Q = 100 \times 1500\ (60\text{–}40) = 3 \times 10^6\,W$$
$$= 50 \times 4200\ (T_{2,out} - 10)$$
$$\underline{T_{2,out} = 24{\cdot}3°C}$$

The flow arrangement is

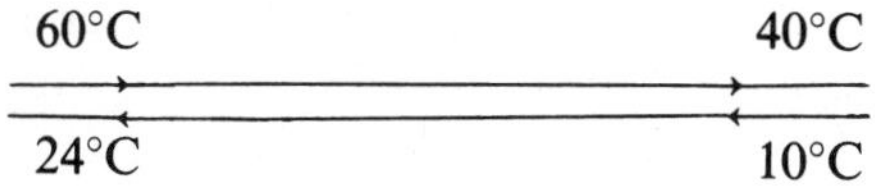

60°C   40°C

24°C   10°C

From equation 7.5.12 the mean temperature difference is

$$\Delta T_{lm} = \frac{(60-40)-(40-10)}{\ln((60-24)/(40-10))} = 33°C$$

The heat transfer area is therefore

$$A = \frac{Q}{U\Delta T_{lm}} = \frac{3 \times 10^6}{500 \times 33} = \underline{182\,m^2}$$

(b) For cocurrent flow the arrangement is

$$60°C \qquad\qquad\qquad 40°C$$
$$\overline{}$$
$$10°C \qquad\qquad\qquad 24°C$$

$$\Delta T_{lm} = \frac{(60 - 10) - (40 - 24)}{\ln((60 - 10)/(40 - 24))} = 30°C$$

$$A = \frac{3 \times 10^6}{500 \times 30} = 200\,\mathrm{m}^2$$

In this example the difference in area is only 10%. However, if the acid were cooled to 32°C the areas would differ by over 70%.

7.6

Separation processes

Separation processes are of central importance in the chemical industry. Appropriate choice and efficient design of separation equipment is essential because it represents a large proportion of the capital and operating costs of a plant. The vinyl acetate flowsheet in Figure 7.2 is almost entirely concerned with separation of the reaction products—this is typical of very many processes. Whenever there is incomplete conversion in a reactor a separation followed by recycle may be necessary. Toxic substances in waste streams must often be removed in order to meet legal requirements. The availability of suitable separation processes is, therefore, crucial when considering the feasibility of any new chemical process.

The subject of separation processes is so extensive that the scope of this section must be extremely limited. We shall therefore concentrate on the general characteristics of separation processes and only examine simple distillation in any detail. Several excellent textbooks on the subject are available.[14-16]

7.6.1 Characteristics of separation processes

A separation process takes one or more feed streams containing mixtures and transforms these into products streams which differ in composition (Figure 7.32). This is achieved by the addition of a *separating agent* which may be energy or another stream of material.

Separation usually involves the formation of more than one phase by the addition of the separating agent. The components of a mixture will be distributed unequally between the phases. Separation of phases of different density is easily accomplished by mechanical means. For example, a liquid feed is passed through a pressure-reducing valve into a vessel (flash drum) where it separates into two phases of different composition. Vapour is taken off at the top of the drum and liquid at the bottom. The separating agent is energy—introduced as work of compression of the feed.

The driving force for interphase mass transfer is always the thermodynamic

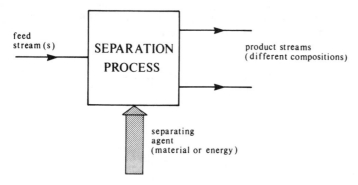

Figure 7.32

chemical potential but in some processes thermodynamic equilibrium is not approached and the degree of separation is dependent on the rate of mass transfer. Some of the more common processes are listed in Table 7.4; all of these are equilibrium separation processes. Mechanical separation processes such as filtration, centrifuging and settling are also widely used.

7.6.2 *Phase equilibria*

A full treatment of the thermodynamics of phase equilibria may be found elsewhere[3]. Only the terminology and some essential concepts will be reviewed in this section.

7.6.2.1 *K-values.* The equilibrium compositions in a vapour-liquid system are related by *equilibrium K-values*:

$$K_i = \frac{y_i}{x_i} \qquad (7.6.1)$$

Table 7.4 Common separation processes

| Process | Phases[a] | | Separating agent[b] |
	Feed	Product	
Flash evaporation	L	L + V	E pressure reduction
Distillation	L/V	L + V	E heat
Stripping	L	L + V	M stripping vapour
Absorption	V	L + V	M liquid absorbent
Extraction	L	$L_1 + L_2$	M liquid solvent
Crystallization	L	L + S + V	E heating or cooling
Evaporation	L	L + V	E heat
Drying	S/L	S + V	E heat
Leaching	S	S + L	M liquid solvent

[a]V, vapour; L, liquid; S, solid
[b]E, Energy-separating agent
  M, material-separating agent

where y_i is the mole fraction of component i in the vapour phase and x_i is the mole fraction in the liquid phase. In general K-values are functions of temperature, pressure and composition but for ideal mixtures they are independent of composition.

The separability of two species i and j is indicated by their *relative volatility*.

$$\alpha_{ij} = \frac{K_i}{K_j} = \frac{y_i x_j}{y_j x_i} \qquad (7.6.2)$$

values of α_{ij} close to unity indicate that separation by methods relying on the formation of vapour and liquid phases will be difficult. For example, at 0·1 MPa α_{ij} for propane:butane is 5·5, whereas for butane:but-2-ene it is 1·05. Separation of the former mixture by simple distillation would be easy but for the latter a very large distillation column would be required.

7.6.2.2 *Ideal mixtures.* Ideal mixing is a good approximation for mixtures of molecules of similar types (e.g. alkanes) at low to moderate pressures.

For ideal mixtures the vapour-liquid equilibrium is described by *Raoult's law*

$$Py_i = P_i^0(T)x_i \qquad (7.6.3)$$

where P is the total pressure and $P_i^0(T)$ is the vapour pressure of component i at the system temperature T. It follows that the K-values in an ideal mixture are given by

$$K_i = P_i^0(T)/P \qquad (7.6.4)$$

and the relative volatility by

$$\alpha_{ij} = P_i^0(T)/P_j^0(T) \qquad (7.6.5)$$

The relative volatility is a function of temperature but not composition or pressure. Over moderate ranges of temperature α_{ij} can often be treated as a constant.

7.6.2.3 *Binary mixtures.* In binary mixtures there is only one independent composition variable in each phase. It is conventional to work in terms of the composition of the *more volatile component* (mvc). The relative volatility of the mvc to the *less volatile component* (lvc) is simply written as α. α is thus always greater than or equal to unity except for azeotropic systems.

The relationship between equilibrium compositions in a binary mixture may be conveniently expressed in terms of the relative volatility

$$\alpha = \frac{y(1-x)}{x(1-y)}$$

Rearranging this equation gives

$$y = \frac{\alpha x}{1 + x(\alpha - 1)} \qquad (7.6.6)$$

Equation 7.6.6 is quite general for binary mixtures but becomes particularly useful if the assumption of constant volatility may be made.

7.6.3 Binary distillation

7.6.3.1 *Distillation columns.* Distillation is one of the most commonly used industrial separation processes. A high degree of separation can be achieved because distillation is a multistage process.

A schematic diagram of a distillation column is shown in Figure 7.33. The column contains a series of *trays* (also called plates or stages) which are usually perforated metal plates. Vapour passes up the column countercurrent to the liquid which flows across each tray and down the downcomers. As vapour bubbles through the liquid on a tray there is mass transfer between phases and

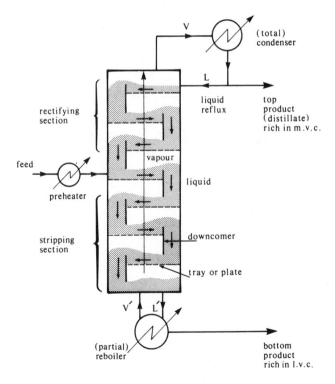

Figure 7.33 Simple distillation column.

their compositions approach equilibrium. Vapour passing up the column becomes enriched in the mvc whilst the liquid is enriched in lvc as it passes downwards. At the top of the column vapour is condensed and part is returned to provide a liquid *reflux*. Similarly, part of the liquid at the bottom is evaporated in the reboiler to provide a vapour phase. The separating agent is the heat supplied in the reboiler.

7.6.3.2 *Material balances*. The notation used to describe the column is defined in Figure 7.34. Vapour leaving a typical plate n has a flowrate V_n and composition y_n, the corresponding liquid flowrate is L_n and the composition x_n:

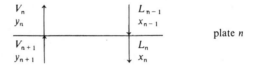

For a *theoretical plate* or stage the vapour and liquid streams leaving are in equilibrium, i.e. $y_n = K_n x_n$. This represents the best possible separation, never obtained in practice. The number of actual plates required to achieve a

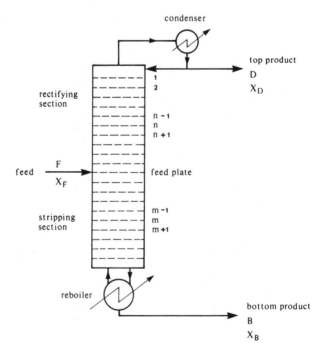

Figure 7.34   Notation for distillation column.

separation may be related to the number of theoretical plates using plate efficiencies.

In order to relate compositions of liquid and vapour streams on successive plates it is necessary to carry out material balances around the top and bottom sections of the column. Consider a balance around the top of the column down to plane n (above the feed plate) as shown in Figure 7.35.

A balance on the mvc gives

$$y_{n+1}V_{n+1} = x_n L_n + x_D D$$

$$\Rightarrow y_{n+1} = \left(\frac{L_n}{V_{n+1}}\right)x_n + \left(\frac{D}{V_{n+1}}\right)x_D$$

This is the equation of the top *operating line* giving the desired relationship between compositions. In general an energy balance is required to determine V_{n+1} and L_n. A major simplification is possible for many problems by assuming *constant molal overflow*, i.e. constant vapour and liquid flowrates. This is approximately correct provided that enthalpy changes on evaporation are about equal for the two components. Constant molal overflow is a good assumption for systems with constant relative volatility. Making the assumption of constant molal overflow

$$L_n = L_{n+1} = L_{n+2} = \cdots = L$$

$$V_n = V_{n+1} = V_{n+2} = \cdots = V$$

Using the overall material balance, $V = L + D$, and writing $r = L/D$ the equation of the top operating line becomes

$$y_{n+1} = \left(\frac{r}{r+1}\right)x_n + \frac{x_D}{r+1} \tag{7.6.7}$$

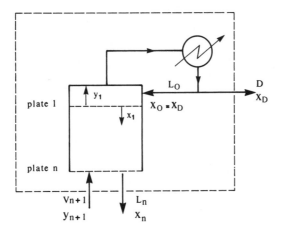

Figure 7.35 Material balance around top of column.

r is the ratio of liquid returned to the column as reflux, to the top product flowrate; usually called the *reflux ratio*.

An analysis similar to the above may be carried out for the bottom of the column below the feed plate. The equation of the bottom operating line, relating compositions on successive plates, is

$$y_m = \left(\frac{s+1}{s}\right)x_{m-1} - \frac{x_B}{s} \tag{7.6.8}$$

where $s = V'/B$. The vapour flowrate in the bottom of the column is V' and the liquid flowrate is L. Because of the feed introduced between top and bottom sections the flowrates in the two sections are not equal.

7.6.3.3 *McCabe–Thiele graphical method.* The equations obtained in the preceding section can be represented graphically on a McCabe–Thiele diagram. The operating lines, equations 7.6.7 and 7.6.8, are plotted on x–y coordinates along with the *equilibrium line*, which is the locus of equilibrium (x, y) compositions. The latter are obtained from tabulations or equations like 7.6.6.

A schematic McCabe–Thiele diagram for the rectifying section of a column is shown in Figure 7.36. We assume that the top product composition x_D and the recycle ration r are known. The composition of the liquid reflux is $x_0 = x_D$. Substituting in equation 7.6.7 gives the composition of vapour leaving plate 1 as $y_1 = x_D$. The top operating line can now be plotted passing through the

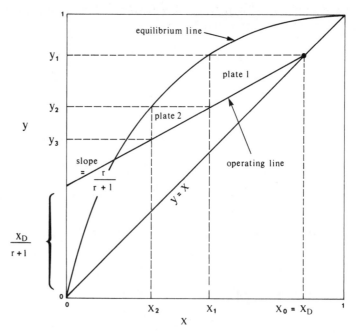

Figure 7.36 Operating line for rectifying section.

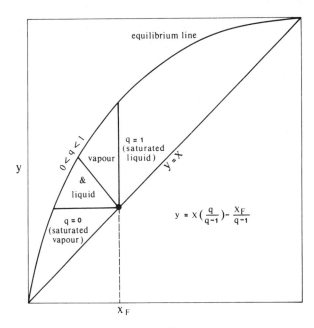

Figure 7.37 q-lines for different feed conditions.

point (x_D, x_D) with a gradient $r/r + 1$. Liquid and vapour phase are in equilibrium leaving each plate. Thus, x_1 is the composition corresponding to $y = y_1$ on the equilibrium line. The vapour composition y_2 can now be obtained from the operating line and the process repeated down the column. Each theoretical plate is represented by a step on the $x-y$ diagram between the operating line and equilibrium line.

The intersection point of top and bottom operating lines is determined by the composition and thermal condition of the feed. A material and energy balance around the feed plate gives the locus of intersection points as

$$y = \left(\frac{q}{q - 1}\right)x - \frac{x_F}{q - 1}$$
(7.6.9)

where $q = \dfrac{\text{enthalpy required to evaporate feed}}{\text{enthalpy required to evaporate liquid of feed composition}}$

The q line (equation 7.6.9) is a straight line passing through the point $(x_F, y = x_F)$ with a gradient $q/q - 1$. The q-lines obtained for different feed conditions are illustrated in Figure 7.37.

A complete procedure for determining the number of theoretical plates in a column is summarized below:

Given x_B, x_D, x_F, r, information on state of feed and equilibrium data;

(i) Draw equilibrium line and $x = y$ line.

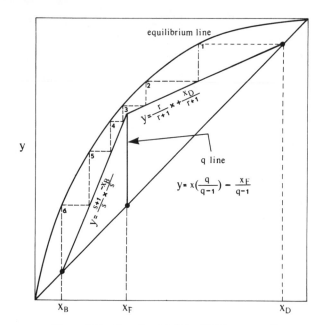

Figure 7.38   Complete McCabe–Thiele construction.

(ii) Draw top operating line through point $(x_D, y = x_D)$ with gradient $r/r + 1$.
(iii) Determine q and draw q line through point $(x_F, y = x_F)$ with gradient $q/q - 1$.
(iv) Draw bottom operating line from point $(x_B, y = x_B)$ to intersection of top operating line and q line.
(v) Step off theoretical plates between operating lines and equilibrium line, transferring from top to bottom operating line after the intersection point.

Figure 7.38 shows a complete McCabe–Thiele diagram. Feed is introduced as liquid at its boiling point $(q = 1)$. The column requires 6 theoretical plates.

7.6.3.4 *Limiting operating conditions.* As the reflux ratio is increased the gradient of the top operating line, equation 7.6.7, approaches unity and the number of theoretical plates required for a given separation is reduced. The *minimum number of theoretical plates* corresponds to *infinite reflux* conditions. At infinite reflux the operating lines are coincident with the $y = x$ line as shown in Figure 7.39. Columns are often tested under infinite reflux conditions by reducing feed and product streams to zero.

The *minimum reflux ratio* is the opposite extreme in operating conditions. As r is reduced the operating lines move towards the equilibrium line and the number of theoretical plates increases. When operating lines intersect on the

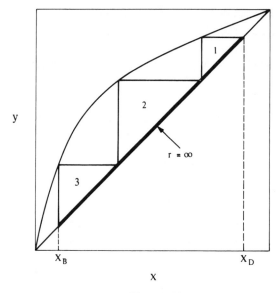

Figure 7.39

equilibrium line the reflux ratio is at its minimum value but the separation requires an infinite number of plates because the driving force for interphase mass transfer becomes zero. Figure 7.40 shows a McCabe–Thiele diagram for this situation. The actual reflux ratio selected for a column must lie between the two extremes. A high ratio reduces the number of plates but increases the

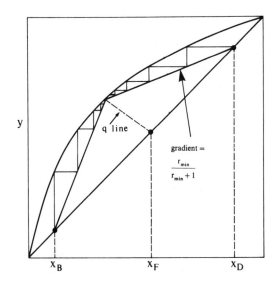

Figure 7.40   McCabe–Thiele diagram for column operating at minimum reflux.

column diameter and energy costs. The optimum reflux ratio depends on physical and cost data and hence varies from column to column. A value of $r = 1.2r_{min}$ is typical.

The following examples illustrates the use of the McCabe–Thiele method.

Example 7.16

A continuous distillation column is required to separate a mixture containing 0·695 mole fraction heptane and 0·305 mole fraction octane to give products of 95% purity. The feed is liquid at its boiling point.
(a) What is the minimum reflux ratio?
(b) Estimate the number of theoretical plates if $r = 2r_{min}$.
Equilibrium data (mole fraction heptane):

x	0·1	0·2	0·3	0·4	0·5	0·6	0·7	0·8	0·9
y	0·188	0·343	0·475	0·588	0·686	0·767	0·840	0·901	0·955

Solution

The equilibrium data are plotted in Figure 7.41.

(a) Minimum reflux corresponds to the top operating line and q-line intersecting on the equilibrium line. For saturated liquid feed the q-line is vertical. The intersection point is at $x = 0.695$, $y \simeq 0.835$.

The gradient of the top operating line is

$$\frac{r_{min}}{r_{min} + 1} = \frac{0.95 - 0.835}{0.95 - 0.695} = 0.451$$

$$r_{min} = 0.82$$

(b) For $r = 2r_{min}$ the top operating line has the equation $y = 0.62x + 0.36$. This line is plotted in Figure 7.41 together with the bottom operating line which intersects it on the q-line. The q-line is vertical because the feed is saturated liquid.

From the construction in Figure 7.41, 11 theoretical plates are required.

7.7
Process control

Automatic control is an essential feature of large-scale continuous processes. Many modern chemical plants are so complex that it would be impractical, unsafe and unprofitable to run them manually.

By its nature process control is concerned with the dynamic behaviour of systems. It is no longer sufficient to make the steady-state assumption. Material and energy balances for unsteady systems must include the accumulation terms so far omitted. Because of the extra mathematical complexity involved in a quantitatve treatment of control this section will, instead, concentrate on general concepts rather than detailed analysis of

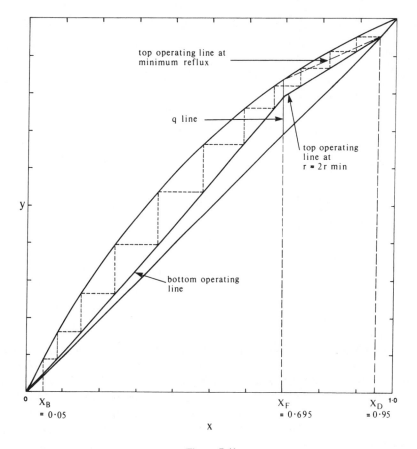

top operating line at
minimum reflux

q line

top operating
line at
r = 2r min

bottom operating
line

X_B
= 0·05

X_F
= 0·695

X_D 1·0
= 0·95

X

y

0

Figure 7.41

specific applications. For a more extensive treatment see the appropriate references[17–20].

7.7.1 *Objectives of process control*

Ensuring safe plant operation is one of the most important tasks of a control system. This is achieved by monitoring process conditions and maintaining variables within safe operating limits. Potentially dangerous situations are signalled by alarms and a plant may even be shut down automatically.

Safe operation is consistent with the economic objective of process control which is to operate the plant in such a way to minimize total costs. This involves maintaining product quality, meeting production targets and making efficient use of utilities such as steam, electricity, cooling water and compressed air.

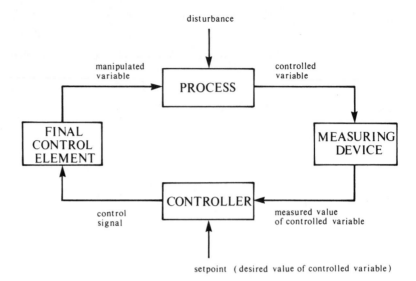

Figure 7.42

7.7.2 *The control loop*

The functional layout of a process control loop is shown in Figure 7.42. A process operates subject to some disturbances e.g. the 'process' may be a preheater and the 'disturbance' a varying feed temperature (Figure 7.43). We identify a variable which needs to be controlled, e.g. the temperature of the stream leaving the preheater. This quantity is measured by a suitable measuring device such as a thermocouple. The output from the measuring device is passed to a *controller* where it is compared with the *setpoint* for the controlled variable. A control signal is generated and activates the *final control element*,

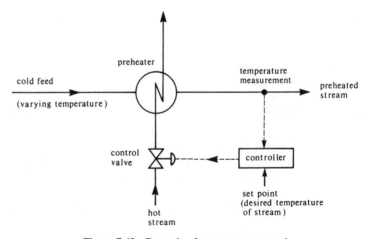

Figure 7.43   Example of temperature control.

usually a control valve. The control valve changes the value of the manipulated variable, e.g. the flowrate of the hot stream to the heat exchanger. A more complex process will typically have a number of control loops each controlling different process variables.

The loop described above is an example of a *feedback* control loop. Other types of control strategy are possible but the feedback controller is the simplest and most widely used. The elements of the control loop will now be examined in turn.

7.7.3 *Measuring devices*

The controlled variable in a process must be amenable to measurement— either directly or indirectly. The most important types of variables controlled in chemical processes are flowrate, temperature, pressure, composition and liquid level.

The characteristics required of a measuring device for process control may differ from those of general purpose instruments. It is necessary to transmit a signal representing the measurement from the measuring element to the controller—at one time this was done pneumatically but now an electrical or optical signal would be used. For example, a mercury-in-glass thermometer may provide a sufficiently accurate temperature measurement but it is difficult to convert it into an electrical signal. By contrast, the output of a thermocouple is naturally an electrical signal and hence this device is a more suitable choice for control applications. Another important characteristic is the dynamic response of the measuring system. For stable and effective control it is important to minimize the time lag between a variable moving off its setpoint and corrective action being taken. A measuring element must therefore respond quickly to changes—this is not always easy to achieve. For example, a thermocouple directly immersed in a flow will respond more quickly than one welded to the outside of a pipe but it may be subject to corrosion.

The literature on measuring systems is extensive, see references 17–20 inclusive plus references cited therein. Table 7.5 lists some of the common measuring techniques used.

7.7.4 *The controller*

Traditionally controllers were complex mechanical devices operated pneumatically and many pneumatic controllers still exist in industry. More recently, analogue electronic controllers have been used but these are now superseded by various digital control devices (see section 7.7.6). Details of controller hardware operation are given by Johnson[20]. The controller compares the signal representing the measured value with the setpoint (i.e. desired) value of the variable. The control action depends on the *control mode* selected for the controller.

Table 7.5   Measuring methods and devices

Variable	Method/principle
Flowrate	Differential pressure devices:
	pitot tube, orifice plate, venturi meter
	Turbine meter
	Rotameter
	Hot wire anemometer
Temperature	Thermistor
	Thermocouple
	Resistance thermometer
	Expansion thermometer
	Optical pyrometer
Pressure	Manometer
	Differential pressure cell (mechanical or
	semi-conductor)
	Bourdon gauge
Composition	Chromatography
	Refractive index
	Density
	Conductivity
	pH
	Infrared/ultraviolet absorption
Liquid Level	Differential pressure
	Float
	Weight of vessel

7.7.4.1 *Proportional control.* The simplest continuous control mode is proportional. The control signal produced by the controller is proportional to the error signal ε, defined as

$$\varepsilon = s - m \qquad (7.7.1)$$

where s is the setpoint and m is the measured value. The controller output is given by

$$c = K_c \varepsilon \qquad (7.7.2)$$

where K_c is the *gain* or *proportional sensitivity*.

The response of a system to a disturbance is shown schematically in Fig. 7.44. The behaviour without control is represented by the solid line. Proportional control (P) reduces the maximum deviation of the controlled variable but gives rise to long-lived oscillations. At the new steady state there is a finite deviation of the controlled variable, known as the *offset*. An offset is characteristic of proportional control which requires an error in order to generate a control signal, equation 7.7.2. If the error (offset) is reduced to zero there is no control action.

7.7.4.2 *Proportional–integral control.* The proportional–integral (or PI) mode is also often referred to as 'proportional plus reset' since this control mode eliminates the offset associated with proportional control alone. The

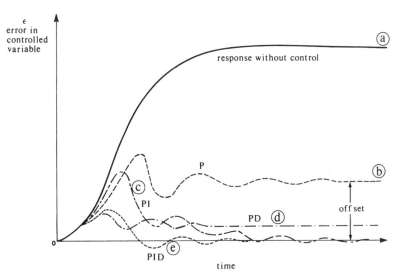

error in controlled variable

response without control

off set

P

PI

PD

PID

time

Figure 7.44 Typical response system with different control modes.

controller output is described by the equation

$$c = K_c \varepsilon + \frac{K_c}{\tau_I} \int_0^t \varepsilon \, dt \qquad (7.7.3)$$

where τ_I is the *integral time*. The output signal is proportional to the error plus the time integral of the error. The values of K_c and τ_I may be set independently to give the best control action. With a PI controller the error itself can be reduced to zero whilst the control action is maintained because the integral of the error remains non-zero. Figure 7.44 shows a response with PI control. The offset is eliminated but oscillatory behaviour still persists.

7.7.4.3 *Proportional–derivative control.* The addition of a derivative mode to a proportional controller results in faster action because the control signal is also proportional to the rate of change of the error:

$$c = K_c \varepsilon + K_c \tau_D \frac{d\varepsilon}{dt} \qquad (7.7.4)$$

where τ_D is the *derivative time*. Derivative action reduces oscillations, thus allowing a higher proportional gain setting and consequently smaller offset.

7.7.4.4 *Proportional–integral–derivative control.* This is a combination of the previous modes and is described by the equation

$$c = K_c \varepsilon + \frac{K_c}{\tau_I} \int_0^t \varepsilon \, dt + K_c \tau_D \frac{d\varepsilon}{dt} \qquad (7.7.5)$$

The integral action eliminates offset and derivative action reduces oscillations but with three modes it is more difficult to select suitable controller settings.

The choice of control mode is influenced by the extent to which an offset or oscillatory behaviour may be tolerated. With pneumatic mechanisms the cost advantage of a simple proportional controller was significant—this is no longer the case.

7.7.5 Final control element

The signal from the controller is used to activate a final control element. In the chemical industry the control element is almost always a valve which controls a flowrate. Changes in flowrate can be used indirectly to change any of the other variables listed in Table 7.5.

Figure 7.45 shows an air-to-close pneumatic control valve. The valve stem is attached to a diaphragm which is held in the up position by a spring. Increasing the air pressure depresses the diaphragm and closes the valve. Different spring arrangements can be used to give an air-to-open valve. The electrical signal from the controller must be converted to a pneumatic signal to activate the valve. Pneumatic valves are still almost universally used.

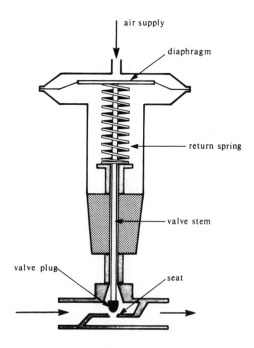

Figure 7.45 Pneumatic control valve.

7.7.6 *Computer control*

Because of the low cost of computer hardware, computer control is now widely used. Computer control schemes can be powerful and flexible. They have the potential to make a process easier to control and more profitable. The different types of computer control are described below.

7.7.6.1 *Supervisory control.* A large plant may have hundreds of PID controllers each operating on a separate control loop. Provided setpoints and controller settings have been well chosen the plant will operate satisfactorily. However, the tasks of supervising the plant, recording data and changing setpoints and controller settings to achieve optimum plant performance are rather complex. In a supervisory control system these jobs are carried out by the computer. Figure 7.46 shows the relationships between the main components in such a system.

As process variables change it may be advantageous to alter setpoints in certain loops. This is accomplished by the supervisory computer which monitors process variables and uses a mathematical model of the process to estimate optimal settings. Sophisticated control strategies are possible because settings of several loops can be altered simultaneously. The task of the plant

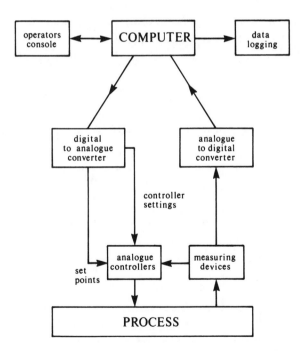

Figure 7.46 Supervisory computer control system.

operator is made easier because all important data relating to the process can be displayed in a compact form at a console. Any abnormal conditions can be quickly identified and acted upon.

7.7.6.2 *Direct digital control.* The conventional controller (as described in section 7.7.4) is a small-scale analogue computer operating on electrical or pneumatic principles. In direct digital control (DDC) measurement of process variables are transmitted directly to a digital computer which compares each signal with an internally stored setpoint. The mathematical operations associated with the desired control mode are carried out and a control signal transmitted to the final control element. Figure 7.47 shows a DDC system.

DDC has most of the advantages of supervisory control but with extra flexibility in choice of control strategies. *Any* control mode can be simulated— the control engineer is not restricted to PID. However, because none of the control loops can function independently of the central computer, elaborate measures may have to be taken to allow for computer failure. In critical applications a standby computer would be provided.

7.7.6.3 *Distributed digital control.* DDC became economically practical with the advent of mini-computers, and the concepts have been extended by the use

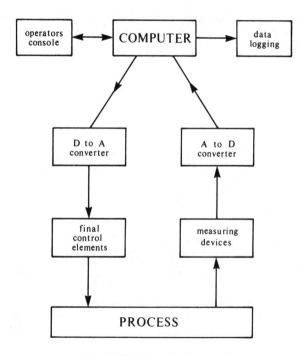

Figure 7.47   Direct digital control.

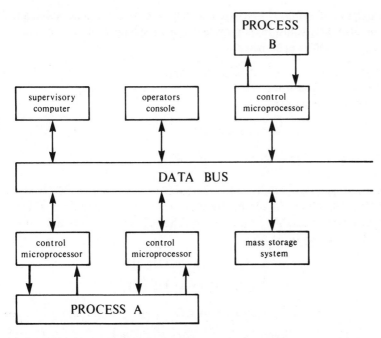

Figure 7.48 Distributed digital control.

of microprocessors. A microprocessor-based control computer can handle a small number of control loops (typically 1–16). Microprocessors are so cheap that, once again, it becomes cost effective to put the control logic in the part of the plant where it is used rather than in a central control room. The costs of cabling a large plant are now significant compared to the cost of the computer.

Distributed control refers to a number of control computers which are linked to each other and to peripherals such as operators' consoles by high speed data channels. A hierarchical structure may be imposed by having certain processors operating in a supervisory mode. Figure 7.48 shows a distributed control system. Such a system is less liable to failure than the central computer based DDC system in Figure 7.47 because each component can operate independently of the others.

Answers are given in parentheses at the end of each problem.   **Appendix**

Problems

1. An evaporative crystallization process is used to produce Na_2SO_4 crystals from aqueous solution. Fresh feed, 22% Na_2SO_4, is mixed with recycle solution and fed to an evaporator where it is concentrated to 45%. This stream is cooled and crystals of Na_2SO_4 are removed in a filter unit as a wet

filter cake containing 95% solids and 5% of a 38% Na_2SO_4 solution. The filtrate, also 38% solution, is recycled. For an evaporator with a maximum capacity of 5000 kg hr^{-1} calculate:

(a) the production rate of solid Na_2SO_4;
(b) the fresh feed rate;
(c) the recycle ratio.

All compositions are on a mass basis.

((a) 1468 kg hr^{-1}, (b) 6468 kg hr^{-1}, (c) 1·68.)

2. Natural gas, (essentially pure methane at 1·5 MPa, 25 °C) and steam (1·5MPa, 450 °C) are mixed in the ratio 4 mol steam: 1 mole methane and contacted with a catalyst to produce synthesis gas for a methanol convertor. The following reactions occur:

$$CH_4 + H_2O \rightarrow CO + 3H_2 \qquad (1)$$

$$CO + H_2O \rightarrow CO_2 + H_2 \qquad (2)$$

90% of the methane is converted to CO by reaction (1) and $\frac{1}{3}$ of the CO produced goes to CO_2 by reaction (2). If 2×10^6 mol hr^{-1} of mixed gases leave the reactor/heat exchanger unit at 600°C, at what rate must heat be supplied?

(19·9 MW—using data from references 2 or 5).

3. 60% sulphuric acid is pumped at a flowrate of 12·6 kg s^{-1} through a smooth pipe to a storage tank. The liquid surface of the storage tank is 15.2 m above the pump outlet. The pipe is in two sections. The first section has an internal diameter of 78 mm and a length of 366 m. The second section has an internal diameter of 63 mm and is 122 m long. Estimate the absolute pressure at the pump outlet. Changes in kinetic energy may be neglected. (Density of acid = 1 530 kg m^{-3}, viscosity = 6·4 $\times$ 10^{-3} Pa s.)

(0·82 MPa)

4. The temperature of oil leaving a cocurrent heat exchanger is to be reduced from 370 K to 350 K by lengthening the cooler. The oil and water flowrates, inlet temperatures and other dimensions of the cooler will remain constant. The water enters at 285 K and the oil at 420 K. The water leaves the original cooler at 310 K. If the original length is 1 m, what must be the new length? Assume that heat transfer coefficients and physical properties do not change.

(1·23 m)

5. A mixture of benzene and toluene containing 25% benzene is to be separated in a distillation column to give a top product containing 90%

benzene and a bottom product containing 10% benzene. The feed enters the column as liquid at its boiling point. A reflux ratio of 8 mol/mol distillate is to be used. Estimate the number of theoretical stages required for the separation. A constant relative volatility of 2·25 may be assumed.

(7)

References

1. *C&R*, vol. 6, ch. 4 (for full details of references denoted by *C&R* see first item in the bibliography).
2. D. M. Himmelblau, *Basic Principles and Calculations in Chemical Engineering*, 4th edn., Prentice-Hall, 1982.
3. K. E. Bett, J. S. Rowlinson and G. Saville, *Thermodynamics for Chemical Engineers*, Athlone, 1975.
4. R. C. Reid, J. M. Prausnitz and B. E. Poling, *The Properties of Gases and Liquids*, 4th edn., McGraw-Hill, 1987.
5. R. M. Felder and R. W. Rousseau, *Elementary Principles of Chemical Processes*, Wiley, 1978.
6. R. B. Bird, W. E. Stewart and E. N. Lightfoot, *Transport Phenomena*, Wiley, 1960.
7. B. S. Massey, *Mechanics of Fluids*, 5th edn., Van Nostrand Reinhold, 1983.
8. M. M. Denn, *Process Fluid Mechanics*, Prentice-Hall, 1980.
9. *C&R*, vol. 1, ch. 1.
10. *C&R*, vol. 1, ch. 7; vol. 3, chs. 1 & 2; vol. 6, ch. 12.
11. D. Q. Kern, *Process Heat Transfer*, McGraw-Hill, 1950.
12. W. L. McCabe and J. C. Smith, *Unit Operations of Chemical Engineering*, 3rd edn., McGraw-Hill, 1976.
13. *C&R*, vol. 6, ch. 12.
14. E. J. Henley and J. D. Seader, *Equilibrium-Stage Separation Operations in Chemical Engineering*, Wiley, 1981.
15. C. J. King, *Separation Processes*, 2nd edn., McGraw-Hill, 1980.
16. *C&R*, vol. 6, ch. 11.
17. *C&R*, vol. 3, ch. 3.
18. Instrument Society of America, *Fundamentals of Process Control Theory*, 1981.
19. D. R. Coughanowr and L. B. Koppel, *Process Systems Analysis and Control*, McGraw-Hill, 1965.
20. C. D. Johnson, *Process Control Instrumentation Technology*, Wiley, 1977.

Bibliography

J. M. Coulson and J. F. Richardson, *Chemical Engineering*, Pergamon.
Vol. 1: *Fluid Flow, Heat Transfer and Mass Transfer*, 3rd edn., 1980.
Vol. 2: *Unit Operations*, 3rd edn., 1980.
Vol 3: *Chemical Reactor Design, Biochemical Reaction Engineering including Computational Techniques and Control*, 2nd edn., 1979.
Vol. 6: *Introduction to Design*, 1982.
A. W. Westerberg, H. P. Hutchison, R. L. Motard and P. Winter, *Process Flowsheeting*, Cambridge University Press, 1979.
R. Aris, *Elementary Chemical Reactor Analysis*, Prentice-Hall, 1969.

The above, together with references 3, 5, 8, 14, and 19, provide a basic coverage of the subject.

8 Energy
Will Bland and Ted Laird

Will Bland and Ted Laird

8.1

Introduction

Energy is usually an important consideration in the manufacture of chemicals. It may be an important cost in the process, e.g. in the manufacture of ammonia energy is required for the high pressures and temperatures used[1]

$$N_2(g) + 3H_2(g) \xrightleftharpoons[]{120\,atm/400°C} 2NH_3(g)$$

Alternatively, energy may give a cost benefit to a process where it is recovered from exothermic reactions, e.g. in the manufacture of sulphuric acid from sulphur a total of 3.26×10^6 kJ per tonne of sulphuric acid is reported to be potentially available from the process[2]:

$$S(g) + O_2(g) \longrightarrow SO_2(g)$$

$$SO_2(g) + \tfrac{1}{2}O_2(g) \longrightarrow SO_3(g)$$

$$SO_3(g) + H_2O(1) \longrightarrow H_2SO_4(aq)$$

This energy can be recovered as steam, e.g. to be used for heating purposes or converted into electricity.

Some aspects of energy were discussed in earlier chapters (6 and 7). In particular general estimates of energy costs in relation to potential profitability were dealt with in sections 6.2 and 6.3. In Chapter 7 the energy factors involved in plant operations, e.g. heat transfer and pumping of liquids were described. These factors are of lesser magnitude than those discussed in Chapter 6, but are still important in the consideration of the details of plant design.

8.1.1 *Energy required by the chemical industry*

Energy is needed in the chemical industry for the following operations:

(i) to drive endothermic reactions
(ii) to provide optimum reaction conditions of temperature and pressure
(iii) to drive separation processes
(iv) to carry out mass transfer operations

(v)   to power control and communication systems

(vi) to provide a satisfactory work environment

Different types of energy are required, e.g. *mechanical energy* used to compress gases for high pressure reactions, *thermal energy* used to raise the temperature of a reaction, *electrical energy* used for electrolytic processes or for driving pumps or just to illuminate the work-place.

8.1.2 *Sources of energy*

Most of the energy used in the chemical industry comes originally from fossil fuels, oil, coal and natural gas, but combustible organic by-products may also be used. Fossil fuels are a high-grade energy source which are especially used for direct heating. For other purposes electricity is more convenient, e.g. for driving pumps and stirrers and in some cases is vital, e.g. temperatures in excess of about 2000°C require heating by electricity and electrolytic processes cannot operate without electricity. Electricity is, however, an inefficient user of fossil fuels. An overall conversion figure of 34% for fossil fuel energy to electrical energy is used in official calculations[3]. Electricity in the U.K. is supplied through the National Grid mainly from two major power generation companies, National Power and PowerGen. In the U.S.A. a large number of private companies supply electrical power. An alternative for chemical companies is to provide their own power generated on the plant and, in some cases, sufficient power is generated for the company to sell some to other consumers.

Because some fossil fuel is converted into electricity care needs to be taken in interpreting statistics of energy supply. Thus the consumption of energy

Table 8.1 UK consumption of energy in 1993

	MTOE[a]
Consumption for energy and non-energy purposes	
Consumption for energy	220·5
Consumption for non-energy (petroleum)	11·7
Consumption for non-energy (natural gas)	2·0
TOTAL consumption of energy	234·2
Breakdown of consumption of energy by source of supply	
Coal	55·1
Petroleum (oil)	79·7
Natural gas	62·4
Nuclear electricity	21·5
Renewable (mainly hydroelectric)	0·4
Imported electricity	1·4
TOTAL	220·5

[a] MTOE = millions of tonnes of oil equivalent.
Source: Reference 3.

Table 8.2 Consumption of energy by different user groups in the U.K. in 1993

	%
Transport	33
Domestic	30
Other industries	19
Other final users	13
Iron and steel industry	5
Total consumption by users = 152.3 mtoe	

Source: Reference 3.

for the U.K. in 1993 shown in Table 8.1 is the total of all energy sources going back to primary fuel. When energy supply is quoted as consumed energy, including electricity, the total is lower because of the inefficiency in converting fossil fuels to electricity. The total consumption broken down by end-user is shown in Table 8.2; other industries, of which the chemical industry is a part, take only 19% of the total and the two major consumer sectors are transport and domestic (particularly heating). Convenient units to use for large-scale energy supply are millions of tonnes of oil equivalent (mtoe). One tonne of oil equivalent is 41.868×10^9 joules (41.868 GJ). Other conversion factors are shown in the appendix at the end of this chapter. An analysis of the data in Table 8.1 reveals that oil together with natural gas is the most important of the fossil fuels with coal now being in third place.

In Table 8.3 the supply of energy to the chemical industry is shown as 5.1 mtoe with petroleum (oil) and natural gas being more important than coal, but electricity being the most important contributor. The total consumption of energy by industry in the U.K. is 36.2 mtoe so the chemical industry takes 14.1% of the energy used in industry. The 5.1 mtoe consumed by the chemical industry is exceeded only by the 7.2 mtoe consumed by the iron and steel industry.

Apart from energy consumption the chemical industry also uses fossil fuels for feedstocks and other materials. In 1993 5.9 mtoe were used as oil based feedstock for chemical plants and of this 3.1 million tonnes were

Table 8.3 U.K. consumption of energy in 1993 for the chemical industry by source of supply

	MTOE[a]
Coal	0·7
Coke	0·07
Natural gas	1·4
Electricity	1·6
Petroleum (oil)	1·4
TOTAL	5·2

[a] MTOE = millions of tonnes of oil equivalent.
Source: Reference 3.

Table 8.4   Sources of world energy supply in 1993

	MTOE[a]
Petroleum (oil)	3121·4
Coal	2141·1
Natural gas	1787·1
Nuclear	557·2
Hydroelectric	197·5
TOTAL	7804·3

[a] MTOE = millions of tonnes of oil equivalent.
Source: Reference 4.

naphtha (see Chapter 12). Another 4.8 million tonnes of non-energy use included 2.5 million tonnes of bitumen, as well as waxes, lubricants, white spirit and industrial spirits[3].

Natural gas and liquefied petroleum gases, propane and butane, are also important non-energy contributors to the chemical industry. 3% of all natural gas comprising 2.0 mtoe is used for non-energy purposes principally to make hydrogen for the production of ammonia, but also for the production of methanol and for hydrogen itself. 2.5 mtoe of liquefied petroleum gases are used for non-energy uses e.g. the dehydrogenation of propane to propene[3].

The sources of supply of energy in the U.K. and usually in the world have changed over time from a predominance of coal in the last century and into this century, followed by the growing importance of oil with its widespread discovery and exploitation, and this is now supplemented by natural gas. The current worldwide pattern is shown in Table 8.4. Oil is dominant with approximately 50% greater consumption than coal, while natural gas is not far behind coal[4].

One other source of energy of importance to the chemical industry is combined heat and power (CHP), which is defined to be an installation where there is simultaneous generation of usable heat and power. This is actively promoted by government since the efficiency may be greater than 70% when the by-product heat from electricity generation is used in a productive way. The chemical industry offers good opportunities for the use of CHP. The chemical industry accounted for 31% of electricity output and 38% of heat output from CHP used in the U.K. in 1993. A measure of the importance of CHP is shown by the fact that CHP accounted for 5% of all electricity generated in the U.K. in 1993.

8.1.3 Cost of energy

The oil crisis of 1974 following the Middle East war resulted in an overnight quadrupling of the oil price from approximately $8 per barrel to $32 per

Table 8.5   Prices (in pence per GJ) of fuels and electricity used by U.K. industry

	1968	1973	1978	1983	1993
Coal	20	32	84	181	156
Fuel oil	21	30	120	294	173
Gas	63	29	111	228	214
Electricity	179	206	528	807	1184

Source: Reference 3 and earlier editions of Reference 3.

barrel. This led to a re-appraisal of energy policy. The second oil crisis in 1979 following the revolution in Iran reinforced the concerns as the price of oil rose to over \$50 per barrel. In the early 1980s coal was priced on a calorific value basis at approximately 60% of the value of fuel oil and this resulted in a partial reversion to the use of coal fired boilers in the chemical industry. This change to coal was promoted from 1981 by a U.K. government-sponsored scheme, but the scheme faltered from 1985 following a coal miners' strike[5]. In 1983 fuel oil was price at £126 per tonne, but in 1993 it had fallen to £69 per tonne[3]. The incentive to convert to coal firing therefore disappeared.

The average prices paid for fuels and electricity are shown in Table 8.5. The prices are quoted on a pence per GJ basis for more easy comparison. Points to notice are the major increases after the oil crises of 1973 and 1979. Also natural gas is a major fuel and has a higher price while the inefficiencies of production of electricity make this the most expensive source of energy.

8.1.4 Environmental factors

In the U.K. in 1992, 3.5 million tonnes of sulphur dioxide were emitted into the atmosphere[6]. Combustion of fossil fuels in electricity generating power stations produced 69% of the emitted sulphur dioxide. Much of this sulphur dioxide is oxidized in the atmosphere, e.g. by ozone, eventually to produce sulphuric acid or sulphates.

$$SO_2(g) + O_3(g) \longrightarrow SO_3(g) + O_2$$
$$SO_3(g) + H_2O(l) \longrightarrow H_2SO_4(aq)$$

Sulphuric acid is one of the main constituents of acid rain. Concern over the effects on the environment of acid rain is shown by action taken in the European Community (EC). The EC Large Combustion Plants Directive 88/609 takes 1980 emissions as a baseline and requires reductions in emissions from existing plants of capacity greater than 50 megawatts thermal of 20% by the end of 1993, 40% by 1998 and 60% by 2003[3].

This focuses the spotlight on the sulphur content of fuels since the combustion of fuels which contain sulphur in any form will result in the formation of the stable combustion product which is sulphur dioxide.

(i) Natural gas from the North Sea is sulphur free although other sources may contain a high proportion of hydrogen sulphide, H_2S. Even where hydrogen sulphide is present it can still be removed by absorption, e.g. in diethanolamine $(HOCH_2CH_2)_2NH$.

(ii) Oil contains sulphur in a variety of forms, e.g. thioethers and thiols. The sulphur can be removed in refinery operations for the production of high grade fuels, but would not normally be removed from fuel oil.

(iii) Coal supplied to power stations in the U.K. has an average sulphur content of about 1.6%[7] of which a half is in the form of metal sulphides, mainly iron pyrites, FeS_2. The iron pyrites content of coal can be lowered by physical methods of treatment, grinding and for example by separation by density differences of the coal and the pyrites. The other 0.8% of sulphur in coal is chemically bound in the macromolecular structure of the coal and it is not economically viable to remove the sulphur before combustion. Flue gas desulphurization (FGD) is used in European power stations, e.g. Drax power station in the U.K.[8], to remove the sulphur dioxide from power station chimneys. The sulphur dioxide is absorbed in a base, eventually to form calcium sulphate which is either dumped or used to manufacture plasterboard:

$$Ca(OH)_2(aq) + SO_2(g) \longrightarrow CaSO_3(aq) + H_2O(l)$$

$$CaSO_3(aq) + \tfrac{1}{2}O_2(g) + 2H_2O \longrightarrow CaSO_4 . 2H_2O(s)$$

The use of FGD and other methods of control of emissions of sulphur dioxide add to the cost of generating electricity especially from coal and this helps to explain the switch-over to the use of natural gas, particularly for the generation of electricity. In the U.K. from 1960 to 1993 the proportion of electricity supplied by natural gas increased from 0.6% to 9.9%[3].

8.1.5 Properties of fuels

The different properties of fuels give rise to different advantages in use. *The calorific value of a fuel* is a term widely used and is the energy released upon combustion per unit mass of a fuel (1 calorie = 4.18 joules). The calorific value of coal varies widely owing to the class of coal available in different locations and to the proportion of contaminating rock in the seam or extracted with the coal. Coal treatment plants remove much of this waste material, but even so typically in the U.K. coal delivered to power stations contains 15% of non-combustible material which ends up as ash[9]. The average value for coal is 26 GJ per tonne. Oil has a narrower range of calorific values than coal; the calorific value is higher due to higher hydrogen content and because oil is effectively fully combustible. An average calorifc value of 42 GJ per tonne can be used.

The use of gross calorific values or net calorific values (which assume that the water of combustion is present as vapour) plus the assumed value of coal leads to different coal to oil ratios being quoted in the literature, e.g. 1.5:1 to 1.7:1 by different authorities[3,4]. The composition of natural gas varies widely in different parts of the world leading to different calorific values. North Sea gas contains about 90% methane and has a typical calorific value of 52 GJ per tonne.

The transport of fuels determines the ease of their use. Gas is easily transported by pipe-line. Some intercontinental trade is conducted by cryogenic ships called liquefied natural gas (LNG) tankers which are used to supply Europe and are of particular importance for exports from Australia to Japan. Oil distribution is carried out principally by ships and road tankers. Coal is more difficult to transport; conveyor belts or pumping of suspensions of coal is only practised over short distances such as are found with giant lignite fields used to supply nearby power stations near to Cologne in Germany. The more usual methods using road or rail transport of coal require much more handling and abrasive wear than occurs in the transport of oil or gas. The transport of fuels on a large scale may be controversial where accidental release occurs as happened with the Exxon Valdez oil tanker, which ran aground releasing approximately 250 000 barrels of oil into a highly environmentally sensitive area, the Prince William Sound off Southern Alaska, creating an oil slick of 3 000 square miles and coating 300 miles of coastline with crude oil[10].

The combustion of fuels varies in its complexity. Gas may be simply burned in a gas jet while fuel oil must be pre-warmed and burnt in a burner head which has provision for atomization of the fuel. For coal either pre-grinding to pulverize material is used with complicated ignition systems or lumps of coal are burnt on less efficient grate stokers. Additional complicating factors with coal are the need to remove clinker from the furnace and ash from the flue gases. With fluids, ignition, the control of flow and metering are all relatively simple so that automatic firing control is more easily achieved than with coal.

The preference for fuels is determined by the combination of their properties and the end-use. The combination of high calorific value, ease of handling and distribution, better controllability, greater reliability, lesser polluting potential during combustion and lower maintenance all make fluid fuels of premium value for combustion with the preferred order being: gas, hydrocarbon liquids, fuel oil and coal. The high energy to weight ratio and ease of vaporization of the liquid fuels derived from crude petroleum (oil) makes them the fuel of preference for use in transport.

8.2

Types of energy

Different types of energy are involved in chemical processing on the production scale. These include chemical energy, thermal energy, electrical energy, mechanical energy, radiation energy, gravitational energy and kinetic energy. The most powerful form, nuclear energy, is usually exploited for the generation of electricity.

Although all forms of energy have important parts to play, there is considerable variation in the magnitude of the different forms. *Chemical energy* is of relatively high magnitude compared, for example, to vaporization because powerful bonds between atoms are being broken and re-formed. The demand for chemicals and the high magnitude of chemical energy affords an explanation of the large proportion of industrial energy use in the U.K. which goes to the chemical industry. Two examples illustrate this. The manufacture of one tonne of aluminium requires 270 GJ, of which 228 GJ are required at the smelting stage when the purified aluminium ore is finally converted to the metal by electrolysis[11]. In the petrochemical industry the manufacture of ammonia from hydrocarbon feedstock in early integrated single stream plants required 14 GJ per tonnes to meet energy requirements[12]. Modern plants are more efficient[13].

Thermal energy is needed for endothermic reactions. This type of energy is also important for some separation and purification processes, e.g. distillation and evaporation. The heat properties of liquids are therefore of importance for these processes. The energy required to heat a liquid is usually much less than the energy needed for vaporization (see Table 8.6).

The main use for electricity is for powering machines, pumps, stirrers and centrifuges. Electric motors are efficient and, most importantly, reliable. Furthermore, the use of flexible armoured cable means that there are no constraints on the positioning of motors. High powered motors are usually associated with gas compression or grinding machines. In contrast pumping of fluids is not usually energy intensive. Electrical energy is also used in electrochemical processes, electrostatic precipitation, drying, area lighting, control and communications.

The transport of liquids (and occasionally solids) between vessels in a batch chemical manufacturing plant is frequently achieved by *gravity*. It is very reliable (the motive power never failing!), but it is not free as liquids have to be pumped initially to the top of the plant. When these fluids pass through pipes they possess *kinetic energy*.

Table 8.6   Heat properties of liquids

	Specific heat capacity (C_p) $/kJ\,kg^{-1}\,K^{-1}$	*Specific latent heat of vaporization at boiling point* (LH) $/kJ\,kg^{-1}$
Benzene	1·70	394
Cyclohexane	1·80	393
Ethanol	2·50	839
Nitrobenzene	1·40	330
Octane	2·20	364
Acetone (propanone)	2·20	522
Water	4·19	2260

8.2.1 *Variation in energy content requirement*

The different order of energy involved in the various aspects of chemical processing operations is illustrated below:

> *Chemical energy:*
> Burning 1 tonne of octane releases   45 000 MJ

> *Thermal energy:*
> Conversion of 1 tonne of octane from
> liquid to vapour at the boiling point
> requires   400 MJ

> *Gravitational energy:*
> Raising 1 tonne of octane through 100 m requires   1 MJ

> *Frictional resistance:*
> Pumping 1 tonne of octane 1 km
> horizontally through typical pipework   0.01 MJ

> *Kinetic energy:*
> 1 tonne of octane flowing at $1\,m\,s^{-1}$ 0.001 MJ

In addition most heat exchangers in the petrochemical industry can transfer 1 to $5\,MJ\,s^{-1}$.

8.3

Use of energy in the chemical industry

More detailed aspects of the use of energy related to specific processes and methods of carrying out manufacture are discussed in this section.

8.3.1 *Batch reactors*

Batch reaction vessels, or batch reactors, are filled with reagents at the start and the reaction mixture is later emptied out or transferred elsewhere at the end of the reaction. Such reactors are commonplace in the dyestuffs, pharmaceutical and speciality chemicals industries. Most of the reaction vessels used have a capacity of 1 to 20 cubic metres and are made of metal perhaps with a glass, ceramic, rubber or resin lining and are equipped with a stirrer and coiled pipes for various heat transfer operations. A flow of steam through a coil might be used to provide energy needed in an endothermic reaction, to distil products from the vessel or perhaps simply to raise the temperature of the contents. A cooling coil might be used to remove energy from an exothermic reaction or perhaps to cool a solution to achieve efficient crystallization of a product. Such reactors are versatile, and some are used to make a succession of different products.

8.3.2 *Continuous reactors*

Large scale processes, as in the petrochemical industry, are generally continuous. Whereas a batch reactor might also be used for distillation or crystallization, this practice is rare in continuous processes. In continuous reactors energy is removed from exothermic processes or added to endothermic processes to control the reaction temperature and possibly to adjust the temperature of the reagents and products. For example, ethylene and other products are made by passing a suitable hydrocarbon feed with steam through steel tubes heated in a furnace. The enormous amount of heat transferred first heats the feed to reaction temperature, then compensates for the endothermic heat of reaction. An example of an exothermic process is the manufacture of phthalic anhydride by passing a stream of naphthalene or ortho-xylene (1,2-dimethylbenzene) vapour with air upwards through a fluidized bed of vanadium(V) oxide catalyst. (A fluidized bed is a suspension of fine particles in gas.) Temperature control is important since if the temperature became too high the system could explode or at least become an expensive way of making carbon dioxide and water! A heat exchanger in the fluidized bed removes heat generated in the exothermic reaction so that the reaction temperature is maintained at the optimum value. The temperature only varies by a few degrees throughout a fluidized bed.

In other cases the heat exchangers are next to the reactors rather than inside them. Thus in several important large scale processes gases react on passing through beds of pelleted or larger aggregates of particles. With these so-called fixed beds a heat exchanger within the bed would be inefficient owing to the lack of movement of the catalyst. An important process in petroleum refining is platforming in which hydrocarbons on passing through a catalyst bed are converted to products suitable for blending into automobile fuels, by increasing the content of aromatic and other desirable hydrocarbons. The process is endothermic. The feed is first passed through tubes in a furnace to heat it to a suitable temperature, t_1. The reactor is essentially adiabatic, so the temperature drops as a result of the endothermic reaction to a temperature, t_2. The reaction is then incomplete, and the fall in temperature is related to the amount of reaction that has occurred. At temperature t_2, the reaction rate has become very slow. The product is reheated to t_1, passed through another reactor where the product cools to t_2, and the whole process is then repeated a third time. Thus the heating is accomplished between stages of the reaction.

A somewhat similar idea applies in the exothermic oxidation of sulphur dioxide by air to sulphur trioxide which is an essential stage in the manufacture of sulphuric acid[2]. Four beds of catalyst with interstage cooling are used. The energy so released generates steam in heat exchangers. This steam can be used for electricity generation as well as process heating.

8.3.3 *Electrochemical reactors*

Electrochemical reactors are usually batch operated. Obviously energy is supplied in the form of electricity. In some cases, such as uranium production, the product can be prepared by both electrochemical and other methods, but other methods may then prove to be more economic in practice. Uranium, for example, is manufactured by heating a mixture of magnesium and uranium tetrafluoride:

$$2Mg + UF_4 \longrightarrow U + 2MgF_2$$

Aluminium production is an intermediate case. It is made by electrolysis of a molten mixture of aluminium oxide and sodium hexafluoro-aluminate(III) (cryolite), but the carbon anodes waste away by reaction with the liberated oxygen and this provides thermal energy to melt the electrolyte, thus lowering the electrical energy input required[14].

8.3.4 *Preparation and separation energy*

Energy demands are heavy in gas compression, size reduction (comminu-tion), and in those separations operations involving vaporization, evap-orative concentration, drying and distillation.

Gas compression requires energy in order to force the gas molecules into a smaller volume. The input energy, which appears as heat must be dissipated to prevent the compression machinery from seizing up due to gross differential expansion or lubrication failure if the temperature becomes too high. High pressure machines are therefore necessarily multi-stage in action with a 4:1 compression ratio at each stage being fairly average.

Comminution of large lumps makes use of flaws within crystal lattices, with rupture along these flaws occurring on compression, e.g. crushing. Reduction in size to fine particles involves abrasion and is very energy intensive partly because of the energy absorbed by the machines themselves.

Distillation involves the vaporization of a liquid mixture and interchange between this vapour rising up a column and condensed liquid descending the column, with contact being ensured by the presence of plates or packing in the column. The degree of separation of the components in the mixture can be improved by using a taller column with more trays or by increasing the reflux ratio, i.e. the ratio of the condensed vapour dropping back down the column to the condensed vapour removed as distillate. For some difficult separations reflux ratios higher than 20:1 may be used. Distillation is an energy intensive process because refluxed vapour has to be revaporized and latent heats of vaporization are much higher than specific heats (see Table 8.6).

Drying is also influenced by energy considerations since latent heat of vaporization is required in order to remove the solvent. As a consequence the maximum mechanical removal of solvent is used e.g. in pressure filters.

8.3.5 *Heat transfer media*

A few operations such as induction heating make use of an energy source directly but most heating (and cooling) processes involve conductive and/or convective transmission via a heating medium. Steam and water are most commonly used for heating and cooling respectively, though steam generation is used as a coolant in high-temperature applications and warm water is used for mild heating. There are occasions when the temperature range of 10°C–250°C normally available with these two media is inadequate.

Cooling water requirements in the chemical industry relative to availability in the locality are such that on large sites the demand cannot be met by once through cooling. Recirculation through cooling towers is needed. These cooling towers are designed to promote droplet formation in a counterflow draught of air with cooling being achieved mainly by increasing the water vapour content of the air. The latent heat required to evaporate 1–2% of the water lowers the temperature of the remainder. The required airflow can be obtained by the use of fans or by the chimney effect, exemplified by cooling towers associated with inland electricity generating stations. Because of the importance of water recovery, direct air-cooling using fans and vaned heat exchangers is becoming more popular.

Steam is useful as a heat transfer medium because:

(i)   it has a high heat content
(ii) it can be easily distributed
(iii) its flow is easily controllable
(iv) it is non-combustible and will not support combustion
(v)   it is non-toxic and relatively non-corrosive
(vi) it is produced from water, which is cheap and abundant

Steam under pressure can be used to heat plant to temperatures well above 100°C (the boiling point of water at atmospheric pressure). Steam gives up its heat by condensing since its latent heat of condensation is 1000 times greater than its heat capacity as gaseous steam. Steam is normally used at three different pressures (10, 15 and 40 bar) to allow for different heating requirements.

One further aspect is that steam can be used as a source of power as well as of heat. Power can be extracted in machine drives and the residual heat in the steam exhausted at lower pressure can be utilized allowing high overall efficiency. If high pressure steam from a boiler plant is passed through a turbine/alternator set to generate electricity, then exhausted and distributed at a lower pressure to supply process heat, most of the energy in the original fuel is usefully used rather than rejected to the atmosphere. This contrasts with conventional power stations where maximum conversion to mechanical rather than electrical energy is the target. Combined heat and power (CHP) is only feasible where there is a steady and continuous demand for thermal energy such as is required for processing operations in a chemical plant. Indeed larger chemical sites have their own electricity generating stations operated on the CHP system. This system using gas turbines is found on sites where ethylene (ethene) is manufactured.

The desirable properties of a high-level heat transfer medium include low cost, non-flammability, nil toxicity, compatibility with common metals, remaining liquid at ambient temperature and most importantly, thermal stability. Materials cannot meet all of these criteria, but some useful ones are discussed below.

Petroleum oils have a low cost and are non-toxic and non-corrosive. They are usable at operating temperatures up to 315°C, but are flammable and subject to oxidative degeneration, which can be countered by using a nitrogen blanket, but thermal cracking will still occur.

'Dowtherm' is a proprietary generic name applied to a range of heat transfer fluids, but in the U.K. normally refers to a mixture of diphenyl (73.5%), and diphenyl oxide (26.5%). The mixture is also marketed as 'Thermex'. It is fluid down to 12°C, leading to a solidification problem in cold weather. High maintenance standards are necessary because of its searching characteristics. For prolonged operation, a temperature restriction to 370°C should apply, but it is usable up to 400°C.

'Hygrotherm' is a trade name for tetra-aryl silicate fluids, which are expensive. However they are non-toxic and non-corrosive. They will burn only at high temperatures, and have good heat transfer coefficients. They can be used at temperatures up to 355°C.

Low temperature heat transfer media are often based on water either in the form of ice or refrigerant brines. *Ice* may be used for direct addition to reaction media where dilution by water is acceptable, e.g. the preparation of diazonium salt solutions as intermediates for the manufacture of dyestuffs. In other cases cooling may be carried out by circulating *sodium chloride brine* (23% NaCl in water) for temperatures down to −21°C. The more popular calcium chloride brine (29%) can be taken down to −40°C. *Ethane-1,2-diol/water* mixtures are sometimes used in intermittent cooling applications where there is no possibility of frost affecting the coolant mixture at any time. For even lower temperatures than those which can be achieved with brines either liquid ethene or liquid propene can be used if they are available. *Ammonia* is commonly used in industrial plants. *Chlorofluoromethanes* should no longer be produced after the end of 1995 because of their effects in depleting the ozone layer; similarly *hydrochlorofluorocarbons* are expected to be controlled[15]. *Hydrofluoro-methanes* and *-ethanes* have been developed as possible alternatives (see section 1.6.6).

It is not unusual to locate a refrigeration plant to serve a number of chemical plants.

Because of the high energy input required for many major chemical processes the chemical industry has been in the forefront of the development and application of efficient energy utilization techniques.

8.4.1 *Exothermic reactions*

Reaction exotherms and endotherms often involve significant amounts of energy, but there are few energy-exporting chemical processes. There are a number of reasons for this. Firstly, the temperature rise associated with many exothermic reactions must be limited since if the temperature rises too much the reaction rate will increase and possibly run away; also by-product formation may be unacceptably high at higher temperatures. Additionally, we know from Le Chatelier's principle that at high temperatures the theoretical yield for an exothermic reaction is lowered. Thus if the reaction temperature is to be kept low ($< 95°C$) then the heat of the reaction must be dissipated into the reaction medium or into cooling coils. This low-grade heat, often in the form of warm water, is of little use although it may be used in some cases to pre-heat another flow.

The question of the grade of the energy is of prime importance. Irrespective of amount available heat will not flow from a cooler to a warmer body. High-grade heat from continuous exothermic chemical processes is always recovered in heat exchangers, either by direct interchange (reactor outlet flow pre-heating reactor feed) or by raising steam, which is subsequently used to drive separation processes or turbine machinery. The success of many modern petrochemical processes is based on the integration of the various chemical processing and separation stages with the heat recovery equipment in order to make the best overall use of the available energy from reaction exotherms and imported fuel. An example is the modern process for the manufacture of ammonia from natural gas (CH_4). The ammonia synthesis state is exothermic, but in order to achieve a satisfactory rate of reaction the temperature must be maintained at $400–500°C$ which gives a conversion of about 15% per pass through the reactor. The quality of the recoverable heat is insufficient to meet the requirements of the steam reforming reaction earlier in the process and which requires temperatures above $700°C$:

$$CH_4 + H_2O \longrightarrow CO + 3H_2$$

Thus primary heat must be added. The heat in the furnace flue gases (and in the process gas stream downstream of the secondary reformer where air is injected to supply the nitrogen) is used to generate steam for the turbine-driven process air compressor, synthesis gas compressor, and gas circulator in the synthesis loop. Exhaust steam from the turbines is used in the regeneration column boiler to strip CO_2 from the liquor used in the absorption system and for other duties. Some plants burn less fuel, recover less heat, and need to import electricity to drive the big machines. Theoretically there is a net evolution of $34\,kJ\,mol^{-1}$ of heat in the manufacture of ammonia from natural gas, steam, and air, but achievement of the necessary reaction conditions[12] and the preparation and separation processes involved requires an energy input of approx. $240\,kJ\,mol^{-1}$. The total improvements which are achievable in a process are well illustrated by ammonia. In the early 1960s ICI introduced a new process based on naphtha rather than coal and with improved engineering design using a single stream rather than the earlier more complicated parallel flow system with interlinks. As a result of this the energy costs for feedstock, fuel and electricity fell from 850 therms per tonne ($79\,GJ\,tonne^{-1}$) to 400 therms per tonne ($42\,GJ\,tonne^{-1}$)[1]. In the early 1980s developments again in engineering design with milder conditions, aided by the use of a new catalyst, led to the energy cost of feed and energy falling to 300 therms per tonne ($32\,GJ\,tonne^{-1}$)[13].

Hydrocarbon partial oxidation reactions are highly exothermic but reaction temperatures are normally kept down to prevent by-product formation. Any energy recovered in the form of steam is usually fully utilized in the distillation train and/or other processes which are used to separate the desired product in sufficient purity from the accompanying spectrum of by-products. One process which is a net exporter of energy is the naphtha oxidation process for the manufacture of acetic (ethanoic) acid (section 12.5). The only two other processes which are significant exporters of energy are the manufacture of sulphuric acid from sulphur and the oxidation of ammonia to nitric acid.

8.4.2 *Separation processes*

Industrial fractional distillation involves the use of large amounts of energy to vaporize the feed material. A proportion of the condensate has to be returned to the top of the column as reflux (see Figure 7.33). The following may be specified in a continuous fractionation: compositions of the feed as well as the top and the bottom and the flow rates (e.g. Figure 7.34). A range of combinations of reflux ratios and number of plates in the column that can achieve this can be calculated (cf. section 7.6.3.3). At a constant distillation rate increasing the reflux ratio is equivalent to increasing the amount of reflux, thereby increasing the cooling water and steam costs. On the other hand fewer plates in the column are required and this can decrease capital costs, although this effect can be negated by the need to increase the diameter to accommodate higher flow rates in the column. All these costs need to be estimated at the design stage, and the combination of reflux ratio, number of plates and diameter giving the cheapest annual costs chosen. As with other items of plant, energy costs can be very important, but total costs are overriding in decision making.

Evaporators may have their efficiency improved by multi-effect operation. In this the evolved vapour from the first stage is condensed in a heat exchanger and gives up its latent heat to the second stage. The necessary temperature differential across each heat exchanger means a progressive reduction in boiling point. This is achieved by having a pressure (vacuum) gradient. Theoretically the heat evolved on condensing 1 kg of vapour equals the heat required to evaporate 1 kg of the same liquid. Efficiencies of 0.85 are available for each stage. Increasing the number of stages improves the thermal economy, but at increased capital expenditure, and therefore each system must be optimized. It should be noted that not all evaporators are muilti-effect, e.g. evaporative concentration of thermally unstable materials requires short residence time and so once-through operation in film evaporators is carried out.

Drying operations also involve the removal of liquids by evaporation. Pre-concentration of slurries by mechanical means, e.g. press filtration and centrifuging, is normal practice in order to reduce the evaporative load in the dryer. The degree of diluent removal is determined by the particle size of the solid and feed characteristics for the selected dryer, e.g. pumpable slurry, preformable paste and hard lumps.

Electrostatic precipitation is another energy intensive separation process, but it is the most efficient method of removing particulate material from off gases. Its use for environmental control purposes in large installations is widespread.

8.4.3 *Restriction of losses*

The cost of energy is of crucial importance in considering energy recovery. Prior to the first oil crisis of 1973 the capital cost of the installation of energy recovery equipment often could not be justified in terms of the possible economic return. If the economic factors are favourable available energy from chemical processes will be recovered. The requirements are that the energy is available continuously at a sufficient temperature differential and quantity to justify the capital expenditure on a heat exchanger. The recovery of heat which is available only intermittently is seldom worth while. The heat input to a process can be kept lower by minimizing the heat losses to the environment. This is achieved in a number of ways, particularly minimizing convection and radiation losses from hot surfaces, and by restricting leaks of heat transfer fluids.

Lagging of interconnecting pipework and of plant, e.g. reactors is often carried out in order mainly to limit heat (or cold) loss but also to protect workers in the plant. Hot pipes are usually lagged with magnesia (MgO) or calcium silicate ($CaSiO_3$) or mineral or mineral wool, whereas expanded polystyrene is frequently used on cold pipework. The thickness of the lagging depends on the duty[16]. Thicker layers are becoming more economical as energy costs increase. The heat loss of $8400 \, W \, m^{-2}$ from a bare surface at a temperature of $325°C$ in still air can be reduced to $200 \, W \, m^{-2}$ by fitting $100 \, mm$ thick lagging[17].

Leaks in heat transfer systems cannot always be completely stopped, e.g. a furnace will contain apertures for air supply, burner guns, feed and withdrawal pipes, peepholes, etc. The apertures and differential expansion of materials of construction, which makes tight sealing difficult, means that it is impossible to prevent leaks of heat. Furnaces are therefore operated at slightly negative pressure which avoids outward loss of hot gases, but gives an inward leak of cold air. The design should be such as to minimize leaks and determined maintenance work should be carried out, e.g. to repair leaks from flanges and valve packings on steam distribution systems. Leaks will tend to increase with time due to thermal cycling (and ageing) and increasing erosion of materials as the leaking material flows out through the leak. Damage to gasket material may then extend to damage to the flange faces necessitating a more complex repair job. Clearly leaks should be repaired as rapidly as possible, but shut-down, e.g. on continuous plant may not be practicable. In such cases temporary repairs may be carried out by fitting a jacket round the leaking flange and filling the space with a special heat setting material. Leaks of steam may cause appreciable loss of energy, e.g. a 3 mm diameter hole in a 20 atmosphere steam pipe[12] will cause a loss of $60 \, kg \, h^{-1}$ or about 10 tonnes per week of high quality steam which is equivalent to 28 GJ per week.

8.5
Conclusions

The energy balance of a chemical process is of prime importance. Continuous processes are more advantageous than batch processes in which the continuous charging and discharging of reactor contents wastes heat. In continuous processes the heat is recoverable all the time, which provides a better economic basis for the investment of capital in energy recovery equipment. Reaction exotherms or endotherms may involve the release or absorption of much energy. In large continuous plants the absolute quantity of energy is very large and so every effort is made to extract this from heat flows where there is a significant temperature differential. This and/or imported energy is used to drive preparation and separation processes, some of which are energy intensive. Successful process design entails the integration of reaction and separation stages as well as heat recovery equipment which makes the best use of energy input. Other important factors are good construction of plant so as to minimize potential leaks and effective insulation to aid heat retention.

For the future the development of less energy intensive processes is a priority on the grounds of the likely future increases in the price of fuel and on the demands to lessen the impact on the environment from the use of fossil fuels. In March 1994 the United Nations Framework Convention on Climate Change came into force following ratification by 50 countries; the requirement of the Convention is that carbon dioxide emissions are returned to 1990 levels by the year 2000[15]. Energy consultants and permanent employees in chemical companies play an important role in attempts to improve energy efficiency. The continual improvements, which can be made to existing processes is well illustrated by the study of ammonia described in section 8.4.1.

Appendix

Units and energy conversion

1 calorie	4.18 joules (J)
1 gigajoule	10^9 joules (GJ)
1 megajoule	10^6 joules (MJ)
1 tonne	1000 kg
British thermal unit	1.055 kJ
1 therm	100 000 Btus
1 kilowatt hour	3600 kJ
1 barrel of oil	42 U.S. gallons
	35 gallons
1 tonne of oil	7.5 barrels

Calorific values

1 tonne of coal	26 GJ
1 tonne of oil	42 GJ
1000 m^3 of natural gas	37 GJ
1 litre of petrol	35 MJ

Note: Further conversion factors are given at the start of this book.

References

1. S. D. Lyon, Developments of the modern ammonia industry, *Chemistry and Industry*, 6th September, 1975, 731.
2. A. Philips, The modern sulphuric acid process, *Chemistry in Britain*, **13**, 471, 1977.
3. *Digest of United Kingdom Energy Statistics 1994*, Dept. of Trade and Industry, Her Majesty's Stationery Office, 1994.
4. *BP Statistical Review of World Energy, 1994*, The British Petroleum Company plc, 1994.
5. *Chemistry and Industry*, 7th January, 1985, 2.
6. *Digest of Environmental Protection and Water Statistics*, Department of the Environment, Her Majesty's Stationery Office, 1994.
7. Technologies for the removal of sulphur dioxide from coal combustion, W. S. Kyte, Desulphurisation in coal combustion systems, The Institute of Chemical Engineers, Hemisphere, 1989, 15.
8. C. Butcher, FGD—the next ten years, *The Chemical Engineer*, 11th April, 1991, 26.
9. J. Longhurst, The British flue gas desulphurisation programme, *Acid magazine*, **8**, 22, 1989.
10. *ENDS Report*, **171**, 11, 1989.
11. P. F. Chapman and F. Roberts, *Metal Resources and Energy*, Butterworth, 199, 1983.

12. C. D. Grant, *Energy Conservation in the Chemical Process Industries*, Institution of Chemical Engineers, 39, 1979.
13. A. Chuter, *ICI magazine*, Autumn 1983, 199.
14. J. E. Fergusson, *Inorganic Chemistry and the Earth*, Pergamon, 1982.
15. *This Common Inheritance, The Third Year Report*, Her Majesty's Stationery Office, 1994.
16. P. M. Goodall, *The Efficient Use of Steam*, Westbury House, 101, 1980.
17. O. Lyle, *The Efficient Use of Steam*, Her Majesty's Stationery Office, 124, 1968.

Bibliography

Principles of Industrial Chemistry, C. A. Clausen and G. Mattson, Wiley, 1978.
Energy Conservation in the Chemical and Process Industries, C. D. Grant, Institution of Chemical Engineers, 1979.
Kirk-Othmer's Encyclopedia of Chemical Technology, 3rd ed., Wiley, 1978–1984.
The Efficient Use of Steam, O. Lyle, HMSO, 1968.
Energy Resources and Supply, 2nd ed., J. T. McMullen *et al.*, Arnold, 1983.
Energy Around the World, J. C. McVeigh, Pergamon, 1984.
Chemical Engineering in Practice, G. Nonhebel, Wykeham, 1973.
Perry's Chemical Engineers' Handbook, 6th ed., eds R. H. Perry and D. Green, McGraw-Hill, 1984.
Ullman's Encyclopedia of Industrial Chemistry, 5th ed. W. Gerhartz, VCLI Verlagsgesellschaft, 1985.
Annual Abstract of Statistics, Central Statistical Office, HMSO, annually.
Digest of U.K. Energy Statistics, Central Statistical Office, HMSO, annually.
Coal and the Environment, Commission on Energy and the Environment, HMSO, 1981.
Development of the Oil and Gas Resources of the U.K., Department of Energy, HMSO, annually.
A Guide to North Sea Oil and Gas Technology, Institute of Petroleum, 1978.

Environmental Impact of 9
the Chemical Industry

Andrew Hursthouse

The earth is a complex and dynamic system, involving the transfer of material on both micro and macro scales. The processes operating on and within the earth can be simply viewed as a series of interlinked compartments with the links allowing material and energy flow in different directions. The general situation, shown in Figure 9.1, summarizes our understanding of the earth's system from a geological viewpoint, in which the fluid flows of rocks or magma within the earth are responsible for the reworking of the earth's crustal materials.

In the context of human behaviour, it is the understanding of surface processes that are of most importance to us. Geological timescales and

9.1
The environment and human interactions

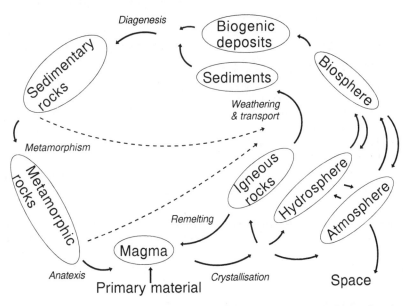

Figure 9.1 A simplified view of earth processes—the Geochemical Cycle. Based on Reference 1.

Table 9.1 Sizes and vertical mixing of various earth reservoirs. Based on References 2 and 3

Sphere	Mass (kg)	Mixing time (years)
Biosphere	$4{\cdot}2 \times 10^{15}$	60
Atmosphere	$5{\cdot}2 \times 10^{18}$	$<0{\cdot}2*$
Hydrosphere	$2{\cdot}4 \times 10^{21}$	1600†
Crust	$2{\cdot}4 \times 10^{22}$	$>3 \times 10^{7}$
Mantle	$4{\cdot}0 \times 10^{24}$	$>10^{8}$

* Surface boundary layer = 1 hour, to tropopause = 1 month, to lower stratosphere 50 years.
† Surface mixed layer = 10 hours, base of thermocline = 2·5 years, Pacific/Atlantic deep ocean 100–1000 years.

major earth movements take place over many millions of years and are only significant when catastrophic events, such as earthquakes and volcanoes occur. Of more immediate concern are those processes operating on the earth's surface that serve to modify or affect systems within our own lifetime (i.e. tens of years). The rates of the processes in operation in the various compartments of the environment are dependent on the nature of the processes themselves. This is highlighted in Table 9.1 which summarizes the sizes, in terms of mass and vertical mixing times, of each of the major earth compartments. From our own everyday observations of climate and other surface processes such as river flow and tidal movement, these time scales can be readily appreciated. Yet within these general compartments, the mixing and movement of materials is not uniform. For example, mixing times of a non-reactive pollutant gas[2] may be of the order of hours within the boundary layer (the atmosphere in contact with the earth's surface to 1 km height), in which flow is affected by surface features; the order of days within the free troposphere (1–10 km from the surface of the earth), within which climatic turbulence occurs to give us weather systems; and the order of years for mixing within the stratosphere (between 10 and 50 km from the surface of the earth).

Human activity on the surface of the earth has had an environmental impact since the first tools and manipulation of resources occurred. But only since the industrial revolution of the 18th and 19th centuries have human activities had a measurable effect on a global scale[4]. To support the maintenance and development of society, humans have developed and refined means to utilize the resources available from within and upon the surface of the earth. This utilization has developed from, in the most part, agricultural practices towards an industrially and technologically dominated society.

The resources available can be classified in a number of ways and are summarized in Figure 9.2, namely:

(i) *Stock resources:* those which are non-renewable and are not re-

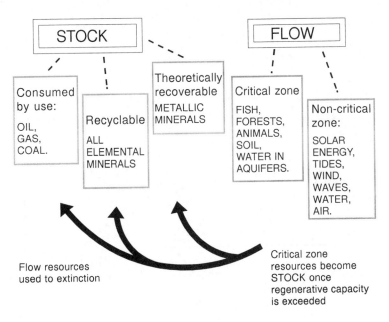

Figure 9.2 The definition and classification of types of resources and their relationship to the sustainability of society. Based on Reference 5.

generated on the human timescale, being consumed by use. However, there exists the possibility for some stock resources to be recycled.

(ii) *Flow resources:* those which are naturally removed in the human timescale. These can be further subdivided into those in the *critical zone* which can be exploited but can also be affected by human activity and those which are *non-critical zone* which can be exploited and are renewable. For some of this latter group one must also add a cautionary note of care that the rates of exploitation and renewal are balanced.

So within any developing society there is an underlying need to capture resources from the earth, process raw materials and utilize the products. The environmental effects of these activities can be severe and can arise at a number of stages[6]:

(i) the removal from a natural location
(ii) handling at the point of removal
(iii) transport and storage
(iv) process handling
(v) the generation and disposal of wastes

An increasing world population and technologically advancing society, increase the net pressure of human activity on the environment. Figure 9.3

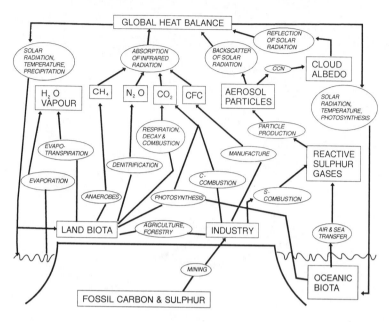

Figure 9.3 The major components of earth surface processes, illustrating human interaction. Based on Reference 6.

shows the types of processes involved in the surface cycling of materials and climate and their effect on observable components of the earth. From this it is easy to see the potential significance of human interactions, through industrial activity and the areas that will be sensitive to increasing activity. This is perhaps more easily appreciated by the more complex schematic in Figure 9.4, which concentrates on the objectives of industrial activity, that is, to produce materials that have a use in society. In so doing, significant amounts of waste of all types are released into the environment.

Waste can be defined as[7]:

(i) any substance which constitutes a scrap material or an effluent or other unwanted surplus substance arising from the application of any process; and

(ii) any substance or article which requires to be disposed of as being broken, worn out, contaminated or otherwise spoiled.

Ultimately, the fate of anything used by humans is that it becomes waste material. If the primary activity of industry is to produce a product, which has a use in society, this activity has the potential to generate waste at many stages in the process. The waste material may be released to the environment at the point of generation or at a later stage. It is important to understand the interaction of any release with the environment and particularly in relation to the biogeochemical cycles, so that any likely environmental

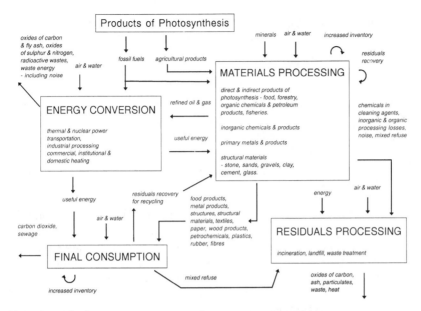

Figure 9.4 The flow of materials in technological society. Based on Reference 1.

impact can be assessed. If that impact is negative, then the release is generally viewed as being a pollution incident.

As a consequence of the complexity of environmental processes, the diversity of industrial activity and inter-relationships between humans and the environment, the term pollution has wide-ranging implications. In a formal sense, a pollutant may be defined as[8]:

> A substance or effect which adversely alters the environment by changing the growth rate of species, interferes with the food chain, is toxic or interferes with the health, comfort, amenities or property values of people.

When this definition is considered with the description of waste, it is apparent that any human activity generates both waste and pollution. Differentiation between the two is difficult. It is generally accepted now, that humans must be more efficient in managing their activities on the surface of the planet if there is to be a sustainable development of society and preservation of the quality of life. As a result of past practices, scientific research and an improving understanding of our own impact on the environment, recent legislative changes have meant that industry, often viewed as the primary polluter, is pressurized to clean up and minimize its environmental impact.

This chapter gives a short, introductory overview of the industrial sources of pollution, present practices, and the constraints of minimizing the environmental impact with particular reference to the chemical industry.

The subject is naturally diverse and rapidly changing, so only a superficial treatment can be given at this stage; in particular the fundamental principles will be emphasized.

9.2

Sources of pollution

From Figure 9.4, it is apparent that through the activity of any industry, a range of pollutants may be released into the environment. The precise composition will vary depending on the nature of industrial activity at a particular site. Most industry, in addition to the extraction of primary raw materials and energy production, requires a considerable amount of water to aid process control. As a consequence, the major environmental impact from industrial activities is the release of material to the atmosphere and to the hydrosphere, through direct discharge to water courses, leaching into the water table and stack emissions. This allows transport of the pollutants away from the site. In addition, the direct contamination of industrial sites has relatively recently become the focus of considerable attention. Changing economic constraints on industrial activity and the rationalization of the industrial base in many developed countries has resulted in the closure of sites, and the redevelopment of land for other domestic or leisure uses has highlighted a significant environmental legacy[9-11].

It is important to remember that the release of pollutants from industrial activity has a geographical control. Overall, the global amounts of naturally occurring materials may be far greater than the industrial releases of certain substances. It is the relative concentration and proximity to sensitive ecosystems and population centres that exaggerates the resulting environmental impact.

9.2.1 *Atmospheric pollution*

The natural composition of air includes the major gases ($N_2 \sim 78\%$; $O_2 = 20.9\%$; $Ar = 0.9\%$; $CO_2 = 0.035\%$; $H_2O = 0.53\%$) and many hundreds of trace components. Some of these can be explained by natural releases or direct industrial activities (*primary pollutants*) and others, that are formed indirectly by chemical processes in the atmosphere (*secondary pollutants*)[12]. It therefore follows that, in minimizing the environmental impact of industrial activity, not only must direct pollutant releases be considered, but also the release of compounds which may have a role in secondary pollutant production.

Natural emissions from biological, geological and meteorological processes include: oxides of sulphur and nitrogen, hydrogen sulphide and ammonia. These compounds, though oxidation in the atmosphere can form acidic and readily soluble compounds, that when dissolved in water, contribute to natural acidity of rainwater, greatly increasing the weathering ability of surface waters. Table 9.2 summarizes the major species of these

Table 9.2 Compounds of sulphur and nitrogen observed in the atmosphere. Based on Reference 13

Species	Concentration	Sources	Sinks
SO_2	0–0·5 ppm v (urban)	Oxidation of fossil fuel S	Direct reaction with Earth surface, oxidation to sulphate
	20–200 ppt v (remote)	Oxidation of S gases	
H_2S	0–40 ppt v	Biological decay of protein in anaerobic water	Oxidation to SO_2
CH_3SH	sub-ppb v	Paper pulping	Oxidation to SO_2
CH_3CH_2SH	sub-ppb v		Oxidation to SO_2
OCS	500 ppt v		Destruction in the stratosphere
CH_3SCH_3	20–200 ppt v	Oceanic phytoplankton and algae	Oxidation to SO_2
CH_3SSCH_3	small		Oxidation to SO_2
CS_2	10–20 ppt v		Destruction in the stratospheric and tropospheric OH
NH_3	0–20 ppb v	Biological	Precipitation
$RNH_2 \ldots R_3N$		Biological	Precipitation
N_2	78·084%	Primitive volatile, denitrification	Biological nitrification
N_2O	0·1–0·4 ppm v	Biological	Photolysis in the stratosphere
(N_2O_3)		Reaction intermediate	
NO	0–0.5 ppm v	Oxidation of N_2 in combustion	HNO_3
NO_2		NO oxidation	HNO_3
HNO_2		OH + NO	Precipitation
HNO_3		OH + NO_2	Precipitation

Units: ppm v = 1 in 16^6 by volume
  ppb v = 1 in 10^9 by volume
  ppt v = 1 in 10^{12} by volume.

elements present in the atmosphere and highlights the main sources. The natural production paths are very numerous and global levels generally exceed anthropogenic releases[12,13]. The main anthropogenic sources being primarily from combustion-related processes. In addition to nitrogen and sulphur-containing species, hydrocarbons (primarily methane) are released naturally from anaerobic digestion in rice paddies, wetlands, tundra and from ruminant animals, and the amounts from these sources appear to be affected by human activities. Other heavier hydrocarbons, such as terpenes, are released through natural degradation processes and greatly exceed anthropogenic emissions.

Natural releases to the atmosphere are geographically variable and superimposed on these are emission to the atmosphere from both stationary and mobile point-sources, of domestic and industrial origin. The routine combustion of fossil fuels for energy production releases significant quantities of CO_2 and H_2O. In addition, CO_2 is released through natural metabolic processes. The increased use of fossil fuels for energy production has increased the total release to the atmosphere of CO_2 and there is some evidence that the natural sinks for atmospheric CO_2, such as photosynthesis, are able to absorb only part of this increase[12,14]. At present there appears to be a net increase of $\sim 0.3\%$ of the atmospheric CO_2 concentration per year[12,15]. The significance of CO_2 in the regulation of global energy balances, means that considerable attention has been given to the control of releases from direct and indirect human activity[16].

Carbon monoxide (CO) can be a significant component of combustion gases from power production, but releases are more significant through the use of internal combustion engines[15]. Through incomplete combustion, flue gases can contain several percent CO. In the production of iron and other metals, oxygen supply is minimized and CO emissions are high[17]. The internal combustion engine, used widely for transportation and mobile energy sources, releases a diverse range of pollutants, including CO. Levels of CO (and hydrocarbons) are higher from petrol engines than from diesel and depend greatly on engine capacity and speed[17,18].

Soot formation generally accompanies incomplete fuel consumption through the polymerization of carbon nuclei[12,17]. Particles of soot have dimensions commonly $< 1 \mu m$ and through effective light scattering properties are easy to see. They also act as nucleating centres for vapour and salt deposition, as gas streams cool. In industrial processes, temperature control is normally adequate to remove the problem of soot production. Domestic combustion, however, is less controlled and with internal combustion engines, the diesel engine is particularly prone to soot and particle release. Size is the most important physical property of soot or particulates[19], as the exposure to humans is dominated by the inhalation pathway. Ash deposits of inert residues from solid fuels are also of concern and particularly problematic with low quality, bituminous fuels. The chemical composition of each can contain high levels of heavy metals, high molecular weight hydrocarbons and halocarbons[12].

High molecular weight hydrocarbons are often produced during incomplete combustion and these are often carcinogenic. Polycyclic aromatic hydrocarbons (PAHs) are a good example of one class that has been relatively well studied[12,15]. These are released from all types of combustion processes and can be particularly hazardous in residues from old coal gas plants[9].

The presence of sulphur in most fossil fuels results in the release of SO_2 and minor amounts of SO_3. The oxides of sulphur react in the atmosphere to form sulphuric acid. Further industrial sources are in the roasting of metallic ores (often found as sulphides). The modification of combustion conditions cannot reduce sulphur released and control is usually by pretreatment of the fuel or desulphurization of post-combustion gases[12,17].

Oxides of nitrogen (primarily NO and NO_2), also have a role in the generation of acid rain at low altitudes and play a crucial role in the ozone balance at higher altitudes[12,20]. Production during combustion involves the high temperature reaction of N_2 and O_2 from air in the combustion chamber and from nitrogenous constituents of the fuel. NO is the main product and variations in burner design are being used to reduce emissions from power stations[17].

Halogen-containing compounds are a major contributor to the acidic component of fuel combustion gases. Hydrogen chloride from chlorides in the fuel or the incineration of chlorinated plastics is corrosive to plant construction materials[17,20]

The most widely known halogen-containing pollutants are the chloro-fluorocarbons (CFCs) used as aerosol propellants, refrigerants, solvents and for foam blowing[20]. Free chlorine in the upper atmosphere, from the photo-dissociation of CFCs, has been postulated as the cause of ozone depletion observed, at an increasing frequency, above the North and South Poles[6,21]. The Montreal Protocol was an international agreement, signed by a number of nations in 1987[20], in an attempt to reduce this effect by banning substances which are thought to contribute. It has met with limited success in reducing emissions, but has helped to stimulate the search for alternative materials[22]. Other fluorine releases at ground level include emissions of hydrogen fluoride from brick kilns, primary aluminium smelting and fluoroapatite fertilizer works[19,20].

All industrial processes release dusts other than soot. There are many sources and forms including fuel ash, metal oxide fumes and silica and releases tend to be localized. Fugitive emissions from outdoor industrial activity or processes associated with the mining and preparation of raw materials can generally be controlled through careful industrial practice. Water is often used to minimize dust raised, resulting in the transfer of an atmospheric pollution problem to one of either aquatic or solid waste.

The final source pollutants, released by industry are volatile organic compounds or VOCs. These encompass solvents and volatile, petroleum based spirits released from paints, petroleum manufacture, distribution and storage[12]. They are often an important fugitive emission and numerous examples exist of methods of control. Simple process control methods, monitoring and tightening of leaking seals[23], removal of volatile components in liquid waste streams[24] are often adequate enough to reduce releases significantly.

With most industrial activities, the high concentration of chemicals, by products or wastes, can result in high levels of associated odour. This is particularly of concern with food processing, tannery processes and the use or production of sulphur-containing compounds such as hydrogen sulphide. It is a particularly difficult class of pollution to deal with as individual senses vary. In addition, the levels at which odorous agents are sensed, whilst a nuisance, may not be toxic[25].

9.2.2 Aquatic pollution

The use of water for drinking, cleaning, power production and as a raw material in many processes coupled with the sensitivity of aquatic ecosystems to environmental disturbances, has resulted in the need for careful water management practices. There exist a number of well documented examples, where uncontrolled discharges have resulted in the destruction of local aquatic ecosystems[26-28]. The two main factors to consider are the transport times and the relative dilution afforded by the receiving water body[28].

The major sources of water pollution are from sewer outfalls or industrial discharges (*point sources*), or directly from air pollution or agricultural and urban runoff (*diffuse sources*). Pollutants are varied in type and effect, and include: inorganic and organic compounds with direct toxicity; nutrients that stimulate aquatic microbial activity and hence oxygen depletion; inert solids that obscure the transmission of light through the water and silt up channels changing water flows; waste heat that creates artificially high temperatures within water courses; radionuclides; and infections agents[29,30].

The identification of point sources of pollution is relatively straight-forward, as is the imposition of remedial action through monitoring, legislation and treatment. Diffuse sources are much more difficult to assess and control, with great difficulty in pinpointing the source and hence apportioning blame. The majority of pollution incidents in water courses come from organic pollution namely organic rich slurries from farms, industrial processes and sewage from sewage treatment plants receiving both industrial and non-industrial liquid wastes[29].

The polluting power of these organic discharges derives from their residual biodegradable components which, if released into water courses would stimulate microbial activity, depleting oxygen levels in the water and killing higher organisms indirectly. As such, this form of pollution is relatively well understood and straightforward to control through engineered systems. The quality of discharges or receiving waters is routinely measured by the chemical (COD) and biochemical (BOD) oxygen demand tests[31]. The COD is the amount of oxygen consumed in the complete oxidation of carbonaceous matter in effluent samples using potassium dichromate. The BOD test involves the determination of dissolved oxygen in a sample before and after generally a 5-day period of incubation in darkness at $20°C$ with a microbial seed (BOD_5). This gives a measure of the amount of microbial nutrient in the sample and a reflection of the waste impact on discharge to a water course.

Additionally, however, both the solid and liquid component of waste discharges may contain chemical substances at sufficient concentrations to cause direct harm[32]. These are commonly: chlorides; sulphates, naturally and from human wastes; nitrogen and phosphorus, in various forms from human wastes; fertilizers and specific chemicals such as phosphorus from detergents; carbonates, bicarbonates, calcium and magnesium salts; toxins, heavy metals such as Cd, Cr, Cu, Hg, Pb, Zn; trace organics, pesticides, polycyclic aromatic hydrocarbons (PAHs), chlorinated hydrocarbons and phenols.

In the U.K., driven by legislative controls, a number of substances have been identified as priority pollutants and are subject to controls on their discharge by either the principle of discharge being allowed to levels that the receiving water can support (environmental quality standards/environmental quality objectives, EQS/EQOs), or a uniform discharge level (uniform emission standards, UES) no matter what the characteristics of the receiving water. A series of priorities has been assigned, which relates to the toxicity of the substances. These have become known as *black, grey* and *red* list substances[29]. The pollutant characteristics and substances involved are summarized in Table 9.3.

Major diffuse sources of pollution are difficult to locate and hence control. However, there are three main types of pollution with major impact on aquatic systems; these are acid rain, nitrate and pesticides.

Acid rain has been shown to be derived from the release of oxides of sulphur and nitrogen from the combustion of fossil fuels (oil, coal and gas), the principal cause being power production. It is a major problem in Europe and eastern and northern America[6,29,33] where the acidity of rainfall is enhanced by the water solubility of these gases. Certain upland regions have soils with a low capacity for the neutralization of an enhanced acidity and

Table 9.3 The classification categories for *black, grey* and *red* list substances which are subject to varying degrees of control in the aquatic environment. Based on Reference 29

(A) List I (*black*)—based on toxicity, persistence, bioaccumulation
1.   Organohalogens, or substances which may form organohalogens
2.   Organophosphorus compounds
3.   Organotin compounds
4.   Compounds exhibiting carcinogenicity in or via the aquatic environment
5.   Mercury and its compounds
6.   Cadmium and its compounds
7.   Persistent mineral oils and petroleum hydrocarbons
8.   Synthetic substances that float, remain in suspension, sink or interfere with the use of waters

(B) List II (*grey*)—based on possible effects, can be confined and depend on the characteristics and location of water body into which they are discharged
1.   The following metalloids and metals and compounds:
     Zn, Cu, Ni, Cr, Pb, Se, As, Sb, Mo, Ti, Sn, Ba, Be, B, U, V, Co, Tl, Te, Ag
2.   Biocides and derivatives, not part of List I
3.   Substances affecting taste and/or smell of products for human consumption derived from the aquatic environment and compounds
4.   Toxic or persistent organic compounds of Si or may give rise to such
5.   Inorganic compounds of phosphorus and elemental phosphorus
6.   Non-persistent mineral oils and petroleum hydrocarbons
7.   Cyanides, fluorides
8.   Substances which have an adverse effect on the oxygen balance, particularly NH_4^+, NO_2^-

(C) UK *red list*—26 initial priority substances identified from carcinogenicity and toxicity studies:
     mercury and cadmium and compounds, gamma-hexachlorocyclohexane, DDT, pentachlorophenol, hexachlorobenzene, hexachlorobutadiene, aldrin, dieldrin, endrin, polychlorinated biphenyls (PCBs), dichlorvos, 1,2-dichloroethane, trichlorobenzene, atrazine, simazine, tributyl- and triphenyl-tin compounds, trifluralin, fenitrothion, azinphos-methyl, malathion, endosulfan.

an increase in lake and river acidity has been observed, with observable toxic effects on both local terrestrial and aquatic ecosystems. In some situations, increased acidity has been shown to mobilize aluminium from surface soils increasing the impact of acidification[33]. The effect can only be controlled by reducing emissions from energy production. In the short term, liming acidified lakes and soils, to restore natural pH, has been found to reverse the effects.

Nitrates released to aquatic systems, from the intensive use of fertilizers to improve crop and livestock yields, has resulted in high nitrate levels in drinking water. Recent evidence suggests that the previously high rates of increase in nitrate contamination of water supplies, is reducing. It is thought that this has been primarily through improvements in land management practices[29,34].

As with nitrates, the increased demands on crop and livestock yields by an ever increasing population has required the intensive use of pesticides. Widespread application in agriculture has resulted in a significant diffuse source to aquatic systems. Many pesticides feature in the lists of priority pollutants (Table 9.3). The nature of environmental residues, behaviour and fate are still uncertain for many compounds[29,35].

In assessing the impact of pollutants in aquatic systems, an overall understanding of the likely pathway and fate must be obtained. The possible reaction paths of a pollutant or any chemical released directly into a water course are summarized in Figure 9.5.

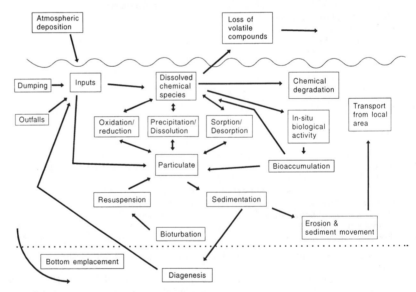

Figure 9.5 A schematic summary of the sources and possible fate of pollutants released into aquatic systems. The pathway followed depends on the physical and chemical properties of the pollutant and the physical, chemical and biological characteristics of the receiving water. Based on References 36 and 37.

The key characteristics of the water body must also be considered. Surface waters differ from ground waters, where in the latter, flow rates are slower and residence times for contaminants are much longer. The subsurface waters are ecologically less diverse and consequently more sensitive to pollution. Being remote from the earth's surface, any contamination is a major problem. A good example is the problem of high levels of nitrate in groundwaters in southern Britain[34]. A high proportion of drinking water is extracted from the ground and in areas where arable land overlies aquifers with no intervening impermeable clay deposits, levels are high enough to require dilution with less contaminated water before use.

9.2.3 *Land contamination*

The surface few metres of the continents consists of an extremely complex covering of disaggregated materials, comprising: minerals, decaying organic matter, water, air and flora and fauna. This is known as soil. It is a valuable resource and supports the growth of crops and provides land for building and transport. Importantly it also acts as a sink, a filter and a reaction matrix and medium for contaminants, and is a crucial component in the discussion of industrial pollution. Most of the world's soils are contaminated to some extent by man's activities. These can again be from local and global, diffuse and point sources. The chemical composition of soil needs to be defined for its compliance with certain 'trigger' concentrations determined by the proposed end use of the land. Major sources of contaminants from industrial sources, posing significant risks are summarized in Table 9.4.

Soil properties and characteristics profoundly influence the impact of pollutants[38]. Soil minerals comprise an intimate mixture which includes clay minerals, quartz, and hydroxides of iron, aluminium and manganese, as well as fragments of primary rocks. Soil organic matter represents the accumulation of organic residues of plants and animals at varying stages of decomposition and as such supports a diverse and significant microbial population. Other soil properties, such as permeability, through water and gas-filled voids or pores, either in solution or the gas phase, is influenced by grain size and compaction through previous use. The flow of material and degree of saturation influences the oxidizing and reducing (redox) status of the soil, which, along with the pH, is important in determining the mobility of pollutants and the degradation of organic contamination. From Table 9.4 it can be seen that the major groups of land contaminants are: heavy metals, organic compounds and sewage sludges.

The heavy metal contaminants are strongly adsorbed by soil constituents such as organic matter, clays and metal hydroxides. Mobility and bioavailability in turn being determined by soil pH, organic matter content and redox conditions. Most metals are more mobile under acid conditions.

Table 9.4 Sources and types of land contaminants. Based on References 9, 10 and 28

Atmospheric fallout
Fossil fuel combustion—S & N oxides and acid radicals
Pb, PAHs from automobile exhausts
Metal smelting—As, Cd, Cu, Cr, Ni, Pb, Sb, Tl, Zn
Chemical industries—organic micropollutants, Hg
Waste incineration—TCDDs, TCDFs
Radioisotopes from reactor accidents—Windscale, UK, 1957; Chernobyl, USSR, 1986
Large fires—soot, PAHs

Agricultural chemicals
Herbicides—2,4-D, 2,4,5-T containing TCDD, B and As compounds
Insecticides—chlorinated hydrocarbons, DDT
Fungicides—Cu, Zn, Hg, organic molecules
Acaricides—tar oil
Fertilizers—Cd, U in phosphates

Waste disposal (controlled and uncontrolled)
Farm manures, sewage sludges, domestic composts—heavy metals, viruses, pathogens, organic
    pollutants
Mine wastes—SO_4^{2-}, Fe, heavy metals
Seepage of landfill leachate
Ash from fossil fuels, incinerators, bonfires—heavy metals, PAHs
Burial of diseased livestock—viruses, pathogens

Incidental accumulation of contaminants
Corrosion of metal in contact with soil—Zn (galvanized), Cu, Pb (roofing)
Wood preservatives—PCP, creosote, As, Cu, Cr
Leakage from storage tanks—petrol, chlorinated solvents
Warfare—organic pollutants from fuels, smoke fires, metals from munitions, vehicles
Sports and leisure—Pb from gunshot and fishing, Cd, Ni, Hg from batteries, hydrocarbons from
    petrol

Derelict industrial sites
Gas works—phenols, tars, cyanides, As, Cd, combustibles
Electrical industries—Cu, Zn, Pb, PCBs, solvents
Tanneries—Cr
Scrapyards—metals, PCBs, hydrocarbons

Organic contaminants are widely varied in type and impact. Pesticides can remain in soils long after application or accidental spillages and there is a significant risk that their useful toxicity may also translate to harm beneficial plants and animals[35] and affect humans. Natural microbial activity can with time degrade many compounds to inactive residues, ultimately CO_2 and water. However, in some situations, degradation products may have significant toxicity or pollutants contain bonds, such as carbon–halogen bonds, which are not found in nature and are slow to degrade[9].

Insecticides include: organo chlorines, organo phosphates and carbamates, and herbicides including phenoxyacetic acids, toluidines, triazines, phenylureas, bipyridyls and glycines. A wide range of chlorinated compounds, such as dioxins and dibenzofurans are released from combustion sources, where they are synthesized at relatively high temperatures, have high stability and are slow to degrade[9]. Polychlorinated

Table 9.5 The hazards and effects of land contamination. Based on References 9 and 10

Hazard	Examples of contaminants
Direct ingestion by children, animals	Heavy metals, cyanides, phenols, coal tars
Inhalation of dusts, vapours, soil	Organic solvents, radon, volatile metals and metalloids (As, Hg)
Plant uptake and food chain transfer	Metals, PAHs
Phytotoxicity	SO_4^{2-}, CH_4, heavy metals
Degradation of construction materials	SO_3^{2-}, SO_4^{2-}, Cl^-, solvents, coal tars, phenols
Fires and explosions	High calorific wastes, organic solvents, CH_4
Contact with site clearance contractors	Coal tars, solvents, phenols, asbestos, radionuclides, PAHs
Aquatic pollution	Solvents, pesticides, cyanides, SO_4^{2-}

biphenyls (PCBs) are stable compounds, manufactured for use in electronic components and plastics and are found in a number of industrial processes. Sewage sludge which contains a wide variety of organic and inorganic compounds, is widely dumped on land as a means of disposal. The heavy metal, PAH and PCB content is of greatest concern. In practice, however, sludge amendment is of little hazard if used to landscape and encourage growth on recreational sites rather than on land used for food production[7,9].

The effects of land contamination are widespread and past industrial activity is the most significant factor. Soil transport and reaction processes are relatively slow compared to air and water, so contamination tends to persist at the point of deposition for a long period. Table 9.5 summarizes common hazards and examples of contaminants. These contaminants can affect humans by absorption into the body through oral, inhalation or skin adsorption pathways. For volatile compounds and dusts, inhalation is the most important pathway. This is of particular concern with young children playing on contaminated land. The re-use of derelict, industrial sites has emphasized many of these problems[9,10].

9.3 Options for the control and treatment of pollution and wastes from industrial sites

Having reviewed the key characteristics of major pollutants in the different compartments of the environment, it is appropriate to consider the options available for the control and for minimization of releases. Within this context, there is a need to consider waste management practices, as often the removal of a potential contaminant from the discharge, will provide a localized concentration of the material, which if not valuable for recycling, will have to be disposed of in another manner. Historically, discharges to air, water or land have only been subject to legislative control in the U.K. since the late 19th century[4].

The disposal of waste is one part of the waste management process. The minimization of waste production in turn reduces the potential for release of pollutants to the environment. Waste minimization can be viewed as a series of logical steps or objectives, which can be applied to any production process. The effects are felt not only in environmental impact, but it also has an impact on the economics of plant productivity. The effects are summarized as[39]:

(i) avoiding the creation of the waste products
(ii) re-using waste products
(iii) if a waste product is not re-usable, recover/reclaim the primary material for new manufactured products
(iv) if primary materials recovery is not practicable, recover, for a secondary material or if combustible use as fuel
(v) if none of the above is practicable, then choose the disposal option with the lowest environmental impact.

This approach can be carried through during the design of chemical processes such that they: afford maximum chemical conversion, have high energy efficiency; use minimum volumes and low hazard solvents; make minimum use of process water; ensure minimum dilution of carrier liquids; and ensure low inventories of liquids. It is outside the scope of this text to include detailed material relating to resource and energy recovery as a method of recycling and waste minimization. The nature of operations involved is dependent on the processes on site and therefore difficult to describe in general terms. However, there are numerous examples in the current literature of recycling routes for specific processes, including the recovery of sulphuric acid in monomer production, antimony recycling from polymer production[24], recovery and recycling of waste oils[40].

At the final stage, where disposal must proceed, there are two basic approaches. Firstly to *contain* the waste or pollutant, immobilized in a controlled manner. The pollutant is then localized and release is subject to the lifetime of the containment barriers, under the storage conditions used. This is relatively straightforward where the lifetime of the hazard is short but a major consideration in the longer term, where containment must perform adequately for many hundred and thousands of years. The management of radioactive waste from the nuclear fuel cycle is perhaps the most appropriate example here of the latter[30].

The second approach, *dilute and disperse*, is to allow discharge of the pollutant in liquid or gaseous forms, diluted to a level such that there is no direct risk to the local environment or that with prolonged releases, there is no likelihood of re-concentration under natural environmental processes close to or remote from the discharge point. This approach (whilst extremely cost effective) is philosophically questionable and the most difficult engineering challenge. Having been routinely used for most disposal solutions, it suffers from the fact that the only feedback to the practice is through the observation of adverse environmental effects. Obviously this is then too late to prevent and the environmental impact has occurred. This practice has driven the increased legislative control on industrial operations, characteristic of the mid to latter part of the 20th century. All waste management practices contain components of both approaches. In the design of containment, it is never possible to ensure complete integrity over time, the costs of this option rapidly escalating with complexity and construction effort and eventual release of the polluted or waste components, at some level, is certain. It is therefore ideal if the release of pollutants can be an integral part of the waste disposal option, which is able to tolerate variations in the physical and chemical characteristics of the waste.

9.3.1 *The control of atmospheric discharges*

Historically, all industrial sites contain exhaust pipes, chimneys or stacks that release volatile gases and entrained particulates directly to the atmosphere. The dilution and natural scavenging and cleaning processes offered by this route take the emissions away from the site and generally ensure wide dispersal. Climatic conditions and stack height are the major factors and, as previously mentioned, dilution is at best a short-term measure, with many infamous examples[12,17,20].

The principles of pollution or waste minimization outlined above, if applied, ensure that minimal discharges occur. Control devices are designed as a secondary approach to destroy, counteract, collect or mask pollutants. Such devices are never effective for both particulate *and* gaseous emissions and are designed for either one or the other[17,41].

9.3.1.1 *Particulate discharges.* A varied range of industries release particulates into the atmosphere. Owing to their visibility, particulate control has received most attention, despite the greater volumes of gaseous components released[17]. Devices can be divided into five groups which, in principle, either alter the flow rate of gas stream or reduce the energy of particulates by impact. The key characteristics, advantages and disadvantages of each approach are summarized in Table 9.6. The devices used are dependent on particle characteristics: size distribution, shape, density, stickiness, hygroscopicity, electrical properties and the carrier gas properties of flow rate and particle loading. Additionally, the desired efficiency, nature of the source, intermittent or continuous, the space available, ultimate waste-disposal method and equipment tolerances to gas stream conditions (corrosion, temperature, pressure) must be compared to the costs of installation and operation. A summary of industrial process examples and control methods is given in Table 9.7. Details of designs for each type of device can be found elsewhere[41].

9.3.1.2 *Gaseous discharges.* From most industrial sites, the main gases of concern are oxides of sulphur, nitrogen and carbon, inorganic and organic

Table 9.6 Control devices available for particulates released from industrial processes. Based on References 17 and 41

Device	Minimum particle size (μm, 90% efficiency)	Efficiency (%, mass)	Advantages	Disadvantages
Gravitational settler	> 50	< 50	Low pressure loss; design simplicity and maintenance	Space; low collection efficiency
Centrifugal collector	5–25	50–90	Low–medium pressure loss; design simplicity and maintenance; space; continuous disposal of collected dusts; copes with large particles and high dust loadings; temperature independent	High headroom requirements; low collection efficiency of small particles; sensitivity to dust loadings and flow
Wet collector			Simultaneous gas and particle removal; cools and cleans high temperature, moisture laden gases; corrosive gas and mist recovery/ neutralization; dust explosion risk minimized; efficiency can be varied	Problems with erosion and corrosion; costs of contaminated waste water treatment; low efficiency for small particles; freezing in cold weather; affects plume rise; can add water vapour to plume
spray tower	> 10	< 80		
cyclonic	> 2·5	< 80		
impingement	> 2·5	< 80		
venturi	> 0·5	< 99		
Electrostatic precipitator	> 1	95–99	> 99% efficiency possible; small particles collected; wet or dry operation; low pressure drop and power requirements; few moving parts; can operate at high temperatures	Initial cost high; sensitive to loading and flow rates; low efficiency for high resistivity materials; electrical hazard; gradual, imperceptible reduction in collection efficiencies
Fabric filtration	< 1	> 99	Dry collection possible; degradation in performance obvious; high efficiencies and small particle collection possible	Sensitive to velocity; high temperature gases must be cooled; chemical degradation of filters; conden- sation effects

Table 9.7 Particle emission characteristics of industrial processes and control options. Based on References 17 and 41

Industry/process	Emission source	Particulate composition	Control options
Iron and steel mills	Blast and steel making furnaces	Oxides, dusts, smoke	cyc.; b.h.; e.p.; w.coll.
Iron foundries	Cupolas, shake-out making	Oxides, smoke, oil, metal fumes, dust	sc.; dry cent.
Nonferrous metallurgy	Smelters, furnaces	Smoke, metal fumes, oil, grease	e.p.; f.f.
Petroleum refineries	Catalytic regenerators, sludge incinerators	Catalyst dust, ash from sludge	cyc.; e.p.; sc.; b.h.
Portland cement	Kilns, driers, material handling	Alkali and process dusts	f.f.; e.p.; venturi sc.
Acid manufacture, phosphoric and sulphuric	Thermal processes, rock acidulating, grinding	Acid mist, dust	e.p.; mesh mist eliminators
Glass and fibreglass	Furnaces, forming/curing, handling	Acid mist, alkaline oxides, dust, aerosols	f.f.; afterburners

Abbreviations: cyc., cyclones; w.coll., wet collectors; b.h., baghouses; sc., scrubbers; cent., centrifuges; f.f., fabric filters; e.p., electrostatic precipitators.

acid gases, and hydrocarbons. Control methods available include: adsorption, absorption, condensation and combustion.

Adsorption of a gas stream passing through a bed of absorbent is by two discrete processes[41]. Physical adsorption involves intermolecular forces (van der Waals) and condensation of gases within the solid materials. The amount of material adsorbed depends on the amount of solid but it is not directly related to the surface area. The process is reversible and desorption can occur by raising temperature or lowering pressure. Chemisorption involves the reaction of the gas with the solid adsorbent to form a bond and is influenced by temperature and pressure. The process is usually irreversible and confined to a single layer of molecules on the solid surface.

Numerous adsorbents exist that are highly efficient at removing gaseous contaminants[42]. Examples are given in Table 9.8. The adsorbents and gas stream can be brought into contact through fixed, moving or fluidized beds of various designs. The adsorbents can be classified as *regenerative* or *non-regenerative* depending on the ease of removal of the collected gas. Chemisorption is naturally the most costly approach; where possible the regeneration processes require extremes of pressure, temperature or chemical treatment. Adsorption methods are an efficient way of material recovery and often valuable by-products are obtained[42].

Absorption (or scrubbing) involves liquid absorbent (solvent) so that one or more of the effluent gases are removed, treated or modified. This may be

Table 9.8 Applications of adsorbents in the treatment of gaseous waste streams. Based on References 41 and 42

Adsorbent	Applications
Activated carbon	Odour removal; gas purification; solvent recovery
Alumina	Drying air, gases and liquids
Bauxite	Treating petroleum fractions, drying gases and liquids
Bone charcoal	Decolorizing sugar solutions
Fuller's earth (mineral)	Refining animal oils, lubrication oils, fats, waxes
Magnesia	Treating petrol and solvents, removing metallic impurities from caustic solutions
Molecular sieves	Controlling and recovering Hg, SO_2 and NO_x
Silica gel	Drying and purifying gases
Strontium sulphate	Removing iron from caustic solutions

through chemical reaction of simple dissolution. The amount of gas removed depends on the gas and solvent properties. A general guide is that solvents with similar properties to effluent components are suitable. So alkaline solvents are suitable for acidic gases and hydrocarbon solvents for organic molecules. Gases commonly controlled in this manner include sulphur dioxide, oxides of nitrogen, hydrogen sulphide, hydrogen chloride, chlorine, ammonia and light hydrocarbons. Other solvent characteristics must include: low freezing point, low toxicity, chemical stability and very low volatility and flammability. Absorption units usually mix a spray of droplets with the effluent gas in a turbulent manner, to maximize gas/ solvent contact area and time.

Condensation methods of air pollution control involve the cooling of the effluent stream by either contact with a cooled solid surface (surface condenser) or by mixing with the coolant itself (contact condenser). Applications of condenser systems are not widespread, being restricted mainly to hydrocarbons, and are usually used as a pretreatment step for other control techniques. Incineration has wide application in the conversion of contaminants such as CO or hydrocarbons to carbon dioxide and water. The main parameter controlling efficiency include: oxygen— concentrations must be such that combustion can continue to completion; temperature—this must be as close to the ignition temperature as possible; turbulence—to allow sufficient mixing of oxygen and combustible components; time—to allow reactions to complete. A range of combustion techniques is available. These include direct flame combustion, in which waste gases are burnt directly in a combustor, with or without the addition of extra fuel. Gas flames in petrochemical plants and refineries are examples. Thermal combustion (after burner), for gas streams with low combustibility, involves preheating the gas stream prior to injection into a combustion zone, containing an independent burner. Design is crucial here, to ensure complete combustion. In catalytic combustion a preheated gas stream is passed over a catalyst, which accelerates the rate of oxidation. This process can be highly efficient and catalytic material from platinum group metals, transition metals and oxides have been used to treat SO_2, NO_xs, hydrocarbons and carbon monoxide. Potential disadvantages include poisoning of the catalyst and the relatively high capital cost of the catalyst. This method has widespread application in automotive emission control, where transportation accounts for $>50\%$ of all major air contaminants[41].

9.3.2 *The control of aquatic discharges*

The range of toxic substances present in industrial waste waters is diverse. Tables 9.3 and 9.4 have indicated examples of the types of potential pollutants present. From the preceding section, aqueous effluents from industry include primary process waters and liquids contaminated by secondary processes to treat atmospheric discharges and other on site operations including cleaning and occasionally fire control. Discharges to water courses during firefighting operations can be catastrophic and result in the release of vast quantities of contaminated water into site and local drainage systems[43]. The major problem being the large water volumes used to combat the fires. The company is liable for any pollution effect. Site design, which involves the equipping of drainage systems to cope with estimated volumes of water and changes in firefighting procedures through the use of containment booms and absorbent pads are being developed.

The most convenient and economical treatment of routine aqueous discharges is to mix industrial wastes with domestic waste waters. This practice is limited only in the situations where: the site is in a rural area with no convenient sewerage system; recycling is feasible (although there will still be a final discharge); domestic effluent is used in irrigation; industrial waste water does not meet the consent conditions for discharge to the sewer[44]. Examples of specific water treatment methods that have been applied to waste streams include the removal of dissolved hydrocarbons with polymer particles[24]; the treatment of mine waters by adding lime to increase pH and the use of reed beds to reduce and precipitate metals[45].

The consequence of the treatment on site allows both the recycling of components and the use of more extreme treatment conditions in addition to the avoidance of contamination of a much larger waste water stream. Toxic contaminants may inhibit routine sewage treatment processes and can be overcome relatively easily with the more specific site treatment processes. In many situations however, on-site treatment may be obviated by simple dilution to acceptable levels prior to discharge. This is especially advantageous if there is not enough space on site, personnel are not experienced enough in treatment or the localization of waste during treatment may cause an odour nuisance[44]. The philosophy of waste minimization has already been described and a number of techniques are available. The decision to treat or not will be a result of the factors outlined above. However, even employing simple physical methods can improve effluent quality con-

Table 9.9 Physical methods of pre-treatment applicable to aqueous waste streams. Based on References 32, 44 and 46

Process	Objective	Typical examples
Screening	Removal of coarse solids	Paper mills, vegetable canneries
Centrifuging	Concentration of solids	Sludge dewatering in chemical plants
Filtration	Concentration of fine solids	Final stages (in above)
Sedimentation	Removal of settleable solids	Coal, clay ore processing
Flotation	Removal of low specific gravity solids	Oil, grease and solids separation in chemicals and food industry
Freezing	Condensation of liquids and sludges	Non-ferrous metal and process liquids recovery
Solvent extraction	Recovery of specific metals and organics	Coal carbonizing, plastics manufacture, metal processing
Ion exchange	(as above)	Metal processing
Adsorption	Trace impurities	Pesticides, dye stuffs
Reverse osmosis	Dissolved solids	Desalination of industrial waters, waste waters with dissolved contaminants

siderably. Generally this involves some form of screening to reduce solid content, balancing of concentration flow, pH and traps to prevent oil and grit escape[32,44,46]. A brief summary of physical methods commonly used is given in Table 9.9. Further pre-treatment may involve the reduction of organic matter content by biological oxidation methods, particularly if the discharge is to go straight to a water course. These methods are varied at the pre-treatment stage and all biological methods are particularly prone to poisoning by toxin concentrations, especially where levels vary. Ultimately, problematic industrial wastes with high organic matter content are usually subject to conventional primary and secondary treatment. Chemical methods are used either to oxidize particular compounds or to aid in the physical treatment process by adjusting pH and/or removing solids and some dissolved species by flocculation and precipitation.

The use of pre-treatment methods is waste stream dependent, as is the resulting effluent quality. Should further processing be required, standard sewage treatment is frequently used[44,47]. The stages involved in this process are summarized in Figure 9.6. Preliminary screening is needed to remove large suspended solids, metals and rags. Grit removal occurs prior to the sedimentation of as much of the suspended solids as possible. Biological (*secondary*) treatment is one of two types: (i) percolating filter, comprising a packed bed of clinker or stones, with a high surface area, that allows aerobic oxidation; and (ii) activated sludge, in which the sewage is aerated in agitated tanks to expose the waste to as much oxygen as possible. Sludge is recycled to seed raw sewage and speed up the process.

Aerobic oxidation should produce low BOD material unless the plant is overloaded. Further settling occurs prior to discharge of the liquid and the disposal or further treatment of the sludge. Tertiary treatment may still be

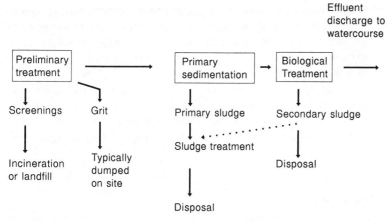

Figure 9.6 A schematic diagram of the sewage treatment process. Most industrial aqueous wastes are introduced to domestic treatment plants. Where the process justifies costs, a purpose built system, with similar operations is used. Based on Reference 47.

required for further solids reduction and various types of filter, lagoons for settling or reverse osmosis processes are used. The treatment of sewage sludge and its subsequent disposal can account for 40% of the operating costs of a sewage treatment plant[47]. Sludge has a water content of 95%, is odorous and a major waste management problem. Four methods are used to deal with sewage sludges depending on relative costs and legislative controls.

(i) Conversion to methane by anaerobic digestion in a variety of tank-based systems. This approach can produce enough gas to burn for energy. It removes pathogens and odours from the sludge and provides a solid residue for use as a soil amendment. The raw sludge must have a high enough organic content, so unless mixed with high organic streams, such as sewage, some industrial effluents may not be suitable. Anaerobic digestion is now routinely used to treat organic rich liquid wastes from a range of industries such as paper mills, breweries, dairies, and distilleries[48]. The technology has been developed from sewage sludge treatment processes and allows relatively rapid processing.

(ii) Incineration can be relatively high cost, requiring additional fuels. Once de-watered, sludge can be burnt (in self-sustaining incinerators at 850°C) in fluidized bed processes, with useful excess heat production[49,50]. The process destroys most toxic organic compounds. Flue gas scrubbing is required to remove acidic gases.

(iii) Disposal directly on land is an option that is attractive where heavy metal, organic pollutants and pathogen levels are low[9,44]. Restrictions on the amounts used, land use and fallow periods are dependent on factors relating to transmission of disease and accumulation of toxins. The sludge can be applied wet—as such it is a valuable source of moisture—or dry. Where agriculture is intensive or soil quality low, the practice can be highly beneficial to crop growth.

(iv) Dumping at sea, for coastal communities, has been a traditional disposal route. Sea dumping can both dilute and treat wastes, provided no re-accumulation occurs under current action and loading rates allow time for processes to complete. In the U.K. this practice has been abandoned under European Union guidelines[49].

9.3.3 *The disposal of solid wastes*

In the discussion of atmospheric and aqueous discharges, the underlying principles of waste minimization, in terms of volume reduction, includes recycling, and treatment. This leads to the removal or minimization of hazards from effluents and results in the production of a solid residue. This must be disposed of along with any other solid material produced on site.

Further treatment options by incineration and biological composting or digestion depend on the total carbon and organic matter content of the solid waste and payback from direct heat and methane generation, which may be used to offset costs of the process[49]. The environmental impact of incineration has recently been reviewed[49,51] and there exists a wide range of technological options in incinerator design that optimize the process, minimizing environmental effects. Incineration produces atmospheric emissions—CO_2, CO, H_2S, HX, NO_xs, SO_xs, dioxins, vapours, heavy metals, inorganic salts; high temperatures; acidity/alkalinity; contaminated waste water and ash. Whilst physically stabilizing the original waste, in some cases it may serve to concentrate a number of potential pollutants. The ash can contain relatively high levels of heavy metals and residual organic compounds, some of high toxicity or carcinogenicity[52].

Biological treatment has been reviewed in section 9.3.2 and depends on the organic content of the waste. It can include aerobic or anaerobic processes. The anaerobic system supplying potentially useful methane is the most difficult to control efficiently, with operational problems that relate to the handling of solids and leachate, and the harnessing of gas yields.

Unless alternative uses can be found (for example as road/construction fill or soil amendment) the solid waste residues require a final disposal route. Ultimately this is through landfilling. The pretreatment methods above serve an important role in stabilizing and reducing the volumes required for landfill. Currently, landfilling is the most economic (and consequently most used) form of disposal in the U.K., U.S.A. and Europe. In the U.S.A. and U.K. this route takes approximately 90% of controlled and 80% of hazardous wastes[7,53,54].

A landfill site should be on inexpensive ground, within economical transport distance from the main waste producers, have year-round access and be a suitable distance from neighbours. The area should be clear, level and well drained, with adequate capacity for the intended use. The surrounding soil should be of low permeability and the site should be isolated from the water table, either through distance and/or design characteristics.

Two approaches are used in landfill design[53,54].

(i) *Containment* in which the waste is isolated from the environment, the generation of aqueous leachate in the waste is minimal, and waste management is more straightforward. Barriers between the landfill waste and the environment are engineered, with designed tolerances. A wide range of materials is available for barrier constructions including polymer membranes and inorganic clays[53,54,58].

(ii) *Disperse and attenuate:* natural sites with materials of known properties are used to alter the composition of leachate and reduce potential environmental damage through physical, biological and chemical processes. There is a net loss of leachate to the groundwater through design[53,54].

A suitable understanding of the site hydrogeology must be gained before operation. Figure 9.7 summarizes many of the key characteristics of a landfill site.

The landfill can be operated in a number of ways to receive both

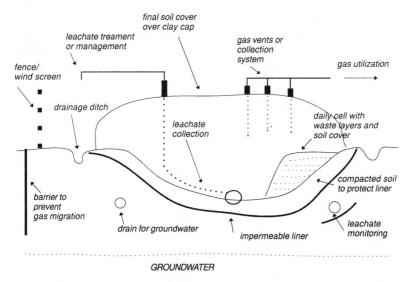

Figure 9.7 A sketch of a typical landfill site, showing leachate and landfill gas management options and cellular filling practices. Based on References 7, 53, 54 and 55.

municipal and industrial wastes, providing waste properties are understood[7]. Co-disposal, in which toxic wastes are added to mature (6 months–2 years) domestic wastes and multi-disposal where, for example, an inert solid may be used to stabilize a slurry, are two options available for industrial wastes[56]. Key factors to consider are waste density on deposition; waste compaction; biodegradability; odour; the variation physical/chemical properties with time, and the compatibility of various wastes (i.e. mixing solid nutrients with biological wastes is unwise!)

The main environmental effects of landfills include:

(i) The generation of landfill gas from the biological degradation of organic components. Landfill gas (LFG) is a general term for a varied mixture of gases, primarily CO_2 and CH_4, that are produced by the wastes commencing immediately after deposition and lasting for many decades, long after the site has been closed[53]. The amount of methane produced can be sufficiently high to 'farm' and be used as a fuel source. The profile of gas composition changes to reflect aerobic and anaerobic activity. They key factors to consider in LFG generation are summarized in Table 9.10. The major LFG hazard is through migration off site and build-up to toxic or explosive levels in enclosed spaces such as residences or site buildings[57,59,60].

(ii) The production of landfill leachate. During the degradation process, moisture from within the waste, the ingress of groundwater, rainfall and surface water runoff can percolate through the site, producing an often noxious liquid, known as leachate[61]. This is a coloured liquid (light yellow, through to red and black) with high levels of organic matter, dissolved and particulate iron and sulphides, and varying levels of heavy metals and organic pollutants[55]. It can be up to 100 times more polluting than raw sewage and contains dissolved LFG. Leachate quality varies greatly between sites and with time. The important factors influencing its production are summarized in Table 9.10.

Table 9.10 Key factors influencing the generation of landfill gas and landfill leachate in solid waste disposal. Based on References 7, 53, 54, 56 and 58

Landfill leachate	Landfill gas
Surface water ingress	Site dimensions
Ground water ingress	Waste composition and variation
Absorbing capacity of waste	Waste input rate
Weight of waste deposited	Site operations
Evaporation losses	Waste density
Cellular filling practice	Moisture content
Capping and landscaping	pH
Waste composition and variation	Temperature
	Ingress of oxygen

The production of LFG and leachate at a landfill, must be minimized, monitored (unless CH_4 production is desired) and controlled if environmental impacts are to be minimized. Some of the uncertainties in landfill operation include: the selection criteria for containment or open site management; consequences of co-disposal; moisture input rates; and gas recovery[53].

9.4.1 *Hazards and historical evidence*

There is a requirement for releases to the environment to be monitored and prevented, where appropriate. By default, the releases will originate on site, in a localized area with potential exposure for site personnel. Industry, worldwide, has a requirement, enforced by various legal systems to provide: clean air at work, safe working practices, well-being at work (which includes occupational health factors) and the control of noise and vibration[62]. Through the manufacture and processing of materials, individuals are exposed to elevated concentrations of chemicals, high energy sources or an elevated risk of injury from physical manipulations. Historically, we have a lot of data relating to the hazards of the modern chemical and process industries[63]. The toxicity of metals and metalloids, extracted from the surface of the earth, the initial development of extraction processes, by smelting, etc., has provided abundant evidence in the history books of detrimental health effects on the exposed workers[63]. Societies have an in-built limitation in this respect, with lessons from the past having a relatively short period of prominence before becoming relegated to the annals of history. As a result of the human perception of risk and other socio-economic factors, such incidents tend to have a habit of repeating themselves. This is particularly true in developing countries, when viewed from a developed, industrial society. Through education and legislation, developed society has evolved a management system to minimize risks at various levels. This system results from past experiences and the effective communication of the results. In developing societies, with poorer communication, higher reliance on primary resource exploitation, this is not the case. A good example of this is the mining of gold in the Amazon Basin, Brazil[64,65]. The extraction of gold has a history of over 20 years, at various levels of activity. Informal sector miners, known as garimpeiros, use vast quantities of liquid mercury to form an amalgam with the gold particles. During the mining process, mercury is sprinkled onto crushed rock, mixed and washed, then separated from the tailings. The mercury is then collected and burnt, releasing the amalgamated gold. The process is summarized in Figure 9.8.

Mercury losses during the process are considerable and the environmental impact of the mining, above and beyond the effects of erosion and deforestation, is enormous. There is evidence of mercury pollution of the

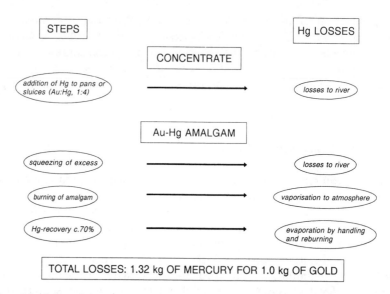

| STEPS | | Hg LOSSES |

CONCENTRATE

addition of Hg to pans or sluices (Au:Hg, 1:4) ⟶ losses to river

Au-Hg AMALGAM

squeezing of excess ⟶ losses to river

burning of amalgam ⟶ vaporisation to atmosphere

Hg-recovery c.70% ⟶ evaporation by handling and reburning

TOTAL LOSSES: 1.32 kg OF MERCURY FOR 1.0 kg OF GOLD

Figure 9.8 A flow diagram showing the losses of mercury during the extraction of gold using the mercury amalgam technique. Based on Reference 66.

local terrestrial and aquatic ecosystems, health effects in the miner and local Indian populations from the high levels of mercury released during amalgam processing and the contamination of local foodstuffs[64]. From a distance this seems quite a remarkable situation. Mercury has long been known to have toxic properties[67-69]. The phrase 'as mad as a hatter' referring to the mercury-induced medical conditions prevalent in the felt-making industry where mercuric nitrate was used to treat furs[68]. More recent examples include the mass poisonings of the fishing community in Minimata Bay, Japan during the 1950s. This occurred through the release of mercury-containing catalysts from a chemical plant manufacturing a vinyl chloride monomer[22].

Further historical lessons include the fact that any detrimental effects from the manufacture and use of chemicals and hazardous materials can only truly be appreciated *after* their introduction into society. There are numerous examples from the pharmaceutical industry, where drugs are produced with the object of providing a beneficial impact on society in treating or alleviating a detrimental aspect of life. With most drugs, there are always additional risks of known 'side effects' and the possibility of unknown effects. In general, there is a degree of general risk to be considered. The effect of a symptom or illness continuing through not taking the drug must be balanced with the known risks from the chemical. One of many tragic examples is the thalidomide episode in the U.K. in the late 1950s[22], where the use of a non-addictive alternative to barbiturates, when administered to pregnant mothers as a sedative, had a major effect on foetus development. This side effect was only appreciated from birth defect statistics, post-administration. The results of such incidents, which still continue today, serve as a salutary reminder of the importance of the holistic approach to the production and use of materials. The testing of toxicity and environmental impact is complex and now statutory for all new chemicals. However, it should never be considered that all hazards have been evaluated. Time, economics and current scientific knowledge can never guarantee future safety.

Turning to the hazards and the impact of industrial processes. Many disasters such as Flixborough, Seveso and Bhopal[22,63] (see section 1.6) have ably illustrated the environmental impact of human errors and the need for rigorous safety controls. The capacities and productivity of chemical plants are driven upwards as are the magnitudes of losses in terms of economic cost and fatalities. Analysis of the major causes of accidents in the U.S. chemical industry in the 1978–1980 period[63] highlighted that:

(i) most frequent and severe losses were from fire and explosion
(ii) explosions were more severe than fires
(iii) accidental and uncontrolled reactions were the main causes of explosions
(iv) most explosions occurred in buildings and with batch processes
(v) vessel and pipe failure was the major contributory factor
(vi) fires were caused by flammable gas and liquid releases

The wide variety of chemicals produced and very poor overall health and safety statistics, make the assessment of occupationally determined disease and ill-health particularly difficult. Taking this a stage further, other than disasters or accidents, where the effect on workers and the environment can be relatively easy to observe, the long-term impact, particularly from low levels of new or poorly studied materials, is a particular problem. Epidemiology, (the study of social groups and patterns of disease as a means of determining causes[70]) is difficult due to the varied exposure conditions and the varied significance of other factors. Exposure to chemicals at work is one hazard that can be controlled.

9.4.2 Toxicity and exposure to chemicals

9.4.2.1 Hazard and toxicity. Any material is a poison if the dose is large enough and the study of the effects of substances on the human body through, dose–response relationships, known as toxicology, is an advanced discipline. It is at the centre of occupational health activities in industry. Exposure can result in a number of effects depending on the levels of substance[63,71].

The major hazards are: *carcinogenicity*, cancer causing or incidence increasing; *teratogenicity*, the induction of non-heriditable birth defects in offspring; and *mutagenicity*, the increase of the risk of hereditary genetic defects. These effects may be *acute*, a rapid response over a matter of hours; or *chronic*, over a matter of months.

Exposure pathways, in particular their relative significance for different substances, are varied[71]. In an industrial context, these are: by swallowing, by absorption through the skin and by breathing. Once in the bloodstream, toxins may attack one or more organs, causing harm.

Dermal (skin) absorption is a significant hazard for a number of volatile substances[72]. These include: aromatic compounds such as benzene and toluene; nitro-compounds such as nitrobenzene, trinitrotoluene and nitro-propane; several insecticides and herbicides; hydrogen cyanide and organo phosphates. Protective clothing can minimize risks except where splashing and adsoprtion into cloth concentrates exposure to the skin.

The inhalation pathway is the major route for chemicals into the body[71]. Gases and vapours can penetrate deep into the lungs where, as the penetration of particles is strongly dependent on size, the general rule being the finer (smaller) the further they penetrate. Deposition in the lungs of very fine ($1-2\,\mu m$) particles is of greatest concern. The action of asbestos is well known[73].

Measures of toxicity are from two routes. Firstly, data from laboratory testing on animals by direct injection, oral or chemical administration and is usually expressed as an LD_{50} value. This is the dose that causes death to 50% of the population after 14 days[71,74], expressed as mg or μg of substance per kg body weight. Tests are performed under standard conditions and subject to rigorous statistical analysis. The lethal concentration (LC_{50}) unit is used for hazardous airborne concentrations and is defined as the 4-hour inhalation of which causes death of 50% of the test group within the 14 day observation period.

The results from this type of study are coupled with the second, very limited, source of toxicity data from previous chemical exposure records of workers. Chemicals are then classified in terms of their inferred, relative toxicity[72]. The two extremes being *practically non-toxic* (probable lethal dose for a 70 kg human $\sim 15\,g/kg$) to *supertoxic* (probable lethal dose for a 70 kg human $<5\,mg/kg$)[72,75,76].

The derivation of these levels and more importantly the dose–response relationships for particular chemicals, incorporates considerable uncertainty when translating experimental data which are often derived from a different species, to the workplace environment[72,75]. Generous safety margins must be incorporated to minimize the risk, and the shape of the dose–response curve is important. The dose–response curve allows the relative effect of increased exposure to a substance to be evaluated. Whether it is linear or non-linear, with or without threshold concentrations has important con-sequences for exposure. The presence or absence of a threshold concentration is critical in the consideration of levels present in the environment and the potential to increase or decrease with time. One situation that receives detailed attention in this respect, is the effect of low levels of ionizing radiation in the environment from nuclear weapons testing, accidents such as the Chernobyl disaster and discharges from the nuclear fuel cycle[30,77]. The effect of small additional radiation doses to a population that has evolved within a naturally radioactive environment is continually debated. Concern centres on the extrapolation of a dose response relationship based on data from high doses to the low environmental doses.

9.4.2.2 *Occupational exposure.* Within the workplace environment, occupational exposure limits (OELs) have been established by several countries[74]. These are the maximum concentrations in air of substances that should not be exceeded in the breathing zone of workers. They apply to single substances only and it must always be considered, that where there are mixtures, effects may be neutralized or enhanced (synergism). As the amount of information on the effects of exposures increases, there is a tendency to reduce these levels. Vapour and gas limits are represented as ppm (volume) or mg per m^3 in air and for solids as mg per m^3.

In the U.K., the Health and Safety Executive (HSE) publish recommended levels for occupational exposure[78] as: maximum exposure limits (MEL), which are reasonably practicable for all work activities; and *occupational exposure standards* (OES) which cover realistic plant design, engineering and control of exposure and can be used to help in the selection of personal protective equipment. Further subdivision of limits occurs into long-term (usually 8-hour) and short-term (10-minute). Short-term limits are of use where brief exposure may cause acute effects. Both levels are expressed as time-weighted average concentrations (TWAs) over the specified period. Studies of OELs in other countries reveal that there are some differences, but generally they fall close to each other in industrial societies[72].

Exposure sources for workers can be divided into two categories:

(i) periodic emissions—short-term releases associated with planned manipulative operations or accidental releases. In these situations, acute exposure effects are common.

(ii) Fugitive emissions—small, but continuous escape of liquids, vapours in general, that tend to arise from dynamic seals and can lead to chronic effects in workers.

Monitoring the working environment is a complex task and advice is available in the U.K. from the HSE, and in the U.S.A. from occupational health authorities, the Occupational Health and Safety Administration (OSHA) and National Institute for Occupational Safety and Health (NIOSH). The primary objectives of monitoring are to estimate personal exposure and the effectiveness of engineering and process control measures, as well as to monitor epidemiological and environmental impacts[72]. The approach is complicated by the varied type and number of emission sources and dispersion processes, but should include the following steps:

(i) Initial assessment, including: the substances present; their physical form; likely health hazards; exposure potential–severity, duration, location; evidence from comparative situations elsewhere.

(ii) Preliminary survey: this will include sampling and analysis. A complex stage, that usually involves the capture of substances by reaction, dissolution or filtration from a known volume of air over a suitable time period prior to chemical, physical or even biological analysis.

(iii) A detailed survey to back up major exposure indications; to evaluate highly variable preliminary results; where previous survey indicated levels that were close to OELs, requiring a more accurate picture of exposure, before the decision to act.

(iv) Routine monitoring to ensure that control measures perform continuously.

Table 9.11   Control techniques of use in chemical plants. Based on Reference 79

Type and explanation	Examples
Substitution—less hazardous chemicals and equipment	Low toxicity solvents, heat transfer through water rather than oil, high flash point chemicals
Attenuation—use chemicals under conditions that are less hazardous	Use vacuum to reduce boiling point, reduce temperature and pressure, dissolve in safe solvents, prevent reactor run away
Isolation—isolate equipment and hazards	Controls distant from operations, strengthen control rooms and tanks
Intensification—reduce quantities of chemicals	Use small continuous reactors, reduce raw material inventories, improve control to minimize hazardous intermediates
Enclosures—enclose equipment/room and use negative pressures	Seal rooms, sewers vents, shielding, remote and continuous monitors
Local ventilation—contain and exhaust hazardous substances	Fume hoods, extractors
Dilution ventilation—control low level toxicity through air dilution	Ensure all work areas adequately ventilated
Wet processes—to control dusts	Water sprays, clean frequently with solvents
Good housekeeping—contain toxic materials, dusts	Place dykes around pumps and tanks, cleaning and flushing lines, sewer systems
Personal protection—last resort	Safety glasses, face masks, protective clothing appropriate to hazards

Throughout this assessment process, the sampling strategy employed requires careful consideration of exposure, and hazards and should be established to cope with the statistical variability of each step.

There exists a wide range of methods for the control of chemical plant hazards which have implications not only for occupational exposure, but by default, the environmental impact. A few basic approaches are summarized in Table 9.11. Personal protection should always be considered as a last line of defence, for use in short-term situations where there is a risk of acute exposure. Environmental control should always be the primary objective.

**9.5
Conclusions—legislative controls affecting the environmental impact of the chemical industry**

The control and management of environmental pollution has become the domain of complex legal frameworks, evolving from a number of specific acts at various governmental levels. The variety and nature of controls varies between countries, but most industrial societies have generally similar structures. This section presents a brief outline of the roles and responsibilities in the U.S.A., E.U. (European Union) and U.K. Ultimately, the objectives of these controls are to minimize environmental impact of any operation.

Within the U.S.A. environmental protection and occupational health has derived from federal administrative agencies through inspection. The U.S. Environmental Protection Agency (EPA), the OSHA and NIOSH establish occupational exposure standards that have the force of law[80]. Ambient standards are defined to minimize risks to the health of humans, animals or the environment. The European approach has been through the issue of Directives and Regulations in a more coherent manner than in the U.K.[62] They concentrate on placing specific duties on employers by laying down quantifiable and minimum standards. Increasingly this will have an impact on U.K. law. Directives have been established to control certain industrial hazards through a central, European Environment Agency (in Copenhagen) and a European Environment Information and Observation Network. The European Union thereby driving member countries to harmonize systems and fill any gaps in procedures. At the time of writing a similar approach to the integrated pollution control principles of the U.K. regulations, is under consideration[81].

Within the U.K., the 1990 Environment Protection Act established a new approach to U.K. environmental law[81,82]. It brings together air framework directives, dangerous substance directives (responsible for the definition of black and grey list substances, see section 9.2.2) and waste disposal directives, as a total emissions management system. Sections are defined that include integrated pollution control and air pollution control. The objectives are: pollution prevention at source; risk minimization; the application of advanced technological solutions appreciating the integrated nature of the environment; a critical loads assessment approach to highlight sensitive environments; and the principle that the polluter pays for the necessary controls.

This has resulted in a system of authorization, the enforcement of pollution control under certain (prescribed) processes and the control and monitoring of the release of certain (prescribed) substances. These are enforced by HM Inspectorate of Pollution (HMIP), river agencies, local authorities and the HSE with a series of registration fees and penalties for meeting authorization.

There are a number of objectives to be met for authorization:

(i) To ensure that the *best available techniques not entailing excessive costs* (BATNEEC) will be used to prevent release of prescribed substances, minimizing release or rendering harmless;

(ii) Compliance with E.U. law;

(iii) Compliance with environmental quality standards (EQSs) or objectives (EQOs) prescribed by the Secretary of State;

(iv) Compliance with emission standards or quotas;

(v) Where releases are likely into one or more environmental media, the principle of BATNEEC will be applied to minimize pollution to the *best practical environmental option* (BEPO). Waste disposal on land is covered in detail and contains improved consideration of the waste disposal process from importing, producing, carrying, keeping, treating and disposing of or controlling waste in a true cradle-to-grave manner. This is through the ability of new Local Authority Waste Disposal Companies to amend contractors' disposal contracts and ensure that the waste contractor is a fit and proper person. An imposed 'Duty of Care' clause proposes to prevent escape of wastes and that the waste is transferred to an authorized person.

The details of the act are widespread and implications to pollution and environmental impact of any industrial operation wide-ranging. This highlights the more recent issues in the environmental impact of industrial processes. In the U.K., past policies on bringing disused/derelict land back into productive use, highlighted the terrestrial environmental effects of industry (see section 9.2.3). In addition a number of disasters, such as Love Canal in the U.S.A., prompted legislative responses such as the Comprehensive Environmental Response and Liability Act 1980 (CERCLA)[83]. This allowed the US-EPA to investigate and clean up contamination and to recover costs from past and present owners, transporters of hazardous waste, previous occupiers and anyone connected with the waste. The UK-EPA 1990 containing the duty of care approach follows similar lines. Today, site investigation techniques[10,84,85], methods for the remediation of contamination[86], which include biological[87], as well as physical and chemical techniques[86], has generated considerable academic and consultancy research, driven primarily by industry. From the above description, worldwide, the drive continually is towards the principle that the polluter pays. Either to operate a plant in which discharges and waste products are minimized, offering safe conditions for the workforce and to ensure that ultimately the releases have minimal environmental impacts. There are economic considerations at all stages, where the cost versus risk analysis of operations (to the workforce and environment) must be balanced by industry and by law.

References

1. Mason, B. and Moore, C. B. (1982) *Principles of Geochemistry*, 4th ed, John Wiley and Sons, Chichester, Chapter 12, pp. 313–328.
2. Harrison, R. M. (1992) Integrative Aspects of Pollutant Cycling in *Understanding Our Environment: An Introduction to Environmental Chemistry and Pollution*, (ed R. M. Harrison), Royal Society of Chemistry, Cambridge, pp. 165–187.
3. Rodhe, H. (1992) Modeling Biogeochemical Cycles, in *Global Biogeochemical Cycles*, (eds S. S. Butcher, R. J. Charlson, G. H. Orians and G. V. Wolfe), Academic Press, London, pp. 55–72.
4. Markham, A. (1994) *A Brief History of Pollution*, Earthscan Publications Ltd, London, Chapter 1, pp. 1–11.
5. Newson, M. (1992) Environmental Economics: Resources and Commerce, in *Managing the Human Impact on the Natural Environment: Patterns and Processes*, (ed M. Newson), Belhaven Press, London, pp. 80–106.
6. Charlson, R. J., Orians, G. H., Wolfe, G. V. and Butcher, S. S. (1992) Human Modification of Global Biogeochemical Cycles, in *Global Biogeochemical Cycles*, (eds S. S. Butcher, R. J. Charlson, G. H. Orians and G. V. Wolfe), Academic Press, London, pp. 353–361.
7. Department of the Environment (1986) Landfilling Wastes, *Waste Management Paper No. 26*, HMSO, London.
8. Porteous, A. (1992) *Dictionary of Environmental Science and Technology*, revised edn, John Wiley & Sons Ltd, Chichester, p. 294.
9. Alloway, B. J. (1992) Land Contamination and Reclamation, in *Understanding Our Environment: An Introduction to Environmental Chemistry and Pollution*, (ed R. M. Harrison), Royal Society of Chemistry, Cambridge, pp. 137–163.
10. Rowley, A. G. (1993) Time to Clean Up The Act? *Chemistry in Britain*, November, 959–970.
11. Martin, I. D. and Bardos, P. (1994) Recent Developments in Contaminated Land Treatment Technology. *Chemistry & Industry*, 6 June, 411–413.
12. Clarke, A. G. (1992) The Atmosphere, in *Understanding Our Environment: An Introduction to Environmental Chemistry and Pollution*, (ed R. M. Harrison), Royal Society of Chemistry, Cambridge, pp. 5–51.
13. Charlson, R. J. (1992) The Atmosphere, in *Global Biogeochemical Cycles*, (eds S. S. Butcher, R. J. Charlson, G. H. Orians and G. V. Wolfe), Academic Press, London, pp. 213–238.
14. Holmén, K. (1992) The Global Carbon Cycle, in *Global Biogeochemical Cycles*, (eds S. S. Butcher, R. J. Charlson, G. H. Orians and G. V. Wolfe), Academic Press, London, pp. 239–262.
15. Harrison R. M. (1993) Important Air Pollutants and Their Chemical Analysis, in *Pollution: Causes, Effects and Control*, 2nd edn (ed R. M. Harrison), The Royal Society of Chemistry, Cambridge, pp. 127–155.
16. O'Neill, P. (1993) *Environmental Chemistry*, 2nd edn, Chapman and Hall, London, Chapter 4, pp. 66–95.
17. Moroz, W. J. (1989) Air Pollution, in *Environmental Science and Engineering*, (eds J. G. Henry and G. W. Heinke), Prentice-Hall Inc., New Jersey, pp. 471–537.
18. Acres, G. J. K. (1993) Catalyst Systems for Emission Control from Motor Vehicles, in *Pollution: Causes, Effects and Control*, 2nd edn (ed R. M. Harrison), The Royal Society of Chemistry, Cambridge, pp. 221–236.

19. Peavy, H. S., Rowe, D. R. and Tchobanoglous, G. (1986) *Environmental Engineering*, International Edition, McGraw-Hill, Singapore, Chapter 7, pp. 417–482.
20. Colbeck, I. and Farman, J. C. (1993) Chemistry and Pollution of the Stratosphere, in *Pollution: Causes, Effects and Control*, 2nd edn (ed R. M. Harrison), The Royal Society of Chemistry, Cambridge, pp. 181–199.
21. Schlesigner, W. H. (1991) *Biochemistry: An Analysis of Global Change*, Academic Press, London, Chapter 3, pp. 40–71.
22. Heaton, A. (1994) Environmental issues, in *The Chemical Industry*, 2nd edn. (ed A. Heaton), Blackie Publishers, Glasgow, Chapter 4, pp. 39–52.
23. Ottewell, S. (1994) Vocalising in Europe. *Supplement to The Chemical Engineer*, 13 October, S14.
24. Nobel, A. (1994) Environmental Technology in the Chemical Industry. *Chemistry & Industry*. 6 June, 414–418.
25. Porteous, A. (1992) *Dictionary of Environmental Science and Technology*, revised edn, John Wiley & Sons Ltd, Chichester, p. 264.
26. Fish, H. (1992) Freshwaters, in *Understanding Our Environment: An Introduction to Environmental Chemistry and Pollution*, (ed R. M. Harrison), Royal Society of Chemistry, Cambridge, pp. 53–91.
27. De Mora, S. J. (1992) The Oceans, in *Understanding Our Environment: An Introduction to Environmental Chemistry and Pollution*. (ed R. M. Harrison), Royal Society of Chemistry, Cambridge, pp. 93–136.
28. Alloway, B. J. and Ayres, D. C. (1993) *Chemical Principles of Environmental Pollution*, Blackie Academic and Professional, Glasgow, Chapter 2, pp. 16–43.
29. Crathorne, B. and Dobbs, A. J. (1993) Chemical Pollution of the Aquatic Environment by Priority Pollutants and its Control, in *Pollution: Causes, Effects and Control*, 2nd edn (ed R. M. Harrison), The Royal Society of Chemistry, Cambridge, pp. 1–18.
30. Hewitt, C. N. (1993) Radioactivity in the Environment, in *Pollution: Causes, Effects and Control*, 2nd edn (ed R. M. Harrison), The Royal Society of Chemistry, Cambridge, pp. 343–366.
31. Porteous, A. (1992) *Dictionary of Environmental Science and Technology*, revised edn, John Wiley & Sons Ltd, Chichester, p. 33.
32. Henry, J. G. (1989) Water Pollution, in *Environmental Science and Engineering*, (eds J. G. Henry and G. W. Heinke), Prentice-Hall Inc., New Jersey, pp. 409–470.
33. Hare, F. K. and Hutchinson, T. C. (1989) Human Environmental Disturbances, in *Environmental Science and Engineering*, (eds J. G. Henry and G. W. Heinke), Prentice-Hall Inc., New Jersey, pp. 114–142.
34. O'Neill, P. (1993) *Environmental Chemistry*, 2nd edn, Chapman and Hall, London, Chapter 5, pp. 96–113.
35. O'Neill, P. (1993) *Environmental Chemistry*, 2nd edn, Chapman and Hall, London, Chapter 17, pp. 229–239.
36. Harris, J. R. W., Gorley, R. N. and Bartlett, C. A. (1993) *ECOS Version 2 User manual*, NERC, Plymouth.
37. Libes, S. M. (1992) *Marine Biogeochemical*, John Wiley & Sons, New York, Chapter 30, pp. 597–647.
38. Ugolini, F. C. and Spaltenstein, H. (1992) Pedosphere, in *Global Biogeochemical Cycles*, (eds S. S. Butcher, R. J. Charlson, G. H. Orians and G. V. Wolfe), Academic Press, London, pp. 123–154.
39. Porteous, A. (1992) *Dictionary of Environmental Science and Technology*, revised edn, John Wiley & Sons Ltd, Chichester, p. 379.
40. Gardner, G. (1994) Waste Not, Want Not. *Supplement to The Chemical Engineer*, 13 October, S11.
41. Peavy, H. S., Rowe, D. R. and Tchobanoglous, G. (1986) *Environmental Engineering*, International Edition, McGraw-Hill, Singapore, Chapter 9, pp. 514–569.
42. Butcher, D. (1993) Vanquishing VOCs. *Supplement to The Chemical Engineer*, 14 October, S19–S22.
43. Williams, G. (1994) In The Soup. *The Chemical Engineer*, 26 May, 17–18.
44. James, A. (1993) The Treatment of Toxic Wastes, in *Pollution: Causes, Effects and Control*, 2nd edn (ed R. M. Harrison), The Royal Society of Chemistry, Cambridge, pp. 63–81.
45. Williams, G. (1994) Mine's a Cornish Nasty. *Supplement to The Chemical Engineer*, 13 October, S12.

46. Peavy, H. S., Rowe, D. R. and Tchobanoglous, G. (1986) *Environmental Engineering*, International Edition, McGraw-Hill, Singapore, Chapter 4, pp. 104–206.
47. Lester, J. N. (1993) Sewage and Sewage Sludge Treatment, in *Pollution: Causes, Effects and Control*, 2nd edn (ed R. M. Harrison), The Royal Society of Chemistry, Cambridge, pp. 33–62.
48. Forster, C. (1994) Anaerobic Digestion and Industrial Wastewater Treatment. *Chemistry & Industry*, 6 June, 404–406.
49. Petts, J. (1994) Incineration as a Waste Management Option, in *Issues in Environmental Science and Technology 2: Waste Incineration and the Environment*, (eds R. E. Hester and R. M. Harrison), Royal Society of Chemistry, Cambridge, pp. 1–25.
50. Eduljee, G. H. (1994) Organic Micropollutant Emissions from Waste Incineration, in *Issues in Environmental Science and Technology 2: Waste Incineration and the Environment*, (eds R. E. Hester and R. M. Harrison), Royal Society of Chemistry, Cambridge, pp. 71–94.
51. Peavy, H. S., Rowe, D. R. and Tchobanoglous, G. (1986) *Environmental Engineering*, International Edition, McGraw Hill, Singapore, Chapter 12, pp. 653–677.
52. Williams, P. T. (1994) Pollutants from Incineration: An Overview, in *Issues in Environmental Technology 2: Waste Incineration and the Environment*, (eds R. E. Hester and R. M. Harrison), Royal Society of Chemistry, Cambridge, pp. 27–52.
53. Henry, J. G. (1989) Solid Wastes, in *Environmental Science and Engineering*, (eds J. G. Henry and G. W. Heinke), Prentice-Hall Inc., New Jersey, pp. 538–581.
54. Peavy, H. S., Rowe, D. R. and Tchobanoglous, G. (1986) *Environmental Engineering*, International Edition, McGraw-Hill, Singapore, Chapter 10 & 11, pp. 573–652.
55. Porteous, A. (1992) *Dictionary of Environmental Science and Technology*, revised edn, John Wiley & Sons Ltd, Chichester, p. 218.
56. Campbell, D. J. V. (1994) Understanding Co-disposal Processes and Practices. *Chemistry & Industry*, 6 June 1994, 407–409.
57. Porteous, A. (1992) *Dictionary of Environmental Science and Technology*, revised edn, John Wiley & Sons Ltd, Chichester, p. 213.
58. Coulson, J. (1994) Contain Your Contamination. *The Chemical Engineer*, 12 May, 25–26.
59. Freestone, N. P., Phillips, P. S. and Hall, R. (1994) Having the last gas. *Chemistry in Britain*, January, 48–50.
60. Her Majesty's Inspectorate of Pollution (1989) *Waste Management Paper No. 27 The Control of Landfill Gas*, HMSO, London.
61. Phillips, P. S., Freestone, N. P. and Hall, R. S. (1994) Dealing with leachate. *Chemistry in Britain*, October, 828–830.
62. Watkins, G. (1992) European Directives Concerning Health and Safety, in *Minerals, Metals and the Environment*, Elsevier Science Publishers, Essex, pp. 34–42.
63. King, R. (1990) *Safety in the Process Industries*, Butterworth-Heinemann Ltd, London, Chapter 1, pp. 6–26.
64. Cleary, D. and Thornton, J. (1994) The Environmental Impact of Gold Mining in the Brazilian Amazon, in *Issues in Environmental Science and Technology 1: Mining and its Environmental Impact*, (eds R. E. Hester and R. M. Harrison), Royal Society of Chemistry, Cambridge, pp. 17–29.
65. Nriagu, J. O. (1994) A Precious Legacy. *Chemistry in Britain*, August, 650–651.
66. Hursthouse, A. S., Patrick, G., *et al.* (1995) Attempting to Generate Student Teamwork and Debate: Case Studies of Environmental and Industrial 'Issues'. *Education in Chemistry*, (in press).
67. Manahan, S. E. (1992) *Toxicological Chemistry*, 2nd ed, Lewis Publishers Inc. Michigan, Chapter 10, pp. 249–268.
68. Rodricks, J. V. (1992) *Calculated Risks: Understanding the toxicity and human health risks of chemicals in our environment*, Cambridge University Press, Cambridge, Chapter 6, pp. 65–107.
69. Levi, P. (1985) *The Periodic Table*, Michael Joseph Ltd, London, pp. 96–108.
70. Porteous, A. (1992) *Dictionary of Environmental Science and Technology*, revised edn, John Wiley & Sons Ltd, Cirencester, p. 147.
71. Rodricks, J. V. (1992) *Calculated Risks: Understanding the toxicity and human health risks of chemicals in our environment*, Cambridge University Press, Cambridge, Chapter 2, pp. 12–24.
72. King, R. (1990) *Safety in the Process Industries*, Butterworth-Heinemann Ltd, London, Chapter 7, pp. 123–155.

73. Rodricks, J. V. (1992) *Calculated Risks: Understanding the toxicity and human health risks of chemicals in our environment*, Cambridge University Press, Cambridge, Chapter 7, pp. 108–144.
74. Porteous, A. (1992) *Dictionary of Environmental Science and Technology*, revised edn, John Wiley & Sons Ltd, Chichester, p. 223.
75. Crowl, D. A. and Louvar, J. F. (1990) *Chemical Processes Safety: Fundamental with Applications*, Prentice-Hall Inc, New Jersey, Chapter 2, pp. 22–46.
76. Manahan, S. E. (1992) *Toxicological Chemistry*, 2nd ed, Lewis Publishers Inc, Michigan, Chapter 8, pp. 193–215.
77. Porteous, A. (1992) *Dictionary of Environmental Science and Technology*, revised edn, John Wiley & Sons Ltd, Chichester, pp. 192–204.
78. Health and Safety Executive (1991) Occupational Exposure Limits EH40, HMSO, London.
79. Crowl, D. A. and Louvar, J. F. (1990) *Chemical Processes Safety: Fundamental with Applications*, Prentice-Hall Inc, New Jersey, Chapter 3, pp. 47–81.
80. King, R. (1990) *Safety in the Process Industries*, Butterworth-Heinemann Ltd, London, Chapter 2, pp. 27–48.
81. Lee, M. (1994) Integrated Pollution Control: The Story So Far. *Chemistry in Britain*, June, 443–445.
82. Wake, B. D. (1992) The Environmental Protection Act 1990, in *Minerals, Metals and the Environment*, Elsevier Science Publishers, Essex, pp. 3–33.
83. Gardner, G. (1994) Remedial Action. *The Chemical Engineer*, 14 July, 14–15.
84. Zayane, S. A., Hursthouse, A. S. and Cooke, D. A. (1994) Site Investigation for Heavy Metal Contamination – An Improved Strategy? *Wastes Management*, January, 33–34.
85. British Standards Institution (1988) Draft for Development: *Code of Practice for the identification of potentially contaminated land and its investigation*, DD 175.
86. Pratt, M. (ed) (1993) *Remedial Processes for Contaminated Land*, Institution of Chemical Engineers, Rugby.
87. Williams, G. (1994) Making a Meal of Pollution. *The Chemical Engineer*, 14 July, 27–28.

Bibliography

Global Biogeochemical Cycles, S. S. Butcher, R. J. Charlson, G. H. Orians and G. V. Wolfe, eds, Academic Press, London, 1992.
Chemical Processes Safety: Fundamental with Applications, D. A. Crowl and J. F. Louvar, Prentice-Hall Inc, New Jersey, 1990.
Understanding Our Environment: An Introduction to Environmental Chemistry and Pollution, R. M. Harrison ed, Royal Society of Chemistry, Cambridge, 1992.
Pollution: Causes, Effects and Control, 2nd edn, R. M. Harrison, ed, The Royal Society of Chemistry, Cambridge, 1993.
Environmental Science and Engineering, J. G. Henry and G. W. Heinke eds, Prentice-Hall Inc, New Jersey, 1989.
Safety in the Process Industries, R. King, Butterworth-Heinemann Ltd, London, 1990.
Environmental Engineering, International Edition, H. S. Peavy, D. R. Rowe and G. Tchobanoglous, McGraw-Hill, Singapore, 1986.
Dictionary of Environmental Science and Technology, revised edn, A. Porteous, John Wiley & Sons Ltd, Chichester, 1992.
Calculated Risks: Understanding the toxicity and human health risks of chemicals in our environment, J. V. Rodricks, Cambridge University Press, Cambridge, 1992.

Chlor-alkali Products 10

Steve Kelham

The production of chlorine and caustic soda is the story of the bulk inorganic commodities chemicals business. It is an example of a large scale manufacturing industry which uses high technology processes to create the basic raw materials used worldwide as the building blocks for industry.

The discovery of chlorine is usually attributed to Scheele in 1774, who referred to it as 'dephlogisticated marine or muriatic acid air' and who described its greenish yellow colour, its bleaching properties and suffocating effects on insects. However, it is certain to have been known earlier through the work of alchemists from the 12th century. The material was assigned the property of an element by Davey in 1810, and he also proposed the name 'chlorine', from the Greek 'chloros', meaning yellow-green.

Use of chlorine developed slowly. Bleaching of textiles with chlorine gas and later with chlorine water was not very efficient. More success was obtained using chlorine as a disinfectant. In 1790, Tennant was granted a patent on the production of solid bleaching powder. Manganese dioxide and hydrochloric acid routes were expensive and inefficient despite the introduction in about 1870 of the Weldon process for the recovery of manganese dioxide. The Deacon process, proposed in 1870, used air and a copper catalyst to oxidize waste HCl from the LeBlanc process to chlorine and gave a cheaper production route. An increasing demand as a bleaching agent for both textiles and paper followed. The first electrochemical production of chlorine was observed by Cruikshank in 1800, and the first patent, on a cell with a porous diaphragm, was granted to Watt in 1851. Both the lack of commercial electricity supplies and of suitable diaphragms slowed development. It was not until 1888 that the Griesheim cell was developed, making potassium hydroxide and chlorine on a batch basis using a diaphragm of porous cement. The forerunners to the modern diaphragm cells were the Hargreaves–Bird cell of 1890, which was first operated by the United Alkali company in Widnes, and the Le Sueur cell, developed in the U.S.A., which was the first to use a percolating diaphragm. In the 40 years that followed, diaphragm cells of all shapes and sizes were developed. Now only a limited number of designs are in service, some of which are described in more detail later in this chapter.

The use of a mercury cathode in the production of chlorine was simultaneously discovered by Castner and by Kellner, in 1892. The inventors joined forces and the first rocking cell was built at Oldbury, with production moving to Runcorn in 1897. Rocking cells have been operated as recently as 1960 but, as with diaphragm cells, development of different types of mercury cell was rapid, with the evolution of one or two basic designs which are now standard within the industry. While diaphragm cells developed mainly in the U.S.A., the mercury cell development was concentrated in the U.K. and Europe.

Soda was known long before chlorine, as naturally occurring sodium carbonate, and was widely used in the growing industrial world. Artificial soda production started in France, as a result of the Napoleonic wars, when lack of natural soda led to a prize of 2400 livres being offered for a practical manufacturing process. In 1783, the prize was promised to LeBlanc, but never paid, and LeBlanc was forced to make his patents public, without payment, in 1793. Leblanc committed suicide in the workhouse in 1806, but his inventions laid the foundations of the chemical industry (and many of its problems) for more than a century afterwards. The LeBlanc process burns coal, sodium sulphate (saltcake, made from salt and sulphuric acid) and limestone together in a ratio 35:100:100, leaches the product and treats the black liquor produced with lime to form sodium hydroxide solution. As the process developed, it became more integrated and the source of many other bulk inorganic materials. Muspratt introduced the process in Liverpool in 1823, and Widnes soon became the centre of the U.K. industry. Even after the introduction of the Solvay process in 1861 for making sodium carbonate directly, the LeBlanc process was still viable for caustic soda production until after world War I, when causticization of carbonate with lime became dominant. This process was superseded by the electrolytic routes which at that time were rapidly becoming established. Crossover of the electrolytic and chemical routes occurred in the mid to late 1930s.

The dominant production routes of mercury and diaphragm cells developed in parallel in the first half of the 20th century, with technological developments leading to lower energy consumptions and capital costs. Both systems used carbon based anodes, and this led to major cost inefficiencies. The use of titanium as a basic electrode material was first proposed in 1956, and platinum based coatings for chlor-alkali electrodes were patented by Beer in 1962. In 1967 titanium-ruthenium oxide catalytic materials were introduced, and these proved highly successful on the large scale. After the early 1970s, all new plants contained metal anodes, and conversion programmes were initiated throughout the industry to change from the inefficient and labour intensive carbon based anodes to the long lasting dimensionally stable metal anodes. This major invention of the 20th century once more revolutionized the chlor-alkali industry.

The principle of the chlor-alkali membrane cell has been known for a considerable time, and patents for this approach to chlor-alkali manufacture were granted in the early 1960s. Membrane cells combine the purity of mercury cell caustic with the power efficiency of diaphragm cells, while

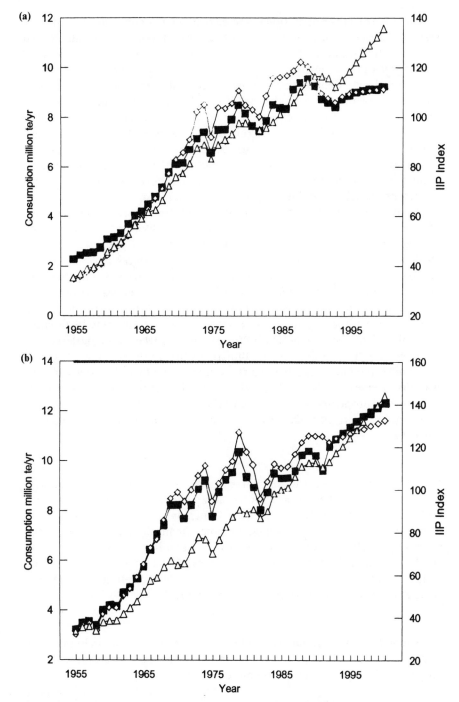

Figure 10.1   Western European chlorine consumption 1955–1983. ■, NaOH; ◇, Cl₂; △, IIP (1985 = 100).

avoiding many of the operational and environmental concerns of both these systems. However, the membranes available at that time were unsuitable, and it was not until the development of new types of membrane by DuPont in 1962 and their use in a chlor-alkali cell in 1964 that the potential of this technology became realized. Membrane cell development followed rapidly, with the first commercial plants on line in the early 1970s, linking membrane, metal anode and cell developments together. Concerns around mercury, particularly in Japan in the late 1960s, were recognized, and the move away from mercury cells gave added impetus to membrane cell development. Today, almost all new plants utilize membrane cell technology, and, as with the introduction of metal anodes in the 1970s, conversion of many older diaphragm and mercury plants to membrane cell technology is starting to take place.

Production of bleaching powder in the U.K. had reached 150 000 tonnes/year by 1900, and the demands for chlorine and caustic soda grew continually at a compound rate of nearly 7·5% until the first oil crisis of 1973. Increased energy prices led to a major world recession and growth paused, being hit by the second energy crisis in 1979. Since then the position has recovered, and there is still a small growth of 1–2% per year. However, as environmental concerns around refrigerants and chlorinated compounds (CFCs) increase, the growth rate may slow even further. Since the 1970s, most of the growth has been in the Third World countries, whereas in the developed nations, growth has been very low, or actually decreased as old, inefficient or environmentally unacceptable plants are shut down. World capacity in 1993 was 40 m tonnes/year (excluding the Communist Bloc), and the growth rate in western Europe and the U.S.A. chlorine/caustic consumption since 1955 is shown in Figure 10.1. Occupacity is currently at around 88% of installed capacity.

10.2
Uses of chlorine

Chlorine consumption is closely linked to standard of living in a country. Initial production of chlorine is for use as a disinfectant, for water treatment and the exploitation of natural resources, such as pulp and paper. Inorganic chemistry applications follow and organic chemicals begin to take large quantities as PVC becomes important in the building industry and for the production of consumer goods. As standards of living increase, so does the usage of solvents, aerosols, dry cleaning materials and refrigerants. Typical market outlets for chlorine are shown in Figure 10.2.

10.3
Uses of caustic soda
(sodium hydroxide)

Caustic soda has similar outlets to chlorine and these are shown in Figure 10.3. Large amounts are used in the organic and inorganic chemicals industries, for soaps and detergents, and for aluminium extraction.

10.4
Uses of hydrogen

Hydrogen, the third product of the electrolysis of brine route, is ideally used as a chemical feedstock, either separately, or integrated with other hydrogen

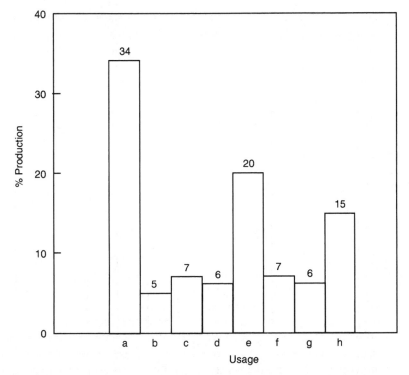

Figure 10.2 Market outlets for chlorine. a, vinyl chloride; b, solvents; c, propylene oxide; d, chloromethanes; e, inorganic chemicals; f, pulp/paper; g, water treatment; h, other inorganic.

plants. Electrolytic hydrogen is very pure, and can be compressed to 200 atm for transportation in cylinders and a wide variety of uses. However, hydrogen disposal is not normally an important consideration and it is often burned as a fuel, or in some cases, even vented to atmosphere, although as a raw material, it has considerable value.

Electrolytic production of chlorine and caustic soda results in fixed ratios of chlorine, caustic and hydrogen being produced. As uses of chlorine and caustic have developed, the balance of production has been critical. Historically, caustic soda demand has tended to be greater than that of chlorine, so processes such as the LeBlanc route and the production of caustic from sodium carbonate have enabled a balance to be maintained. Chlorine and caustic demands are currently essentially in balance, and although it has long been expected that chlorine demand will eventually exceed caustic demand this has not proved to be the case to date, and markets can be balanced by stocking and pricing policies. As caustic soda prices rise, alternative sources of alkali are found, and the market restabilizes. While caustic is a world scale commodity chemical, and is bought and sold freely, chlorine is not readily transported or stored in quantity. Large chlorine users are generally tied closely to the chlor-alkali plant. If chlorine has to be transported it is usually

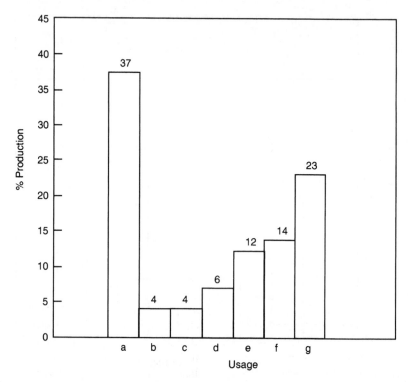

Figure 10.3   Market outlets for caustic soda. a, vinyl chloride; b, solvents; c, propylene oxide; d, chloromethanes; e, inorganic chemicals; f, paper/pulp; g, miscellaneous.

converted into ethylene dichloride which contains over 70% w/w chlorine and is an important intermediate in the manufacture of PVC. The high power consumptions and opportunities for process integration tend to move the chlor-alkali industry to areas of cheap power, natural gas or hydroelectricity, and in many cases, chlor-alkali production forms part of an integrated petrochemicals and plastics complex.

10.5

Types of cell

10.5.1 *Mercury cell process*

The mercury cell process has developed extensively since its inception in 1892. A number of distinct phases can be identified, from the rocking cell to the 400 000 A modern mercury cell unit. The cell chemistry is identical in all cases

   Anode reaction (titanium or graphite)

$Cl^- - e = Cl^•$ chloride ion loses an electron to form a chlorine atom

$2Cl^• = Cl_2$ chlorine atoms combine to escape as chlorine molecules

Cathode reaction (thin film of mercury)

$Na^+ + e = Na$ sodium ion gains an electron to form a sodium atom

$Na + Hg = Na/Hg$ sodium atom immediately dissolves in the mercury film electrode to form sodium amalgam which passes into the decomposer where it reacts to form sodium hydroxide solution and hydrogen, leaving mercury which is recycled.

Decomposer reaction (essentially a short circuited cell)

$$2Na/Hg + 2H_2O = 2NaOH + H_2 + 2Hg \quad \text{(iron/graphite electrodes)}$$

The caustic soda is produced at up to 50% concentration. The reversible voltage of the cell overall is 3.1 V, to which has to be added the operating electrode overvoltages and resistance effects of the cell components and liquids within the electrolysis compartment. The design of the cell has centred on reducing these resistances, and increasing cell efficiencies. The major inefficiency of the mercury cell is the premature decomposition of the amalgam and the evolution of hydrogen within the brine cell rather than in the decomposer.

The first operating membrane cell was the rocking cell, largely developed by Baker, who was Castner's chief chemist, at Oldbury, and later at Runcorn. The history of the Castner Kellner plant at Runcorn gives the history of the development of the mercury cell, and indicates the way in which a technology has developed in the drive to increase production quantities and efficiencies while minimizing capital costs.

The 1896 rocking cell plant at Runcorn was of 1000 HP and operated at 575 A. The cells consisted of two outer brine compartments of $0.9\,m^2$ with graphite anodes and a central amalgam decomposing chamber of approximately $0.6\,m^2$ containing iron grids. The cell was rocked periodically from side to side so that the mercury flowed from one compartment to the other. Only one anode compartment was in use at a given time. While rocking cells continued in service until 1938, the slow movements and mass transfer limited performance and current densities, and the long cell concept was developed.

The long cell was first introduced by Baker in 1902 and has been the basis of all subsequent mercury cells. The cell was 50 ft long and consisted of a brine channel, with mercury flowing along the base, an amalgam channel where decomposition of amalgam occurred, returning the mercury to the front end of the cell, and a pump to return the mercury to its starting point. The brine cells were of concrete, with steel mushrooms in the base to provide electrical contact with the mercury. The slope was 1 in 300 to give a high mercury flow rate. The denuder was half the width of the cell and initially made of concrete, although this was later replaced with steel. The denuder contained graphite grids. An Archimedean screw pump was used, and the current rating eventually achieved 11 000 A with a base area of $7.3\,m^2$. Anodes were of graphite rods impregnated with wax which were sealed into the covers, and as these suffered appreciable wear, anode–mercury gaps increased and electrical efficiency deteriorated, leading to the need for frequent replacement. Concrete cell covers placed on wooden laths, which were progressively removed to lower the anodes, were used as a primitive adjustment system.

Ebonite (hard rubber) lined steel based cells were introduced in 1937, although concrete cells continued to run until 1967. Ebonite lasted several times as long as concrete and introduced fewer impurities into the cell. The retention of the steel mushrooms still limited current loadings, but these cells still continued to operate until 1971 at Runcorn. The current limitations were overcome by the use of steel baseplates, which became widespread after 1943, with the base area increased to $12 \cdot 5\,m^2$. The denuder was mounted under the cell instead of alongside to conserve space and more efficient mercury pumps developed, leading to centrifugal rather than screw designs. Anode design and anode adjustment developed, with more emphasis on the optimum gaps and the ability to release gas bubbles. Individual anode adjustment was introduced. Vertical denuders, packed with iron impregnated graphite balls, were introduced in 1966 and by 1975 all the remaining cells at Runcorn had been converted to vertical denuders. Introduction of metal anodes in 1971 led to further improvements, and with the limitations on current density removed, currents increased, reaching a maximum of 90 000 A on the final versions.

The modern generation of mercury cells began in 1966 with the introduction of the $25\,m^3$ cell, still 48 ft long but double the width and with a slope of 1 in 100. A sketch of such a cell is shown in Fig. 10.4. Water cooled vertical

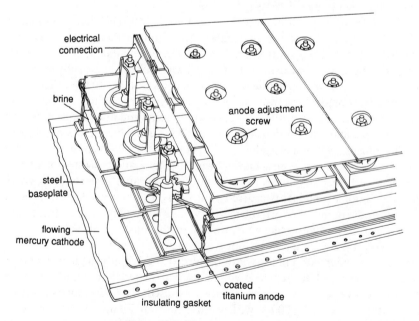

Figure 10.4 Mercury cell diagram.

denuders, centrifugal pumps and later, metal anodes were installed. Currents were raised to 225 000 A. The major design change came from the cellroom layout, with cells on a single level instead of several floors, and a much greater attention given to working conditions and mercury hygiene. Adiabatic denuders were installed in later designs, giving more efficient and compact units operating at higher temperatures, and aluminium started to replace copper for electrical connections. In 1976, the 33 m^2 cell was introduced. This was the first unit designed specifically to take advantage of metal anodes, with automatic anode adjustment and currents of up to 400 000 A.

As currents increase and power consumptions and efficiencies become more critical, the problems of brine quality become more apparent. Brine purity is controlled at 20–40 ppm of calcium and less than 1 ppm of magnesium. Brine can be supplied on a resaturation circuit or on a once through basis. Poor brine quality can lead to high hydrogen levels in the chlorine and the formation of 'thick mercury', a buttery compound which has to be removed from the cell from time to time, and to denuder deactivation. As the mercury cell technology has reached its peak, chemistry, rather than engineering problems have started to become dominant, and the fundamentals of mercury cell operation have been increasingly investigated.

Of particular interest has been the relationship of electrical power consumption and current density to development activities. Current densities have increased continuously, giving greater production capability per unit area and hence for a given capital cost. Cell efficiencies, in the form of voltage

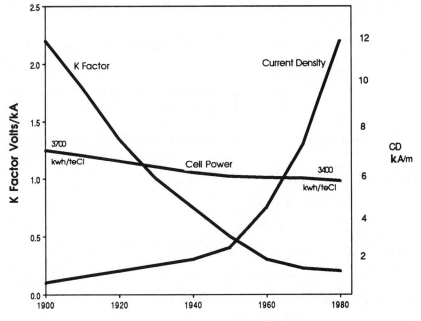

Figure 10.5   Mercury cell performance 1900–1980.

loss reduction have increased, but the relationship between the two has led to a virtually constant power consumption, of about 3400 kW/tonne chlorine as shown in Figure 10.5.

With the introduction of membrane cell technology, mercury cell development has virtually ceased, and it is highly unlikely that any significant new mercury cell capacity will ever be installed. However the modern mercury cell is still a highly effective way of producing high quality caustic soda, and provided mercury discharge levels can be maintained at sufficiently low levels mercury cells will remain viable until well into the 21st century.

10.5.2 *Diaphragm cell process*

The diaphragm cell principles have been in use commercially since 1888, when the Greisheim cell was introduced in Germany. The principle is to reduce reverse flow and reaction of hydroxyl ions with chlorine by use of a porous diaphragm, across which a hydraulic gradient is set up, and the efficiency of the cell depends on the design of this separator.

The principal reactions are

Anode reaction (titanium or graphite)

$NaCl = Na^+ + Cl^-$ equilibrium of sodium and hydroxyl ions

$Cl^- - e = Cl^\bullet$ chloride ion loses an electron to form a chlorine atom

$2Cl^\bullet = Cl_2$ chlorine atoms combine to escape as chlorine molecules

Liquor containing ionic mixture flows through the diaphragm

Cathode reaction (steel)

$2H_2O + 2e = H_2 + 2OH^-$ water decomposition to form hydrogen and hydroxyl ions

$OH^- + Na^+ = NaOH$ hydroxyl ion combines with sodium ion to form caustic soda

The reverse reactions are

$$Cl_2 + OH^- = Cl^- + HOCl$$

$$HOCl = H^+ + OCl^-$$

$$2HOCl + OCl^- = ClO_3^- + 2Cl^- + 2H^+$$

$$4OH^- = O_2 + 2H_2O + 4e$$

With graphite electrodes, there is in addition

$$4OH^- + C = CO_2 + 2H_2O + 4e$$

It can be seen that in a commercial cell, the major byproducts are sodium hypochlorite, sodium chlorate, oxygen and carbon dioxide, which lead to overall current efficiencies in the range 90–96%. The electrolysis current only converts about half the 25% w/w sodium chloride feed to caustic soda, which leads to a maximum of about 12% w/w caustic in the product sodium chloride/sodium hydroxide liquor.

The early cells used a porous diaphragm made from concrete which included soluble salts leached out after setting. The anodes were carbon or magnetite and a steel pot was used as the cathode. The system was operated on a batch basis, initially with KCl. The first continuously percolating diaphragm cell was the Le Sueur cell in 1897 which used a concrete trough with a porous base to separate the liquids, and cells using this principle were still in operation in the U.S.A. until the mid 1960s. Cell design development rapidly moved towards more effective diaphragms and improved anodes in the form of graphite rather than carbon. Most of these had vertical diaphragms. Major breakthroughs occurred with the Marsh cell, with finger cathodes and side entering anodes, with an asbestos paper diaphragm, and the Hooker deposited diaphragm cell, the forerunner of all modern diaphragm cells, which was first introduced in about 1928. Current ratings were raised from 5000 to 30 000 amp as the technology developed. Filter press bipolar designs were developed by Dow and later by PPG. Some of these units are still in operation.

Hooker S type cells and the Diamond Alkali D type cells became the standard units for the industry, and had many similar features. The anodes consisted of vertical plates of graphite set in a shallow steel or concrete pan and attached to copper connection bars by pouring in molten lead which was then covered with pitch to protect it from the corrosive conditions in the cell. A steel gauze finger cathode with a deposited asbestos diaphragm was placed over the anode blades and insulated from the anode base by a rubber gasket. A concrete cover was then located over the cathode box. The asbestos diaphragm was deposited by sucking a slurry of asbestos fibres onto the gauze in a depositing bath. In operation, brine is admitted to the top of the cell, and percolates through the diaphragm and cathode gauze into the catholyte compartment. Chlorine is evolved from the anodes and is collected from inside the concrete cover. Hydrogen is collected from inside the cathode box, from where the caustic/brine mixture is also removed. As the percolation rates through the diaphragm vary, so does the level of brine in the cover, and the art is to produce a diaphragm which gives a long lifetime before the brine level in the cell becomes too high. Not surprisingly, as most of the components have limited lifetimes, the cells are relatively expensive to rebuild, and handling asbestos and lead can lead to health and environmental concerns.

As with the mercury cell, the introduction of metal anode technology led to a considerable review of cell design and the opportunity to optimize the anode–cathode gap, which as with the mercury cell, increased with time as the anode deteriorated, but unlike the mercury cell, could not be adjusted with the cell in service. Diaphragm technology improved, with the replacement of wet

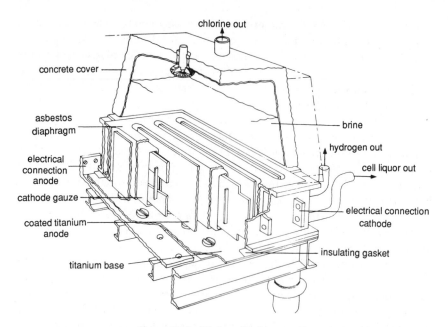

Figure 10.6   Modern diaphragm cell.

asbestos diaphragms with heat treated and polymer modified diaphragms. Lead bases became unnecessary, and the ICI DMT cell is typical of a modern diaphragm cell unit and shown in Figure 10.6. This cell has a current capability of 120 000 A and a cathode area of 43 m^3, giving a current density of 2.8 kA/m^2. The titanium anodes are permanently welded onto a titanium sheet which is explosion bonded onto a massive steel base to minimize electrical resistance losses. The cathode box and base has explosion bonded copper plates to provide low resistance intercell connections and the anode–cathode gaps can be minimized by expanding the electrode structures into the cathode fingers once the box is in place. The asbestos diaphragm is heat treated to improve stability and lifetimes, and the cell cover is of GRP lined concrete to maintain hydraulic heads. Current leakage is minimized through flow interrupter devices, and the cell can operate without maintenance for in excess of a year, depending mainly on brine qualities. Asbestos is a magnesium rich material and, in operation, the pH changes in a diaphragm lead to deposition and leaching of magnesium. For long lifetimes and good permeability, the magnesium balance must be maintained by keeping the feed concentration of magnesium at 1–2 ppm in the brine.

Recently a number of companies have offered synthetic fibre diaphragms, made primarily from PTFE, which mimic the performance of asbestos and are applied in a similar way. If these are shown to be successful commercially, the operational life of diaphragm cell units will be significantly extended.

Reversible voltages are of order 2.3 V, which leads to intrinsically significantly less electrical energy than for a mercury cell where the decomposition voltage is 3.1 V. However, the actual resistance losses (known as the k factor) and therefore added voltage, are significantly higher—typically 0.43 V per kA/m^2 compared to 0.1 V per kA/m^2 for a mercury cell. This is offset by the lower current density, typically 2.8 kA/m^2 compared to 12 kA/m^2. Total energy requirements must, however, include the need to purify the product caustic soda to remove the salt and increase the strength to the typically 50% sales specification. This is usually carried out in three or four effect evaporators. As the caustic concentration increases, the sodium chloride crystallizes out and is removed by filters and/or centrifuges. However, solubility levels are such that most diaphragm cell products still contain up to 1% w/w salt, and this can restrict its use, e.g. in the rayon industry. While processes have been devised for further purification of the caustic, they have never been a commercial success.

10.5.3 *Membrane cell process*

The membrane cell concept has been known for many years, but early work failed as a result of non availability of suitable ion exchange membranes which would resist the very demanding conditions within the chlor-alkali cell. It was also realized that the graphite anodes, which wore away in use, would present major difficulties from the engineering side, so it was not until metal anodes became available that interest revived. However, much of the early work and patents of the 1950s set the scene for the way in which this technology was to develop.

The membrane cell operates in a similar way to the diaphragm cell, with the same basic reactions. However, instead of the hydraulic gradient preventing reverse flow of hydroxyl ions, the cation exchange membrane will only allow sodium ions to pass through in the direction anode–cathode, and will inhibit the reverse flow of hydroxyl ions. The effectiveness of the system depends on the selectivity of the membrane, coupled with the electrical resistance of the cell itself. Instead of the hydrogen-producing reaction taking place in a brine solution, the catholyte liquor consists of a recirculating caustic system, in which the strength is increased on each pass through the cell, and the concentrations maintained by purge and water addition.

The first commercial membranes were fabricated by DuPont under the Nafion tradename. These had a backbone of carbon and fluorine atoms, to which were appended sidechains containing active sulphonic acid ion exchange groups (see Figure 10.7). These end groups gave high conductivities and therefore low voltage drops, but had relatively poor current efficiencies. The major breakthrough came in 1978 when Asahi Glass introduced a carboxylic acid grouping to the sidechains. This gave high current efficiency but relatively poor voltage, and was attacked by the anolyte side conditions. The

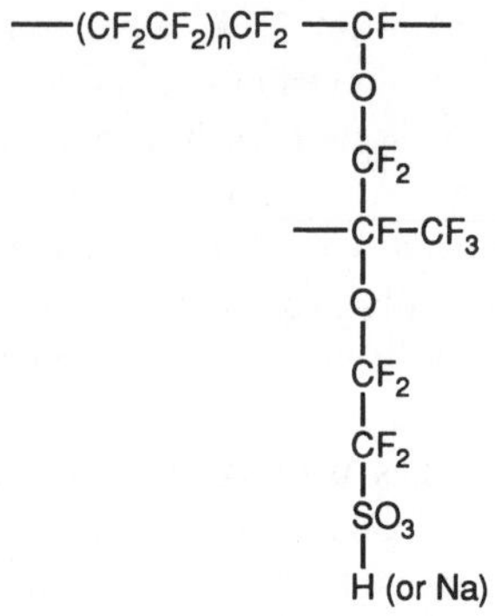

$$-(CF_2CF_2)_nCF_2-CF-$$
$$|$$
$$O$$
$$|$$
$$CF_2$$
$$|$$
$$-CF-CF_3$$
$$|$$
$$O$$
$$|$$
$$CF_2$$
$$|$$
$$CF_2$$
$$|$$
$$SO_3$$
$$|$$
$$H \text{ (or Na)}$$

Figure 10.7 Nafion structure.

compromise was to produce a laminated membrane with a thick anolyte side of sulphonic acid structure to give low voltage and physical integrity, and a thin catholyte face containing carboxylic structures which provided a good current efficiency. These membranes are now manufactured by both DuPont and Asahi Glass. From time to time other manufacturers have produced membranes, but these are largely based on Nafion membranes with minor chemistry modifications. Further enhancement of membrane performance has been by addition of surface coatings to promote bubble release, reducing thicknesses, and incorporating reinforcing meshes. Membranes have typically a 96% initial current efficiency and voltage drop of about 250 mV. Lifetimes are of the order of 3 years. Caustic concentration is typically 30–35% so a simple evaporation system, to bring concentrations up to 50%, is still required. However, the product material contains only a few ppm of salt.

Early membrane cell development was dominated by the Japanese drive to eliminate mercury cell activities. Initial conversion was to diaphragm cells, but it was soon realized that the inferior grade of caustic would not fulfil all needs, and the early membrane systems were rapidly brought up to full commercial scale. Early cells were of the bipolar rather than the monopolar design. In a bipolar cell, the voltage is applied across a stack of membranes and electrodes such that all the individual components are running in series. The current passes through each compartment in turn, going in sequence anode, membrane, cathode; anode, membrane, cathode etc., and the design of the cell means that the anodes and cathodes are fabricated as a single unit. Monopolar cells, which are more akin to the conventional mercury and diaphragm cell layouts have the electrodes arranged in the cell in parallel. Each cell has the same voltage across it, and the current is spread between the cells in the unit. The sequence would be anode, membrane, cathode, membrane, etc., with the electrode structure being either wholly anode or wholly cathode based. Bipolar cells show voltage savings over monopolar equivalents as the voltage drop within the composite electrodes can be reduced to very low levels, but there are potential problems with materials, due to hydrogen diffusion and embrittlement and current leakage. For this reason, special precautions have to be taken in feeding liquors into the cell and removing them from it, using long plastic tubes to maximize electrical resistances and hence reduce current bypassing which, with a 300–400 V stack of plates, can be considerable. Asahi Chemical, in Japan, developed a successful bipolar cell design, using welded titanium anodes and steel cathodes approximately $2·5\,m^2$ area and incorporating their own variety of membrane to make 22% caustic soda. This type of cell is closed with hydraulic rams, and a typical 50 000 tonnes/year plant would contain 5 of these cell units, operating at 350 V, 10 000 A and about $4\,kA/m^2$. An advantage of this design of cell is that the rectifiers required are high voltage, low current, and are therefore relatively cheap. A disadvantage is that the loss of a single membrane can lead to shutdown of the whole cell, whereas for the monopolar cell, this is not a major hindrance.

Monopolar cells have been developed more extensively, and are more appropriate for larger installations. The ICI FM21 cell provides a radical departure from the traditional heavy engineering of electrochemical cells as shown in Figure 10.8. The electrodes are pressed from sheets of nickel for the cathode and titanium for the anode, each with an active area of $0·21\,m^2$. The electrodes are coated with electroactive materials, and the cell is assembled, sandwiching the membrane between alternate anode and cathode plates. Sealing is by rubber gaskets and compression is maintained by disc springs. As the voltage drop across the unit is only of order 3.5 V, current leakage is not a problem and liquor inlets and liquor-gas outlets are through internal porting systems. A typical cell will operate at up to 100 000 A and $4\,kA/m^2$, using an option of membrane types. A 50 000 tonnes/year plant would contain 50 such cells.

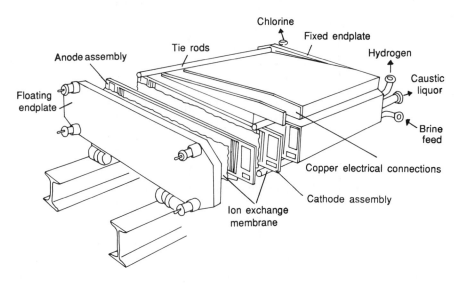

Figure 10.8 Membrane cell diagram.

Electroactive coatings for membrane cell anodes are essentially the same as for mercury or diaphragm cells. The cathodes, unlike those on diaphragm or mercury cell systems, can also be coated to reduce the hydrogen evolution overvoltage. These coatings are usually platinum or nickel based, and although capable of reducing overvoltages to as low as 50 mV, tend either to be expensive or have short lives, and so the economic justification for these has to be critically examined. Recoating operations have to be carried out, and the style of fabrication of the cell can significantly change the cost and complexity of this operation. Generally, the lifetime of an anode coating can be defined as 24/current density, i.e. typically 2 years for a modern mercury cell at 12 kA/m^2, 8 years for a diaphragm cell at 3 kA/m^2 and 6 years for a membrane cell at 4 kA/m^2. The cost of electrode coatings is therefore virtually the same per tonne of chlorine whatever technology is utilized. The actual cost of this operation can vary considerably, depending on the cell design, the cost of dismantling and the loss of production incurred through the shutdowns involved.

Other manufacturers such as Asahi Glass, Tokuyama Soda, De Nora, Oxytech, Chlorine Engineers and Uhde offer a variety of either monopolar or bipolar cells and most of these technologies now have representative plants. Membranes are almost all Nafion based.

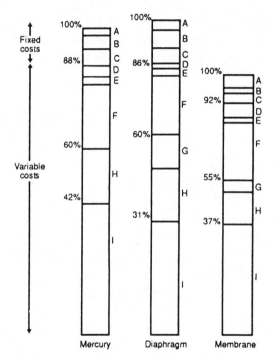

Figure 10.9 Production costs. A, operation; B, maintenance; C, others; D, cell renewal; E, others; F, salt; G, steam; H, resistance power; I, reaction power.

Table 10.1   FM21 cell voltages at $3 \, kA/m^2$

Anode assembly and over voltages	210 mV
Cathode assembly and over voltages	140 mV
Anolyte	130 mV
Catholyte	100 mV
Membrane	250 mV
Reversible anode voltage	1220 mV
Reversible cathode voltage	1000 mV
Total	3050 mV

The emphasis on improving power performance of membrane cells has led to a continual drive to reduce voltages and resistance losses in the system. Power consumptions are now less than 2400 kW/tonne of chlorine, and a comparison of costs and energy consumptions for mercury, diaphragm and membrane cell systems is shown in Figure 10.9. Reversible voltages are fixed although there are areas where some changes could potentially still be made in the long term, for instance by the use of air or oxygen depolarized cathodes which produce water instead of hydrogen gas at the cathode. These developments are becoming more widely used and a major part of the cathode voltage can now be recovered. However, most of the other achievable improvements are based on the fine tuning of the cell design itself. The voltage breakdown components of the monopolar cell would be typically as shown in Table 10.1.

It can be seen that there are finite limits as to how far voltage performance can be decreased as internal resistances tend to smaller and smaller values. Further improvements can be made by reducing resistance losses within liquid films by bringing electrodes closer to the membrane and most modern membrane cells can accommodate essentially zero gap configurations so that the electrodes virtually touch across the membrane. However, this has been shown to create secondary problems within the cell, mainly resulting from high local brine depletion rates and deterioration of membranes, and the optimum is now viewed as being 1–2 mm. Brine circulation and degassing is important and again there is an optimum to be achieved between a tall cell for good liquor circulation through gas lift effects and a short cell to minimize bubble voidages and electrode voltage losses. Uniformity of operating conditions between compartments is essential. Overall the whole concept of cell design is one of optimization to ensure that the minimum operating costs in terms of membrane, power consumption and anode lives are achieved, together with the minimum installation costs for the cell itself. Membranes are the highest cost item, but membrane lives are increasing, with 3–4 year lives being achieved in many plants. It is therefore vital that there should be no requirement to take the cell off-line to recoat electrodes or change gaskets while the membrane is still usable. The drive in membrane cell design has therefore been to extend component lives to ensure that a membrane cell rebuild is governed by membrane life and not for other reasons.

The actual time when a membrane is changed depends very much on local power costs, the type of membrane and cell systems used. In the economic assessments for membrane installation high priority must be placed on obtaining the right system overall, including the costs of all the replacement operations and other maintenance activities and also the correct choice of current density. As current densities increase so do power consumptions, while capital costs reduce. However, higher current densities will tend to lead to shorter operating lives, so optimization studies are essential in any project assessment.

10.5.3.1 *Installation of membrane cells*   The electrolytic cell itself comprises only a small part of the total chlorine plant, perhaps 15–20% of the capital for a new installation. A typical schematic layout is shown in Figure 10.10. The system of providing the brine as a salt source and handling the product chlorine, hydrogen and caustic soda adds significantly to the cost of the total installation, and rectification and power distribution systems for handling the tens or hundreds of megawatts involved lead to considerable capital cost. Ion exchange membrane cells require high purity brine with less than 20 ppb of total hardness and this is achieved by using ion exchange systems. The brine is circulated around the anolyte side of the cell and is normally resaturated to return the feed concentration to near saturation. The catholyte circuit usually consists of 30–32% caustic which is concentrated within the cell to about 34–35% liquor strength. Externally to the cell this liquor is further evaporated to give a 50% sales specification, although there are now developments in membranes, that make 50% caustic directly from the cell a possibility. Hydrogen treatment is relatively straightforward with cooling, possibly filtration and compression before the product is used either as a fuel or for further chemicals manufacture. The chlorine product is cooled and filtered and then dried, usually with concentrated sulphuric acid before being

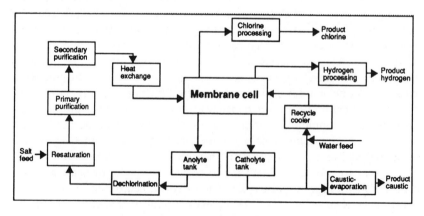

Figure 10.10 Chlorine production from membrane cells.

compressed for use either as a chlorine gas stream or subsequent liquefaction. The major by-product from a membrane cell is oxygen in the chlorine from the back-migration and reaction of hydroxyl ions. Achieving a high current efficiency is important, not just from the electrical energy point of view but also to minimize this level of oxygen in the cell gas product. Energy conservation within the membrane cell circuit is important and various systems are devised to minimize heating and cooling duties so that a well-designed circuit should be essentially energy neutral. However, as the operation of membrane cells becomes more efficient the heat input due to resistance losses decreases and at low loads membrane cell systems often require additional heat input to maintain thermal equilibrium.

Conversion of an existing diaphragm or mercury cell plant to membrane technology is usually cheaper but not necessarily less straightforward than in a greenfield site plant. Much of the existing brine system for both mercury and diaphragm cell plants can be re-used for membrane cells but the purity levels have to be increased by addition of appropriate ion exchange systems. A diaphragm cell plant has a once through brine system in that all the brine is consumed by the cell and there is no liquid recycle. Salt recycle is through the evaporator stage and so extensive modifications may have to be done in this area. Chlorine and hydrogen systems are usually very little changed but the caustic system is usually entirely different for a membrane cell plant. Rectification systems in existing mercury and diaphragm cell plants are normally related to monopolar cells and so the conversion of an existing plant to membrane cells is much easier if monopolar membrane cells are to be used rather than the bipolar variety. However, in many cases rectifiers currently in use are old and there is a good financial case for replacement of these units with more efficient and up-to-date systems which could be used with either monopolar or bipolar geometries, although for large plants the monopolar system is usually preferred.

10.6 Future developments

Membrane cells present many advantages over the mercury and diaphragm alternatives. However, many of the existing mercury and diaphragm cell plants are relatively new. After metal anodes were introduced there was a major rebuilding programme in the 1970s and early 80s and nearly all producers upgraded their old graphite anode plants or built new plants to use the new technology. As a result, the majority of chlor-alkali plants which were built at that time are still in relatively good condition and unless there are major environmental or economic reasons for conversion then there is a high likelihood of these plants continuing to operate for some time to come. What has happened is that over the last few years there has been a significant increase in the number of small plants, particularly in the developing part of the world where on-site production of chlorine and caustic soda is essential to local industries. In addition the transportation problems associated with liquid chlorine are leading to more on-site generation in the developed countries and again a proliferation of relatively small plants away from the traditional centres of production. All these new plants are almost without exception ion exchange membrane cell systems. The drive to convert large plants is starting to take off and it requires a trigger mechanism to set the ball rolling. In Japan this was the Minamata mercury scare in the early 1970s which led to a Government requirement that all mercury cells were phased out. This initially led to conversion to diaphragm cells, but diaphragm cells were not able to produce the quality of caustic required for consuming industries and so membrane cell technology really became the only possible alternative to shutting down and as referred to earlier, this led to many of the major technological advances.

Mercury cells account for about 14 million tonnes/year of chlorine or 31% of capacity, with 180 plants worldwide; diaphragm plants for 23 million tonnes/year or 51%, with 300 plants; and membrane cells for 8 million tonnes/year or 18%, with 170 plants. The average membrane cell plant is typically only half the capacity of a mercury or diaphragm cell unit. However, the rate of growth of membrane cell capacity has far outstripped the growth rate in the industry, suggesting that many of the other technologies have shut down plants to give way for the new technology introduction. If the Japanese indigenous capacity, which is now virtually all ion exchange membrane, is excluded it can still be seen that the growth rate in the rest of the world is significant and is likely to escalate rapidly during the next decade. As energy costs and environmental pressures continue to increase there is no doubt that membrane cells will become dominant within the industry and the major producers will be adopting this technology.

Bibliography

History of the Chemical Industry in Widnes, D. W. F. Hardie, ICI, 1950.
Modern Chlor-Alkali Technology, Vol. 2, C. Jackson (ed.), Ellis Horwood, 1983.
Modern Chlor-Alkali Technology, Vol. 1, Coulter (ed.), Ellis Horwood, 1980.
Diaphragm Cells for Chlorine Production, Soc. Chem. Ind., 1977.
Chlorine–Its Manufacture, Property and Uses, Sconce (ed.), Reinhold/Chapman & Hall, 1962.
Developments and Trends in the Chlor-Alkali Industry, C. Jackson and S. F. Kelham, *Chemistry and Industry*, 1984, 397–402.
Modern Chlor-Alkali Technology, Vol. 3, K. Wall (ed.), Ellis Horwood, 1986.
Chlorine Bicentennial Symposium, T. C. Jeffrey et al., Electrochemical Soc. Inc., 1974.

Catalysts and Catalysis 11
John Pennington

A young research chemist, particularly in the field of organic chemistry, could possibly be forgiven for lacking any 'feel' as to whether a newly proposed reaction is feasible. The laws of thermodynamics provide signposts, but these are often obscured by the inherent stability of covalent bonding, which presents a barrier to change at the molecular level. Of course, were such kinetic barriers absent, no natural or synthetic organic matter could survive in an oxygen-containing atmosphere. This general lack of reactivity, with a few notable exceptions, also implies that very high temperatures would be required for many reactions, providing little hope for selective conversions; reactions which become thermodynamically unfavourable with increasing temperature, such as methanol synthesis and olefin (alkene) hydration, would be impracticable.

11.1
Introduction

However, the existence of life forms shows that such constraints need not limit the scope of organic chemistry. Photosynthesis, the controlled stepwise oxidation of carbohydrates and many more highly specific chemical transformations all proceed at near ambient temperatures.

Early in the 19th century, Berzelius began to recognize a common feature within a number of isolated observations, such as the behaviour of hydrolytic enzyme concentrates *in vitro* and the effects of acids, platinum metal and other materials on simple chemical reactions. In 1835, he introduced the terms 'catalytic force' and 'catalysis' to describe the property by which some materials, soluble or insoluble, could effect or accelerate a reaction, to which the catalytic material itself remained 'indifferent'.

The complexity of enzyme systems limited the interaction between biochemical and chemical interpretations of catalytic mechanisms until the middle of the present century. In the meantime, industrial catalysis progressed through empirical developments, with homogeneous catalysis by sulphuric acid, alkalis and transition metal compounds. But the historical landmarks, the Contact Process for sulphuric acid (1890s) and the Haber–Bosch Process for ammonia (1913), established a very strong tradition for predominantly heterogeneous catalysis by metals and metal oxides. Organometallic chemistry has introduced some important new applications of homogeneous catalysis, but has by no means changed the face of the petrochemical industry as yet.

Not all petrochemical processes are catalytic—the steam cracking of hydrocarbons to lower olefins is a thermal process at 700 to 800°C or more. However, excluding free-radical polymerization processes, this is a rare example, though severe conditions may still be required in some catalysed processes on thermodynamic grounds or to achieve acceptable rates (several $mol\,h^{-1}$ per litre of reaction volume). As we shall see in this and the following chapter, the major impact of catalysis is to provide a remarkably wide range of products from a small number of building blocks.

11.2 Definitions and constraints

11.2.1 Essential features

A catalyst brings about or accelerates a specific reaction or reaction type, which proceeds in *parallel* with any other existing thermal or catalytic reaction within the particular system. Hence, useful catalysis requires that the rate of the desired reaction considerably exceeds the rates of all other possible reactions.

The requirement that the catalyst remains 'indifferent', as quoted in the original definition, must be reinterpreted in the light of our increased knowledge of catalytic mechanisms. Mineral acid catalysts used in esterification, etherification, etc., may be partially converted to esters. Soluble salts and complexes of metals may be converted into a variety of forms, often in different oxidation states. Heterogeneous catalysts will become covered in adsorbed and chemisorbed reactants, intermediates and products, often again with changes in the oxidation states of catalytically active metal centres. In continuous processes we should see a constant compositional state, reaction rate and yield pattern develop. In batch operations, compositional changes during a reaction may cause changes in the proportions of different catalytic species, but recovery and re-use of the catalytic material should thereafter provide indefinite reproduction of rates and yields.

Even with our modified definition of 'indifferent', we still require that the catalytic material should act indefinitely once introduced. This requirement is also fulfilled by a number of essential materials added to some catalytic processes, and often referred to as co-catalysts or promoters. For example, the copper (I)-copper (II) chloride redox system used in Wacker's palladium-catalysed oxidation of ethylene to acetaldehyde[1] (section 11.7.3) behaves in a true catalytic manner in the single-reactor variant of the process (ethylene and O_2 introduced into the same reaction vessel).

$$C_2H_4 + \tfrac{1}{2}O_2 \xrightarrow{\text{PdCl}_2/\text{CuCl}_2} CH_3CHO$$

However, when the copper (II) chloride is used as a stoichiometric oxidant in one reaction vessel, and is then re-oxidized in a second (another commercialized variant of the process), it should possibly be regarded as a regenerable or catalytic reagent.

$$C_2H_4 + 2CuCl_2 + H_2O \xrightarrow{PdCl_2} CH_3CHO + 2CuCl + 2HCl$$
$$2CuCl + 2HCl + \tfrac{1}{2}O_2 \rightarrow 2CuCl_2 + H_2O$$

Nonetheless, it may still be treated as an integral part of the 'catalyst package' in the overall process.

11.2.2 *Initiators*

A distinction must be made between catalysis and 'initiation' of a reaction. Initiators usually decompose spontaneously, effectively introducing packages of energy into a small number of reactant molecules to generate reactive species, most frequently free radicals. These propagate the reaction by a chain mechanism, and the chemical nature of the initiator is no longer relevant; moreover, the number of reactive species can change with time. (In many oxidation processes, individual steps in initiation and the chain reaction may also be catalysed by metallic ions, leading to very complex mechanisms.)

11.2.3 *Co-reactants*

In the oxidation of *p*-xylene (in acetic acid solution) to terephthalic acid, catalysed by cobalt and manganese cations, bromide ions are generally introduced as co-catalysts to accelerate the oxidation of the second methyl group (section 11.7.7.1). Now halide ions in carboxylic acid solution are extremely corrosive even to high-grade stainless steels. However bromides can be omitted if materials such as acetaldehyde or 2-butanone, which are oxidized rapidly to acetic acid via peroxide intermediates, are added continuously to the reaction mixture in relatively large quantities (0·25–1 mole per mole of *p*-xylene). The peroxides generate high concentrations of Co(III) and Mn(III) ions to initiate attack on the methylaromatic substrate. Such added substances are referred to as co-oxidants, or more generally co-reactants. They are always consumed in the process.

11.2.4 *Inhibition*

No simple, single-step chemical reaction can be inhibited, in the sense that the rate constant cannot be reduced. The occurrence of inhibition implies that

(a) a reactant or important intermediate has been removed by a competing reaction with, or catalysed by, the inhibitor (occasionally by complexation in a way which does not affect the analytically determined quantity of that reactant), or

(b) a catalytic species required for the reaction, whether previously recognized or not, has been destroyed or removed from the sphere of activity.

11.3 Thermodynamic relationships

11.3.1 Application

A true catalyst cannot change the thermodynamic properties of reactants and products. Hence, a catalyst can only effect, or accelerate, a reaction which is thermodynamically feasible; in a reversible process, the back reaction is also accelerated and thermodynamic calculations remain essential for the selection of preferred reaction conditions.

If we consider any reversible reaction of the form

$$aA + bB + \ldots \rightleftharpoons lL + mM + \ldots \tag{11.1}$$

we can relate the thermodynamic properties of the reactants and products to the equilibrium constant expressed in terms of *activities* of the elements or compounds, A, B, etc. Activities of materials may be defined with respect to one of several 'standard states', the most common for organic compounds being the pure compound in its natural form or the ideal gas state. Standard thermodynamic data are usually quoted at 25°C (298·15 K) and 1 atmosphere (1 atm = 1·01325 bar; 1 bar = 10^5 Pa), though a standard S.I. pressure (1 bar or 1 Pa) has been adopted in some recent tabulations.

Liquid mixtures of organic compounds are almost invariably non-ideal, and the activity of an individual component with respect to the pure compound is usually determined by measuring the partial pressure of each component to provide 'vapour–liquid equilibrium data' (used extensively in the design of distillation systems). Hence, it is quite logical to use the ideal gas state as a 'standard state' for both liquid-phase reactions and the many gas-phase chemical processes of industrial interest. Several texts present thermodynamic data for a range of organic compounds in the ideal gas state, and provide guidance in their use. Furthermore, a number of methods can be used to predict approximate ideal gas-phase data for new organic compounds from their structures (and spectroscopic data, when available).

In the gas phase, the term 'fugacity' (f_A, etc.) usually replaces the term activity; but at modest total pressures (up to several atmospheres) this is approximately equal to the ratio of the partial pressure (p_A, etc.) to the standard pressure $p^\ominus$ in consistent units. Hence

$$K_p^\ominus = \frac{(p_L/p^\ominus)^l (p_M/p^\ominus)^m}{(p_A/p^\ominus)^a (p_B/p^\ominus)^b} \tag{11.2A}$$

where $p^\ominus$ is a standard pressure.

If $p^\ominus = 1$ atm and p_A, etc., are expressed in atmospheres, this reduces to

$$K_p^\ominus = \frac{p_L^l p_M^m}{p_A^a p_B^b} \tag{11.2}$$

If ΔG° is the Gibbs free energy change from reactants to products (individually at 1 atm. and at a temperature T in kelvin) we have

$$\ln K_p^\circ = -\Delta G^\circ / RT. \qquad (11.3)$$

(However, even at low partial pressures, carboxylic acids may be extensively dimerized in the vapour phase, and corrections are necessary; at higher pressures the fugacities of components must be obtained from published data or by applying a 'fugacity coefficient', $\phi = f/p$, derived from a generalized or compound-specific chart.)

11.3.2 *Effect of total pressure*

In equation 11.2, each partial pressure can be represented by the product of the mole fraction (y_A, etc.) and the total pressure (P in atm.)

hence

$$K_p = \frac{(y_L^l y_M^m \ldots) P^{(l+m\ldots)}}{(y_A^a y_B^b \ldots) P^{(a+b\ldots)}}$$

If $r = a + b + \cdots$ the number of moles of reactants
and $q = l + m + \cdots$ the number of moles of products

$$K_p P^{r-q} = \frac{y_L^l y_M^m \cdots}{y_A^a y_B^b \cdots}$$

Hence, the mole fractions of products will increase with increasing pressure if $r > q$, a relationship of considerable importance in seeking high conversions in methanol synthesis and ethylene hydration for example.

11.3.3 *Rough calculations*

The free energy of reaction, ΔG°, is derived from the change in enthalpy, ΔH°, and the change in entropy, ΔS°, by the equation:

$$\Delta G^\circ = \Delta H^\circ - T\Delta S^\circ$$

and $\ln K_p = \Delta S^\circ / R - \Delta H^\circ / RT$.

For *rough* purposes we can take ΔS° and ΔH° as constant for gas-phase reactions, while for many organic reactions, values of $\Delta S^\circ / R$ are of the order of $15(q-r)$, less 10 for C—C bond formation with cyclization (at a standard pressure of 1 atm. or 1 bar). We now have a very crude method for assessing feasibility based solely on (gas-phase) heats of formation.

Thus, the value of $\Delta H^\circ / R$ for the hydrogenation of an olefin (alkene),

$$RCH = CH_2 + H_2 \rightleftharpoons RCH_2CH_3$$

is typically of the order $-15\,000\,K$. (ΔH° is approximately -120 to $-130\,kJ/mol$.)

Therefore $\qquad\qquad \ln K_p \sim -15 + 15\,000/T$

At low temperatures, $\ln K_p$ will be positive, but will fall to zero at about 1 000 K, above which the reverse reaction becomes possible. In contrast, for the hydration of an olefin ($\Delta H^\ominus/R \sim -5\,500\,K$) $\ln K_p$ becomes negative ($K_p < 1$) at about 370 K, and the reverse reaction will be significant under all practical conditions.

11.3.4 *Thermodynamic traps*

In some well-known reactions, the quantity of 'catalyst' (or catalytic reagent) required is so large that it does affect the thermodynamic properties of reactants and products.

A particularly misleading situation can occur when a reaction product forms a complex with the catalytic material. Thus the Gatterman–Koch reaction of benzene and carbon monoxide to produce benzaldehyde is thermodynamically unfavourable, but reaction proceeds under pressure in the presence of at least 1 mole $AlCl_3$/mole benzene. However, the benzaldehyde–$AlCl_3$ complex formed must be cleaved *chemically*, with water for example, to liberate the product. Surprisingly, the similar conversion of toluene to *p*-tolualdehyde (*p*-methylbenzaldehyde) is favourable, and separation from the catalyst system can be achieved by simple distillation; workers with Mitsubishi in Japan have used BF_3/HF[2].

The above examples are from an area of traditional chemistry. However, in seeking to translate organometallic chemistry into catalytic processes, we may often find products or intermediates in the desired process scheme as ligands on the metal. In such circumstances, it is important to recognize that their formation confirms that a mechanistic pathway exists, but does not imply that the formation of the free product or intermediate is necessarily feasible under the conditions used.

11.4
Homogeneous catalysis

11.4.1 *General features*

In homogeneous catalysis, the catalyst is usually dissolved in a liquid reaction mixture, though some or all of the reactants may be introduced as gases, or even as solids. A small number of examples exist in which the reactants and catalyst are all in the vapour or gaseous state; one example is the 'cracking' of acetic (ethanoic) acid to ketene (ethenone) and water at about 700°C, with diethyl phosphate vapour as an acid catalyst.

In solution chemistry, considerable scope exists for accurate kinetic investigation, to throw light on mechanistic aspects. Further information can be derived from the kinetic behaviour of isotopically labelled or otherwise substituted reactants, catalysts or intermediates[3].

Changes in the catalytic species can often be followed by conventional analytical and spectroscopic methods, NMR, ESR, etc. Interest in the nature of catalytic species under actual working conditions has prompted the development of high-temperature and high-pressure cells for such techniques. It may also be possible to isolate and characterize the catalytic species, though the true catalyst may be a minor component or changes may occur during the isolation procedure. Nevertheless, such characterized species and their stoichiometric reactions can often make a major contribution to understanding the catalytic mechanism.

11.4.2 *Catalyst life and poisons*

In seeking to apply any new catalytic material the useful life of the catalyst is important. Loss of activity may be inherent i.e. an inevitable consequence of the chemical nature of the catalytic material, reactants and products, or a result of 'poisoning'.

Figure 11.1 shows some typical modes of physical loss and chemical inactivation of soluble catalysts within a reaction system (with the possible introduction of organic impurities into the products). Many such reactions are more readily identifiable by extending studies to extreme conditions. Rather more difficult to identify are reactions between catalytic materials and by-products which accumulate only after extended operation with recycle streams.

By catalyst 'poisons' we normally mean materials which do not form part of the defined process chemistry, but which gain entry into the reaction mixture and lead to permanent or temporary catalyst deactivation. In general, poisons are impurities in the chemical feedstocks or corrosion products from materials of construction. Many chemical feedstocks derived from petroleum contain traces of sulphur compounds, whilst chloride ion is almost ubiquitous (in the air at seaboard factory sites). These impurities, together with traces of organic impurities such as dienes and alkynes in alkenes, can again displace preferred ligands, destroy ligands (e.g. oxidation by traces of air) or compete with the reactant. Metal ions, in feedstocks or introduced by corrosion, may lead to progressive neutralization of an acid catalyst, or otherwise interfere with the catalyst performance; metal ions may also catalyse undesirable side reactions.

11.4.3 *Limitations*

In practice, there are a number of serious limitations to the application of homogeneous catalysis.

Firstly, no suitable catalyst may have been found in the past. For example, organometallic chemistry has introduced soluble and stable zero- or low-valency metal complexes, and derived hydrido-complexes, which catalyse hydrogenations in the liquid-phase. But with the enormous amount of accumulated general and company-specific information and experience in the

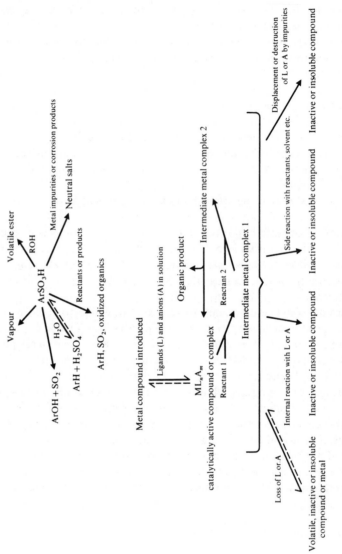

Figure 11.1   Some modes of deactivation of homogeneous catalysts during reaction or product separation.

heterogeneous catalysis of hydrogenation processes, there is little incentive for industrial organizations to evaluate novel (and expensive) homogeneous catalysts, except in a very small number of specific applications. We still have no homogeneous catalysts for many of the industrially important lower olefin (alkene) oxidation processes (e.g. acrylonitrile (propenonitrile) from propylene (propene), butadiene from butenes).

In other cases, homogeneous catalysis or mechanisms have been demonstrated, but rate, or more often thermodynamic, requirements can only be met at temperatures too high for liquid-phase operation. Gas-phase reaction over a heterogeneous catalyst permits a far wider range of reaction conditions.

However, the major problem encountered in homogeneous catalysis, particularly when the catalytic material is expensive, lies in the separation of reaction products from the catalyst such that the recovery process is efficient and does not impair catalytic activity. Thus we can afford to lose small, catalytic quantities of mineral acid, alkalis or base metal compounds. But only minute losses can be tolerated if a noble metal catalyst costs over £12 000 (say $20 000) per kg, and the product sells for about 40–70 pence (say 70–120 cents) per kg.

11.5
Heterogenization of homogeneous catalytic systems

A number of methods have been used to convert catalytic species into a heterogeneous form, and thereby simplify separation from the reaction mixture, but without changing drastically the nature of the catalytic species, or its immediate environment. To ensure that the original mechanistic pathway is unaffected, this category must exclude the substitution of a supported metal for a soluble metal compound or complex. The original modes of decay and interference by poisons are then similarly unaffected.

Two distinct approaches have been adopted: (a) the chemical attachment or tethering of the original catalytic species to a support material (applied mainly to reactions in the liquid phase), and (b) the physical absorption of the original soluble catalyst/solvent system into an inert porous material (applicable solely to gas-phase reactants and products).

Possibly the simplest example of the first approach would be the substitution of a sulphonated polystyrene-divinylbenzene polymer ion-exchange resin (in free acid form) for p-toluene sulphonic acid as the catalyst in an esterification or other acid-catalysed chemical reaction (see section 11.7.2). The ion-exchange resin may be used in bead form and packed into a tubular reactor, or may be introduced as a powder and subsequently filtered off. Furthermore, the solid resin may prove less corrosive to the metal walls of an industrial reaction vessel.

The physical nature of such resins, whereby the majority of sulphonic acid groups lie within the resin beads or particles, and are accessible only via narrow pores, introduces considerable resistance to the flow of reactants to, and products from, these catalytic 'sites'. Hence, in general, reaction rates are reduced compared with those attainable with an equivalent quantity of a soluble sulphonic acid, though 'macroreticular' resins with improved access for relatively large organic molecules are now available. As an ion-exchange material, the sulphonic acid resin efficiently scavenges the reaction mixture of traces of corrosion metal ions, thereby leading to slow neutralization of the sulphonic acid groups. Further, at high temperatures we encounter the typical modes of thermal and hydrolytic decomposition of aromatic sulphonic acids, together with the possibility of degradation of the polymeric structure itself. Nonetheless, several processes utilizing sulphonic acid ion-exchange resins as catalysts have now been commercialized in the petrochemical field.

A further example of the first approach to the heterogenization of organometallic complexes entails the chemical attachment of a ligand to an inorganic or organic support[4]. BP workers utilized the reaction of a silyl alkyl phosphine (amine, thiol) with the hydroxylic surface of a porous silica (alumina, etc.). Rhodium, for example, can then be attached by liganding to the phosphine groups, giving materials capable of catalysing hydrogenation and hydroformylation of alkenes. Finnish workers are seeking to prepare improved styrene polymers functionalized with groups designed to ligand catalytic metals[5]. Again, diffusional limitations can affect performance. At higher temperatures, slow loss of the catalytic metal may occur, by dissociation, displacement or cleavage of the ligand, and such materials have not found commercial use as yet.

To illustrate the second approach to heterogenization, we will consider the hydration of ethylene. Ethylene can be converted into ethanol by absorption into concentrated sulphuric acid, to form ethyl hydrogen sulphate, followed by dilution with water; the sulphuric acid is a regenerable reagent, rather than a catalyst, in this scheme. In contrast, phosphoric acid is less corrosive, non-oxidizing and can be used as a catalyst for hydration, though its lower acidity necessitates temperatures of over $250°C$ and pressures of 60–70 atm. to achieve acceptable rates and conversions. If a controlled quantity of phosphoric acid is absorbed into a siliceous material of appropriate pore-structure, the acid spreads over the internal surface of the support material and leaves an appreciable part of the pore volume unoccupied, permitting access to most of the acid solution by gaseous reactants. The material can be packed into a simple tubular reaction vessel. Obviously, the use of such a system is limited to operations in which reactants are introduced, and products removed, in a gaseous form; if liquid were to condense within the reaction vessel, a part of the phosphoric acid solution would be washed off the support material.

This type of catalytic material and mode of use provided the basis for the Shell process[6], the first direct hydration route to ethanol. The approach remains the basis for the manufacture of all synthetic ethanol and most of the isopropanol (2-propanol) produced today.

A similar approach has been adopted in a number of other reaction systems. A common example is the use of supported melts of copper chloride (with possibly other metal chlorides incorporated to reduce volatility or provide a lower melting-point eutectic mixture) for 'oxychlorination' processes, in which hydrogen chloride introduced into the reaction mixture, or formed in it, is reoxidized to chlorine *in situ* (section 11.7.7.6).

The maintenance of a supported liquid layer in gas-phase reactions is also important in other heterogeneous catalytic applications, such as the Bayer/Hoechst process for vinyl acetate manufacture[7]. However, in these systems, the catalytic metal is reduced to the metallic state, leading to significant mechanistic differences from the formally related homogeneous Wacker-type alkene oxidation/acetoxylation processes (section 11.7.7.3).

11.6.1 *Introduction*

11.6
Heterogeneous catalysis

A heterogeneous-catalyst is a solid composition which can effect or accelerate reaction by contact between its surface and either a liquid-phase reaction mixture (in which the catalytic material must be essentially insoluble) or gaseous reactants. In liquid-phase systems, one or more of the reactants may be introduced as a gas, but access of such reactants to the (fully wetted) surface of the catalyst is almost invariably by dissolution in the reaction medium and subsequent diffusion.

Although heterogeneous catalysis has been recognized for over 150 years and practical applications increased dramatically from the beginning of the present century, mechanisms were poorly understood and selection was based on trial-and-error or empirical rules. In the last 25 years or so many new techniques have been applied to characterize the materials and throw light on mechanistic aspects, though no overall theory has found universal acceptance.

We will start, therefore, with some of the more pragmatic aspects of heterogeneous catalysis, for which little theoretical background knowledge is required.

11.6.2 *Major (primary) and minor (secondary) components*

Many heterogeneous catalysts in commercial use contain several components, often referred to as the major (or primary) component, minor (or secondary) components and the support, with which we shall deal later.

While all components contribute to the overall performance of the catalytic material, the major component is essential for any activity in the type of reaction required. If several components show individual activity, the major component is often taken to be the most catalytically active material. (Two major components may be essential in bifunctional catalysts, section 11.7.4.)

With metal catalysts, classification is usually reasonably straightforward. However, with oxide catalysts none of the components may be individually active. In most cases of this type, a true mixed oxide, containing two metals (or metalloids) within the crystal structure, is active when used alone, and is regarded as the major component. Such a component may be named by joining the names of the two constituent oxides, e.g. silica-alumina or tin-antimony oxide, but often names of the type bismuth molybdate, cobalt tungstate, copper chromite are used. Such names should not be taken literally, as the two metals may not be present in a simple stoichiometric ratio or the composition may contain microcrystals of different structures each containing differing ratios of the two metals.

Minor (or secondary) components are introduced to modify the crystal structure or electronic properties of the major component, to improve activity, selectivity or thermal stability. The situation is highly complex, but some examples of secondary components, and their effects, will be included in subsequent sections.

11.6.3 *Operational modes*

In liquid-phase reactions the heterogeneous catalyst may be introduced in powder form into the reaction mixture, and maintained in suspension by stirring or other means. The powdered catalyst must be recovered sub-sequently by settling, filtration, centrifuging etc. The efficiency of this recovery process usually dictates the minimum particle size usable.

Alternatively, the catalytic material may be used in a coarser form (0·5 to 8 mm or more in diameter), and packed as a *fixed bed* into a tower or tubular reaction vessel. Practical considerations would be concerned with the means of retaining the catalytic material, while avoiding high pressure drops which might cause physical damage (crushing) to the catalyst beads, granules, pellets etc. It is sometimes possible to pass the liquid upwards through the catalytic bed so that the particles are slightly lifted and separated, but are not carried away in the supernatant liquid; such *expanded-bed* operation requires careful choice of particle size and liquid velocity for successful operation. When one of the reactants is gaseous, a *trickle-bed* configuration may be adopted, in which the liquid reaction medium trickles down over the surface of the catalytic material but the gas occupies most of the space between the granules (up or down flow).

When all the reactants and products are gaseous under reaction conditions, the fixed bed configuration is frequently adopted. However, if the chemical reaction is either strongly exothermic or strongly endothermic, the diameter of a tubular fixed bed may have to be limited to 25–50 mm (1–2 inches), to permit heat transfer through the wall and avoid excessive changes in temperature along the bed. Thus, a 'reactor' for the gas phase oxidation of *o*-xylene (1,2-dimethylbenzene) to phthalic (benzene-1,2-dicarboxylic) anhydride will actually comprise 10000 to 15000 parallel tubes of about 30 mm diameter, each containing coarse catalytic material, all surrounded by a circulating

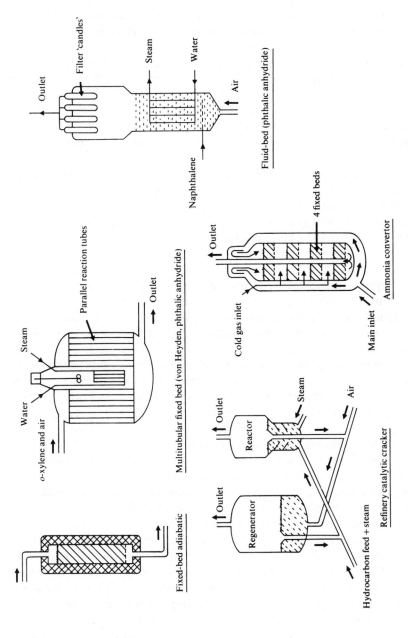

Figure 11.2 Various reactor types (schematic) for heterogeneous catalysts and gaseous reactants.

molten salt mixture to remove the heat (enthalpy) of reaction[8]. To simplify engineering design, a number of other modes of operation have been adopted—see Figure 11.2.

The most common form of *fluidized bed*[9] is similar to the expanded bed referred to above. Gaseous reactants pass upwards through a bed of small catalyst particles at a velocity which induces turbulence in the bed, to given good heat transfer to the walls and immersed cooling tubes, but does not carry away excessive quantities of the catalyst. (Filters may be fitted.) A further variant, used for the catalytic cracking of petroleum fractions with catalyst regeneration by oxidation, employs a transported bed. The catalyst is fluidized, but transported through the main reactor and separated from the emerging product stream. The catalytic material is again transported (in air) through a second reactor (regenerator) if required, before its return to the main reactor. (Compare this operation with the pneumatic transport of grain, powdered coal etc.)

When we move to very high-pressure exothermic processes (such as methanol and ammonia syntheses), different engineering approaches have resulted in a variety of other reactor types[10].

11.6.4 *Chemisorption and active sites*

Heterogeneous catalysis occurs on the surfaces of solid materials. These surfaces are almost invariably covered by various species—water, carbon dioxide, and other materials (including ions) from the preparative mixture or atmosphere. The majority of these species are very weakly adsorbed, or physisorbed, via van der Waals' forces. However, for a chemical reaction involving an otherwise stable molecule to be initiated, a significant electronic disturbance in that molecule, or even bond scission, must occur at the catalytic surface.

The adsorption process that brings about such a chemical modification, which may occur in several steps, is usually referred to as *chemisorption*. Furthermore, when two or more molecules are involved in a reaction on a catalytic surface, as in hydrogenation or oxidation processes, we usually find evidence that the major reactions occur between chemisorbed or surface species derived from each of the reactant molecules.

It was recognized early in the history of heterogeneous catalysis that, in many instances, only a relatively small proportion of the surface was catalytically active. The pre-exponential rate factor was seen to be very small in relation to likely collision frequencies of molecules adsorbed on the surface, taking into account steric requirements, while poisoning (inhibition of the catalysed reaction) could result from surprisingly low levels of specific impurities (see below). Hence the term *active sites* was coined to describe those localities on the surface which would induce the desired chemical reaction.

11.6.5 *Physical forms and their preparation*

Solid catalytic materials can be divided into two main groups:

(a) Bulk catalytic materials, in which the gross composition does not change significantly throughout the material, such as a silver wire mesh or a compressed pellet of 'bismuth molybdate' powder.
(b) Supported catalysts, in which the active catalytic material is dispersed over the surface of a porous solid, such as carbon or silica.

Bulk metals can be used in traditional engineering forms, more particularly as fine wire woven into gauzes. Such forms are generally used only in high-temperature processes, such as the partially oxidative dehydrogenation of methanol (over Ag) or ammonia oxidation (over Pt–Rh) at about 500–600°C and 850–900°C respectively. Mechanical stability is of greater importance than high surface area.

However, for high activity at modest temperatures, forms which present a much higher surface area to the reactants are highly desirable. Finely divided metal powders often show very high catalytic activity, but may present separation problems, whilst agglomeration often leads to a progressive loss in activity. Hence, we find the development of methods to produce coarser particles of metals in porous form, such as platinum sponge or, far more commonly, Raney nickel. (The Raney technique has been applied to other metals such as cobalt and copper.) However, the scope for producing physical forms with a high proportion of accessible metal atoms is seriously limited.

In the case of oxides of metals and metalloids, even small particles (0·1 mm diameter or less), whether naturally occurring or produced by precipitation, gelling, crystallization or powder reactions, are usually polycrystalline or agglomerates, with appreciable porosity (after driving off water and other weakly adsorbed materials). Quite often, by allowing precipitates or micro-crystals to stand for appreciable periods, further agglomeration will provide relatively large, porous granules, several millimetres across. Alternatively, larger *formed* catalysts in the shape of cylinders, spheres, rings, etc. can be produced from powder, optionally incorporating a small amount of a 'cement' or 'binder', by compression into moulds or extrusion of a slurry and drying. In all cases, preparative conditions affect surface properties. Most oxide catalytic materials provide (specific) surface areas from about 0·5 up to 700 m^2 per gram, more usually 10 to 300 m^2/g. The zeolites (molecular sieves), of course, also have very small channels of 0·4–1 nm (4–10 Å) diameter extending throughout the crystal structure[11].

There are two major reasons for the use of supports, to provide a stable extended surface over which an active component can be dispersed, and to confer mechanical strength. In either case, the desired property may well have to be maintained to a high temperture.

Synthetic silicas, aluminas and carbon blacks—there are many variations in form, surface area, impurity levels and crystal structure (for aluminas)—are probably the most common supports for metals. Their use with noble metals can provide a high *dispersion* (the proportion of metal atoms on to which a reactant can be adsorbed), and thereby provide a very considerable economic benefit. However, these materials, together with calcined clays and diatomaceous earths, are also commonly used as supports for less expensive metals, such as nickel and copper, for use at higher temperatures where sintering of base metals can occur.

Preparation most frequently involves absorption of a solution of a suitable metal compound into the support, followed by drying, thermal decomposition and reduction. This treatment, and the reductant employed (such as hydrazine in solution, or hydrogen gas), often has a significant effect on the physical form and dispersion of the resulting metal. There are few rules, and this area is far too complex for discussion here.

When we turn to metal-oxide catalysts, a few of low porosity, such as zinc oxide, benefit from supporting the finely divided material. However, a great many metal-oxide systems of acceptable porosity prove too soft for commercial handling and use; often only small additions (15–30%) of a hard 'support' will confer adequate strength. Preparative techniques include absorption of salt solutions into the support (followed by calcination to oxides), precipitation of hydroxides in the presence of support material, mixed powder compression, slurry extrusion and coating or spraying on to low-porosity support materials (quartz, glazed ceramic). The preparative method and conditions again significantly affect performances.

11.6.6 *Support interactions*

While the practical reasons referred to above represented the main driving force for the use of supports, this is by no means the whole story. As all the surface is rarely active, the support material can act as a heat sink or source for active sites or zones in exothermic and endothermic reactions, while increasing the total area for heat transfer between solid material and the gas (or liquid) phase.

The support material may have catalytic properties in its own right, either useful or otherwise; hence exposed areas of the support can influence the selectivity of the overall process, and special pretreatments may be required to enhance or inhibit such behaviour, as appropriate. More importantly, however, the support may induce crystallographic modification of a thin layer of overlying metal or oxide, or electron transfers may occur between the support and the supported catalytic material, leading to significant changes in the structure and electronic (acidic/basic or electrophilic/nucleophilic) properties of the latter material, and hence its catalytic behaviour. When the primary component is a metal, the term strong metal-support interaction, often abbreviated to SMSI, has recently appeared in the literature.

Such chemical interactions have prompted extensive investigations of the influence of support materials on overall catalytic behaviour in many systems, and have extended considerably the range of mainly oxide materials used as supports in commercial catalysts. In many systems the so-called support material is thus an important part of the complete catalyst formulation, fulfilling a dual role, though usually retaining crystallographic identity except at boundaries with other components.

11.6.7 Catalyst structure

The surface area, pore structure and chemical composition of the surface are important parameters of any support material or solid catalyst. Even when mechanistic interpretation is not a primary aim, many techniques are now applied in industrial laboratories to establish correlations between these parameters and the performance of specific catalysts. Such techniques may also be used on a routine basis to monitor the reproducibility of purchased materials and catalyst preparation methods.

The *pore volume* of a solid material can be estimated from the actual density and the true density of the constituents. More frequently, the absorption of a suitable liquid into a (pre-evacuated) sample is used as a guide.

Surface areas are usually obtained by the *BET Method*[12]. The adsorption isotherm (at $-196°C$) for nitrogen onto the thoroughly de-gassed surface is measured; a 'knee' at monolayer coverage provides a value for the *specific surface area*, while the complete curve gives information on the *pore size distribution*, particularly for micropores of less than 50 nm diameter. Low-temperature gas-solid chromatographic (GSC) techniques have also been applied[13]. Information on the size distribution of larger pores can be obtained by forcing mercury into the pores, a technique known as *mercury porosimetry*. The *tortuosity* of larger pores through formed or shaped catalysts (particularly cylinders) can be estimated from the pressure differential required to establish a gas flow through a suitably mounted sample.

With a supported catalyst system, the information derived by the above methods relates to the composition as a whole. A number of methods have been applied to measure the sizes of metal crystallites on the support surface, and hence estimate the dispersion or surface area of the metal itself. These include electron microscopy, X-ray diffraction and Mössbauer spectroscopy. More direct measurements of metal surface area have been obtained from adsorption isotherms for hydrogen, carbon monoxide and other adsorbates, by assuming that all the exposed metal surface adsorbs such gases. GSC methods can again be applied.

A host of spectroscopic techniques have now been applied to obtain compositional, structural and valency information for the topmost layers of catalyst compositions. These include traditional methods, such as UV/visible reflectance, IR and Raman spectroscopy. X-ray fluorescence spectroscopy

(XRF) has been applied to catalytic samples or surface rubbings. A potentially more informative group of modern techniques includes low-energy electron diffraction (LEED), X-ray photoelectron spectroscopy (also known as electron spectroscopy for chemical analysis, ESCA) and Auger electron spectroscopy (AES)[14]. LEED provides lattice information, and led to the discovery that the surfaces of some metals and most semiconductor oxides have structures which differ from those projected by simple extension of bulk geometry. ESCA and AES provide moderately quantitative compositional information, and indications of valencies or bonding states. The former has shown that some components may be present in excess, or be deficient, at the catalyst surface. However, *no* single technique can be regarded as definitive.

11.6.8 *General kinetic behaviour*

Rate data have often been approximated in terms of simple powers of reactant concentrations or partial pressures, but more widely applicable rate equations were derived by Hinshelwood and others to take into account competition between reactants, and possibly intermediates and products, for active sites[3]. The form of the best equation may provide some mechanistic guidance, but when the process is multistep, with desorption/readsorption of intermediates, the kinetics may be extremely complex.

Normally, the dependence of the rate constant on temperature takes the traditional Arrhenius form, $k = A \exp(-E/RT)$. However, some systems, particularly oxide catalysts in gas-phase processes, can show a change in activation energy with temperature; this may reflect a multistage process or changes in the composition (or oxidation state) of the catalytic surface, which may also be time-dependent. If the chemical process is strongly exothermic, such as the oxidation of o-xylene (1, 2-dimethylbenzene) over vanadia catalysts, heat-transfer limitations lead to appreciable variations in the catalyst and gas temperatures along the catalytic bed. Considerable effort has been devoted to the mathematical modelling of interactions between kinetic and heat-transfer parameters for such systems, aimed at improved control and reactor designs.

A common feature of heterogeneous catalysis is an increase in rate with increasing subdivision of the catalytic material. This arises from increasing accessibility of the surface, and reduction in diffusional constraints between reactants and catalytic sites. In general, continued subdivision will eventually lead to a levelling off in the reaction rate at a value dictated solely by adsorption and chemical processes. For simple, selective reactions (olefin hydrogenation, oxidation of sulphur dioxide), the ratio of the reaction rate with a practical catalyst form to the maximum rate attainable by mechanical subdivision is often referred to as the *effectiveness factor*.

When parallel reactions or further reaction of the desired product occur, as in most oxidation processes, larger catalyst particles can also affect selectivity.

Differences in diffusion rates of individual reactants lead to changes in their ratios within deep pores, which also favour further reaction of a product in the pore network. Therefore, some catalysts have been developed with a 'shell' structure, in which only the outermost zone of a porous support is loaded with active material or an impervious central core is coated with the catalytic components. All such effects are, of course, very important in scale-up for commercial processes, for which relatively large granules may be desirable.

Not all pore diffusional effects are necessarily undesirable. The molecular sieving action of zeolites, as catalysts or supports, can lead to preferential reaction of small or linear molecules (such as *n*-paraffins) in complex mixtures, or modify the product distribution in other ways (cage effects). In the latter context, the zeolite ZSM-5 shows exceptionally high activity for the cyclization and aromatization of hydrocarbons[15].

11.6.9 *Catalyst deactivation and life*

In general the activity of any catalyst falls off with time. Ultimately the conversion of reactants or production rate reaches an unacceptable level, or a compensatory temperature increase degrades selectivity. Long life is especially important for expensive catalytic materials; deactivated precious metals may be recoverable, but the credit will only partially offset the cost of new material. With fixed beds, even of cheap catalysts, considerable labour costs are incurred in discharging and recharging the reaction vessel, while loss of production during these operations will mean a loss of income (and possibly customers).

Loss of catalytic activity can occur in several ways, but firstly we will consider simple physical loss. Particulate catalysts used in agitated liquid phases or fluidized beds are liable to both wear and fracture through collisions of the particles with each other, the vessel walls and fittings. The process is usually referred to as *attrition*. The finest particles formed tend to escape the main separation or filtration equipment, and continual make-up of the catalyst charge is required.

Losses can also arise from dissolution or volatilization of catalytic components. Phosphoric acid is slightly volatile in the gas phase hydration of ethylene, leading to its depletion in the catalytic material at the reactor inlet and deposition in downstream equipment. Similar problems occur with alkali metal acetates and, in the presence of steam, alkali metal hydroxides. Some compensation may be possible by introducing a small quantity of the appropriate material with the reactants. Nickel and palladium catalysts are subject to slow dissolution in unsaturated fatty acids if the hydrogen concentration falls too low locally, for example within deep pores. Surprisingly, even the precious metals in platinum-rhodium gauzes used for ammonia oxidation slowly volatilize, owing to the formation of uncharacterized gaseous oxides at high temperatures.

A far more common mechanism for catalyst deactivation in high-temperature gas phase reactions is *sintering* of the support or active material, leading to a reduction in the effective surface area. Many metals and oxides begin to sinter at a temperature equal to about half the melting point in kelvin (the *Tamman* temperature), though surface diffusion of atoms can occur at still lower temperatures. Thus silica should be stable at temperatures up to about 500°C, but structural changes still occur at lower temperatures under high partial pressures of water vapour. The presence of certain components in the gas phase (often oxygen or water) can again accelerate sintering or growth of metal crystallites. Fortunately, alloys of the principal metal with a second, often catalytically inactive, metal may provide more thermally stable dispersions; examples include supported Pt–Ir, Pt–Re, Pd–Au and Ni–Cu. Small changes in catalytic behaviour may also result.

As in the case of homogeneous catalysis, poisons can also lead to deactivation of heterogeneous catalysts. Soluble or volatile metal or nitrogen compounds can destroy acid sites, while carbon monoxide and sulphur compounds almost invariably poison nickel and noble metal hydrogenation catalysts by bonding strongly with surface metal atoms. These considerations often lead to the selection of less active, but more poison-resistant, catalysts for industrial use.

Slow catalyst deactivation may also result if certain impurities, present in commercial grades of chemicals used for catalyst preparation, migrate to the catalyst surface in use.

Finally, the formation of deposits on the surface and pore blockage may introduce a physical barrier between reactants and catalytic sites. In acid-catalysed reactions of olefins at modest temperatures, insoluble or involatile polymers, derived from the reactant olefin or impurities, may accumulate slowly. However, *coke laydown* is a much faster process in the high temperature, gas-phase reactions of hydrocarbons over acid- and metal-acid catalysts. The coke can be burned off, but only at higher temperatures; hence the adoption of continuous regeneration in catalytic cracking, the incorporation of additives to accelerate burn-off, and the advantages of using the more thermally stable zeolites as catalysts or supports.

11.6.10 *Studies on surface chemistry*

More specific information on the nature of active sites can often be deduced from adsorption isotherms and heat of adsorption distribution measurements (by calorimetry or GSC)[13] with individual reactant gases or vapours. These can be backed up with many of the spectroscopic techniques referred to above, to characterize the nature of the adsorbed layer and chemisorbed species, and their effects on the structure and oxidation state of the surface itself[14].

Acidity is frequently an important parameter in the chemisorption process or general catalytic behaviour. The extent (and heat) of adsorption of various bases, such as ammonia, pyridine and hydrocarbons (e.g. triphenylmethane and perylene), again backed up with spectroscopic techniques, and a titrimetric (H_0 indicator) method for n-butylamine adsorption, have all proved informative. Similarly, basic sites can be quantified by adsorption of acidic materials. (Total surface hydroxyl can often be assessed by NMR and deuterium displacement for comparison). Kinetic studies on partially neutralized samples can point to the acid/base strength required to effect a particular reaction. These are specific examples of a general approach, in which the effects of selected poisons can throw light on the useful catalytic area and mechanism.

Isotopic labelling of reactants, or a switch to labelled reactant during reaction, can give information on kinetically limiting steps, symmetry patterns in intermediates, the involvement of lattice oxygen, etc. Microreactors and pulse techniques are usually most suitable for such work.

The co-adsorption of reactants at below normal reaction temperature, with temperature-programmed desorption of reactants, intermediates and products, has also received increasing attention in recent years. GSC techniques again find use in this context.

Simulation of catalytic reactions on metal films allows closer spectroscopic examination of the reacting system. A more recent step from convention is the use of molecular beams.

11.6.11 *Theoretical approaches*

Early ideas on heterogeneous catalysis assumed the need for appropriate spacing of surface atoms to effect bond breaking and making, the *geometric concept*. However, chemical bonds in liquids and gases are by no means rigid, while the use of crystal data for the solid can be misleading as the surface structure may differ or show some ionic mobility.

The *electronic, semiconductor*, or *band theory*, as developed during the 1950s and early 60s, treats the catalytic surface of an oxide as a general entity. Reactions on the surface are effected by transfer of electrons into and out of the acceptor bands of a *p*-type semiconductor (or vice-versa for *n*-type donor bands). The theory showed modest success in rationalizing the hydrogenation and oxidation performance of some oxides, but is now considered inadequate for general use. However, as electron transfers *are* involved in such processes, the idea of acceptor bands warrants incorporation into many mechanistic interpretations, and semiconductor theory can provide useful guidelines for the selection of minor components to improve performance.

A more sophisticated approach starts from molecular orbital theory, which in itself proves too tedious for all but simple molecules and complexes. However, a combination of *electrostatic theory* and *crystal field theory* has proved valuable in relating changes in the stabilization energy of *d*-orbitals with changes in coordination patterns and the number of *d*-electrons. In particular, no changes in energy are predicted for metal ions with zero, five or ten d-electrons, pointing to a twin-peaked pattern of activity. Such a pattern has indeed been observed in many catalytic reactions over single oxides, but little real progress can be made to define the chemical reactions on the surface.

Table 11.1 Some industrial catalytic systems

Reaction	Homogeneous	Heterogeneous (Major components only)
Acid-catalysed		
Esterification	$ArSO_3H$;$Ti(OR)_4$, etc.	
Cumene (isopropybenzene) hydroperoxide rearrangement	H_2SO_4	
Methanol to gasoline		Zeolite (ZSM-5)
Hydration of olefins	H_2SO_4	Acid ion-exchange resin
Alkylation of benzene	$AlCl_3$-HCl	Supported H_3PO_4
Alkylation of isobutane	H_2SO_4 or HF	Supported H_3PO_4
Catalytic cracking		Silica-alumina; Zeolites
Hydrogenation		
General olefin, aromatic		Supported Ni or noble metal
Desulphurization		Co-Mo oxides
Dehydrogenation		
Butane → butenes		Chromia-alumina
Ethylbenzene → styrene (phenylethene)		Fe_2O_3-Cr_2O_3-K_2CO_3
Isopropanol → acetone (propanone)		ZnO
Dual-function (acid-catalysed reactions with hydrogen transfer)		
Isomerization of alkanes		Pt-acidified Al_2O_3
Catalytic reforming		Pt-Re-acidified Al_2O_3
Other metal-catalysed		
Oligomerization of C_2H_4	$AlEt_3$; Ni(O) complexes	
Polymerization of C_2H_4		Supported chromium (II) oxide; supported Ziegler
Oxidations		
p-Xylene (1,4-dimethylbenzene) → terephthalic (benzene-1,4-dicarboxylic) acid	Co-Mn bromides	
C_2H_4 → acetaldehyde (ethanal)	$PdCl_2$-$CuCl_2$	
C_2H_4 + HOAc → vinyl acetate (ethenyl ethanoate)		Supported Pd(-Au)-KOAc
C_2H_4 → ethylene oxide (oxiran)		Supported Ag(-Cl)
CH_3OH → HCHO		Ag (partial ox.); Fe-Mo oxides
C_3H_6 + NH_3 → acrylonitrile (propenonitrile)		Bi-Mo; U-Sb; Sn-Sb oxides
C_4H_8 → butadiene		Ferrite spinels
o-Xylene (1,2-dimethyl benzene) → phthalic (benzene-1,2-dicarboxylic) anhydride		Supported V_2O_5-TiO_2
C_2H_4 + 2HCl → $ClCH_2CH_2Cl$		Supported $CuCl_2$ melt
Carbon monoxide chemistry		
$CO + 3H_2 \rightleftharpoons CH_4 + 3H_2O$		Supported Ni
$CO + H_2$ → paraffins (alkanes), etc.		Iron oxide (promoted)
$CO + 2H_2O \rightarrow CH_3OH$		Cu-ZnO (-Cr_2O_3, etc.)
$CO + H_2O \rightleftharpoons CO_2 + H_2$		Fe_3O_4; Cu-ZnO
Olefin + CO/H_2 → aldehyde	Co; Rh-phosphine	
$CH_3OH + CO$ → acetic (ethanoic) acid	Rh-CH_3I	
$CH_3OAc + CO$ → acetic (ethanoic) anhydride	Rh(Ni)-I-base	
$(CH_3OAc + CO/H_2 \rightarrow CH_3CH(OAc)_2$	Rh(Ni)-I-base)	

Furthermore, the majority of practical oxide catalysts contain several metal species.

11.7.1 Introduction

The following sections are by no means comprehensive, but are intended to illustrate the scope of catalysis. Table 11.1 presents a partial summary of petrochemical applications.

These highlight some chemical advantages of heterogeneous catalysis.

(a) A very wide range of reaction temperatures is usable in gas phase reactions, allowing many kinetically difficult reactions to be effected and thermodynamic boundaries to be crossed.
(b) Heterogeneous catalysts can often be tailored empirically to give a wider range of products from an individual feedstock.

However, extensive use of heterogeneous catalysts has shown up a number of problem areas.

(a) Where a reaction can be effected with both homogeneous and heterogeneous catalysts, the latter are often less active, and the compensatory use of higher temperatures may degrade selectivity.
(b) Heterogeneous catalyst systems are less well defined, and the structural and electronic properties are subject to disturbance by trace impurities. Hence reproducibility may prove problematical.

The main emphasis here is on mechanisms, and few process details are included (see Chapter 12 for the latter). Whenever possible, related homogeneous and heterogeneous catalytic processes have been brought together for comparison. However, many of the mechanistic equations, diagrammatic representations and catalytic cycles have been greatly simplified, and details often continue to attract considerable debate.

The absence of a section devoted to polymerization processes may be noted. However, the majority of these, including the production of low density polyethylene and the polymerization of vinyl chloride, vinyl acetate, acrylonitrile, butadiene and styrene for example, involve initiated free-radical chain reactions which are considered to lie outside the scope of the present chapter. Despite their growing importance, only brief reference is made to olefin polymerizations by metal complexes, as kinetic and other characteristics may depart significantly from those associated with traditional catalytic processes.

11.7.2 Acid catalysis

Acid catalysis finds very wide industrial application in refinery operations and chemicals production. Both Brönsted acids (proton donors) and Lewis acids (electron acceptors) are used; the latter are often precursors for Brönsted acids or related species (traditional Friedel–Crafts' systems);

$$AlCl_3 + H_2O \rightleftharpoons H^+[HOAlCl_3]^-$$
$$AlCl_3 + EtCl \rightleftharpoons Et^+[AlCl_4]^- (\rightleftharpoons C_2H_4 + H^+[AlCl_4]^-)$$
$$BF_3 + HF \rightleftharpoons H^+BF_4^-$$

The author has adopted the negative of the H_0 scale as a convenient means of comparing the acidity of widely different acid systems, though rates of acid-catalysed reactions are rarely simply related to acidity on this $-H_0$ scale. Some values for strong (anhydrous) acid systems are listed below.

System	$-H_0$
HF	10·2
H_2SO_4	12·0
CF_3SO_3H	13·1
FSO_3H	15·0
$AlCl_3/HCl$ or $GaCl_3/HCl$	14 to 15
SbF_5/HF	up to 20

The value of $-H_0$ associated with the hydrated proton is about 1·7, and water acts as a base in systems of higher acidity. Alcohols and ethers, ketones and esters also act as bases at progressively higher values of $-H_0$.

Earlier reference has been made to sulphonic acid resins and supported mineral acids as heterogeneous catalysts. The chlorination of alumina provides a strongly acidic surface, while the performance of a number of 'solid superacid catalysts' ($-H_0$ values up to 16) has recently been reported[16]. However a number of mixed oxides of metals and metalloids also show useful acidic properties.

Differences in coordination number become significant in mixed oxide systems, where we find many examples of enhanced Lewis or Brönsted acidity. even for binaries such as SiO_2–MgO[17]. However, the best known and possibly most remarkable system of this type is silica–alumina. Typical materials have acid titres of the order 1 milliequivalent/g, but only a small proportion of the acid sites show $-H_0$ values of 3 or more.

A reduction in alumina content enhances acid strength up to a point, but a reduction in physical strength and thermal stability limits the utility of such materials. A breakthrough came with the synthesis of the low-alumina-content zeolites, by the growth of highly regular silica–alumina crystals around bulky quaternary ammonium ions. These materials provide nearly 1 milliequivalent/g of sites of $-H_0$ from 3 to 10 or more, while showing markedly improved thermal stability[11].

11.7.2.1 *Esterification and ester hydrolysis.* A considerable number of commercial esterifications are still carried out with conventional acid catalysis (for which a kinetic equation based on proton sharing is most appropriate). However, acid catalysts promote side reactions, and 'uncatalysed' esterifications become very slow at high conversions. Hence, for the production of high-quality esters, a large family of metal compounds are now used as esterification catalysts. These include the soluble alkoxides and carboxylates, and insoluble oxides and oxalates of titanium, aluminium, zinc, cadmium, antimony, tin and many other metals. The phenomenon is often referred to as amphoteric catalysis, and is related to the coordinating capability, i.e. Lewis acidity, of the metal towards the acyl $C=O$ group; electron withdrawal permits nucleophilic attack by alcohol (or alkoxide) in a similar, but less effective, manner to protonation. Temperatures required are of the order $200°C$ or more; essentially all such processes are carried out in the liquid phase, under pressure if necessary.

11.7.2.2 *Rearrangement of oxonium ions*. In the acid-catalysed cleavage of cumene hydroperoxide (to phenol and acetone), an important step is aryl transfer from carbon to oxygen in the intermediate oxonium ion:

$$Ph-\underset{\underset{Me}{|}}{\overset{\overset{Me}{|}}{C}}-O\overset{+}{O}H_2 \xrightarrow{-H_2O} Ph-\underset{\underset{Me}{|}}{\overset{\overset{Me}{|}}{C}}-\overset{+}{O} \longrightarrow \underset{/}{\overset{Me}{\diagdown}}C-OPh$$

Alkyl transfers from O to C (Stevens rearrangement), carbenes and methyl carbonium ions have all been postulated to explain the formation of lower olefins from methanol and dimethyl ether over heterogeneous acid catalysts[18]; the reaction is autocatalytic, e.g.

$$Me\overset{+}{O}H_2 + MeOH \rightleftharpoons Me_2\overset{+}{O}H + H_2O \rightleftharpoons Me_2O + H_3\overset{+}{O}$$

$$\left.\begin{array}{l} Me\overset{+}{O}H_2 + Me_2O \rightarrow Me_3\overset{+}{O} + H_2O \\ Me_3\overset{+}{O} \rightarrow EtMe\overset{+}{O}H \rightleftharpoons C_2H_4 + Me\overset{+}{O}H_2 \end{array}\right\} \text{ slow}$$

$$Me\overset{+}{O}H_2 + EtOMe \rightarrow Et\overset{+}{O}Me_2 \rightarrow PrMe\overset{+}{O}H \text{ etc. (faster)}$$

$$(\rightarrow \text{propylene})$$

These reactions represent the first steps in the conversion of methanol to hydrocarbon fuels over Mobil's ZSM-5 catalyst; further reactions of the olefins are described in section 11.7.2.5.

11.7.2.3 *Formation of carbonium ions from olefins (alkenes)*. Many industrial reactions of olefins involve protonation to give a carbonium ion, which is subject to nucleophilic attack, followed by proton transfer from the product to olefin. The ease of protonation follows the stability of the carbonium ion formed in the sequence tertiary > secondary > primary. Additional proton exchanges can occur at any stage in the overall process, leading to double-bond shifts in the olefinic feedstock and mixed products in some cases. (At high temperatures, products with terminal substituents may also be detectable).

11.7.2.4 *Hydration and etherification.* The direct hydration of ethylene has been discussed (section 11.5). Propylene can also be hydrated in the gas phase over supported phosphoric acid (*c.* 180°C), or with an ion-exchange resin catalyst at about 140°C, with liquid water and gaseous propylene. The use of an ion-exchange resin as a catalyst has also been commercialized for the hydration of *n*-butenes, though the sulphuric acid two-stage process still predominates. The use of very weak acid systems at much higher temperatures ($> 250°C$) has also been studied.

In all the above processes, the corresponding symmetrical ethers are co-produced by the reaction of product alcohol with the carbonium ion. Ion-exchange resins protonate *iso*-butene in methanol (at below $100°C$) to produce methyl *tert*-butyl ether (MTBE).

Despite the reversibility of these reactions, many acidic oxides catalyse the dehydration of alcohols but show no significant activity for olefin hydration. The high partial pressure of water thermodynamically necessary leads to excessive surface coverage by water, with a marked fall in effective acidity.

11.7.2.5 *Carbon–carbon bond formation and cleavage.* In olefin hydration processes, minor amounts of dimers and polymers are produced by the mechanism

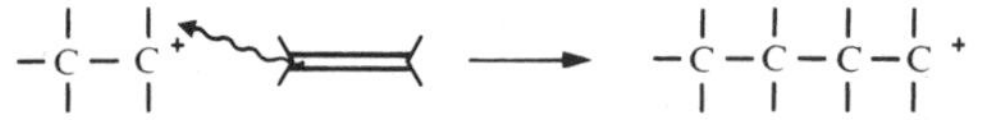

Under anhydrous conditions, traditional mineral acid and Friedel–Crafts' systems (liquid phase), as well as supported phosphoric acid (gas phase), can be used to produce dimers and trimers through to relatively high molecular weight viscous liquid polymers from C_3 and C_4 olefins. These same catalyst systems are also used in the alkylation of aromatic hydrocarbons.

Branched alkanes are also alkylated by lower olefins in the presence of concentrated sulphuric acid or anhydrous HF at near-ambient temperatures; an additional reaction, *hydride transfer*, is involved. If we consider the reaction of propylene and *iso*-butane (in excess), the chain reaction sequence is as follows:

$$MeCH=CH_2 + H^+ \rightarrow Me_2CH^+ \qquad \text{(protonation)}$$

$$Me_2CH^+ + Me_3CH \rightarrow Me_2CH_2 + Me_3C^+ \qquad \text{(hydride transfer)}$$

$$Me_3C^+ + CH_2=CHMe \rightarrow Me_3C \cdot CH_2\overset{+}{C}HMe$$

$$Me_3CCH_2\overset{+}{C}HMe + Me_3CH \rightarrow Me_3CCH_2CH_2Me + Me_3C^+$$

(alkylation)

All the above reactions are reversible. Hence, at higher temperatures with zeolites, cleavages of olefins and isomerization and trans-alkylation of alkylaromatics can occur; in the presence of alkenes and alkylaromatics as hydride acceptors, alkanes can also take part.

If we generate a hydrocarbon of appropriate structure with both unsaturation and a protonated centre, then intermolecular addition,

cyclization, can occur. C_5 and C_6 ring systems are most favoured thermodynamically. Further hydride transfers to olefins can lead to aromatization of
C_6 ring systems. Thus, under appropriate conditions, lower olefins can be
converted to mixtures of alkanes and aromatic hydrocarbons by acid catalysis
alone. Such reactions form the basis for aromatics production from intermediate olefins in the homologation of methanol over Mobil's ZSM-5
zeolite (section 11.7.2.2). Further hydride transfers from polyaromatic
hydrocarbons lead to 'coke'.

11.7.2.6 *Koch reaction.* Carbonium ions react with carbon monoxide (under
pressure) to form acyl cations; the overall reaction from isobutene gives 2, 2-
dimethylpropionic (2, 2-dimethyl propanoic) acid, for example.

$$R^+ + CO \rightleftharpoons RCO^+ \xrightarrow{H_2O} RCO_2H + H^+$$

Traditionally, concentrated sulphuric acid has been used as the reaction
medium, necessitating dilution with water to give the carboxylic acid, and
reconcentration of the mineral acid. The more direct reaction of olefin, water
and carbon monoxide in the presence of BF_3 requires much higher temperatures and pressures, but has possibly become the preferred system.

11.7.2.7 *Carbonium ion rearrangements.* At low temperatures, strong acids
($-H_0$ about 10) induce methyl shifts in branched alkanes (a hydride acceptor
must be present to form the carbonium ion).

Similarly, methyl shifts on aromatic rings are relatively facile. Under the
severe conditions used in refinery processes, more dramatic rearrangements
occur towards thermodynamically favoured highly branched products.

11.7.3 *Hydrogenation*

The two critical steps in alkene hydrogenation by metal complexes are now
moderately well defined; recent reports of alkane activation essentially
confirm that the final step is also reversible[19]. (Di-hydrogen addition can
occur at various stages to complete the cycle).

There have been few reports of carbonyl hydrogenation with metal complexes.

However, alkoxy-metal intermediates have been proposed in hydrogen transfers from alcohols.

Similar mechanisms are postulated for commercial alkene/arene, carbonyl and nitrile hydrogenations on metal surfaces; in particular, individual metal atoms are involved. In contrast hydrogenolysis, the cleavage of C—C or C—O (N, S, etc.) bonds, appears to need two or more adjacent sites and can sometimes be reduced by alloying the main component (addition of copper to nickel, for example). The stability of supported metal (especially platinum) catalysts permits their use at high temperatures, to promote hydrogen transfers between alkanes, alkenes and arenes or dehydrogenation processes.

However, a number of metal oxides also show high hydrogenation/dehydrogenation activity at higher temperatures, and find extensive commercial use in dehydrogenation processes. The oxides of chromium, iron and zinc, among others, are common catalyst components for the dehydrogenation of alkanes, alkenes (e.g. butane or butene to butadiene), ethylbenzene and secondary alcohols (for ketones). Alkali 'promoters' are often added to eliminate side reactions caused by Brönsted acid sites, but the slightly reduced surfaces show Lewis acid behaviour and semiconductor properties. A possible scheme for the dehydrogenation of isopropanol over zinc oxide involves proton transfers, thus:

$$Me_2CH - O - H \qquad Me_2C=O \qquad\qquad H_2$$
$$O^{2-} - M - O^{2-} \longrightarrow \bar{O}H - M - O\bar{H} \longrightarrow O^{2-} - M - O^{2-}$$
$$\qquad\qquad\qquad 2e \text{ to acceptor band} \qquad 2e$$

(effective reaction is $2OH^- + 2e \rightarrow 2O^{2-} + H_2$)

Metal-hydride bonding may also be involved. More readily reducible mixed oxides, such as copper chromite, possibly present a more metallic surface, but with an electron acceptor band of the semiconductor type.

Metal sulphides are also used in hydrogenation processes. Thus nickel sulphide (Ni_3S_2) permits reduction of alkynes and dienes to alkenes. Various tungstates and molybdates (particularly of cobalt) are used in sulphided form for the hydrogenolysis of sulphur compounds, and the saturation of sulphur-containing refinery streams.

11.7.4 Dual-function catalysis

For hydrocarbon reactions, metals (particularly platinum and its alloys) are frequently applied to acidic supports to catalyse hydrogen transfers. Thus platinum on a chlorinated alumina support accelerates the acid catalysed isomerization of n-alkanes (at about 150°C). In hydrocracking, the metal catalyses hydrogenation of heavy aromatic and polyaromatic components; the resulting cycloparaffins (cycloalkanes) undergo zeolitic cracking, with olefin hydrogenation, to give paraffinic naphthas (mainly C_5 to C_8). In the *catalytic reforming* of naphthas the presence of a Pt–Ir or Pt–Re alloy on an acidic alumina (with halogen in the feed), leads to fast dehydrogenation, cyclization and aromatization of the paraffinic hydrocarbons without the cage effect which promotes such reactions on the ZSM-5 zeolite.

Metallic components have also been added to a variety of heterogeneous oxide catalysts, to introduce additional hydrogenation, dehydrogenation and hydrogen transfer processes during aldolization, ketonization or Tishchenko reactions. Examples include acetone (propanone) to 4-methyl-pentan-2-one, ethanol to acetone and methanol to methyl formate (methyl methanoate), e.g.

$$CH_3OH \xrightarrow{Cu(ZrO_2)} HCHO + H_2 \qquad 2HCHO \xrightarrow{(Cu)ZrO_2} HCO_2CH_3$$

11.7.5 Olefin (alkene) polymerization and dismutation on metals

Although the addition of olefins to aluminium hydrides is not readily reversible, exchange occurs between aluminium alkyl groups and olefins. Olefin insertions into the metal–alkyl bond extend the alkyl chain, followed by displacement by the reactant olefin to give relatively low molecular weight polymers (up to $c.$ C_{20} linear alpha-olefins from ethylene).

$$\rangle AlEt + nC_2H_4 \longrightarrow \rangle Al(C_2H_4)_nEt \xrightarrow{C_2H_4} \rangle AlEt + CH_2 = CH(C_2H_4)_{n-1}Et$$

Phillips' supported chromium (II) catalyst, the most commonly used for high density polyethylene (HDPE) manufacture, possibly behaves in a similar manner, but the olefin insertion reaction is faster by several orders of magnitude. In the original Zeigler catalyst systems for HDPE, an aluminium alkyl is used to reductively alkylate the primary component, most frequently a titanium compound, to give the true catalytic species.

$$AlR_3 + \equiv Ti\text{-}X \rightarrow XAlR_2 + \equiv TiR \xrightarrow{nC_2H_4} \equiv Ti(C_2H_4)_nR (X = Cl \text{ or } OR')$$

A similar scheme, albeit by activating supported complexes of titanium, zirconium, etc., appears to describe the behaviour of the new 'metallocene catalysts', claimed to allow greater control and a potentially exciting new range of polymer properties[20].

Aluminium and other metal alkyls also activate tungsten and molybdenum compounds (particularly oxychlorides), to generate homogeneous or supported olefin dismutation catalysts. It is now believed that an initial M-alkyl (M = W, Mo) group is converted to a metal alkylidene group by α-hydrogen abstraction. Coordinated olefin now gives a metallocyclobutane (isolable in some cases)[21]:

$$
\begin{array}{ccccc}
M{=}CR_2 & & M{-}CR_2 & & M & CR_2 \\
+ & \rightleftharpoons & |\quad| & \rightleftharpoons & \| + \| \\
R_2''C{=}CR_2' & & R_2''C{-}CR_2' & & R_2''C & CR_2
\end{array}
$$

However, some oxides, such as supported Re_2O_7 (at 50–120°) or molybdenum oxide (at higher temperatures), show dismutation activity without alkyl treatment. Presumably, partial reduction of the surface by olefin exposes suitably liganded $(M—O—M=CR_2)$ centres. Further reduction (slow anion migration to the surface) leads to deactivation, but activity is restored by re-oxidation.

Finally, organo–nickel chemistry has provided a number of industrial applications[22]. Nickel(0) complexes and supported nickel show somewhat similar properties to aluminium alkyls in ethylene oligomerization to alpha-olefins. However, sparsely liganded (phosphine) nickel (0) complexes form dually bonded σ, π (allyl) C_4 ligands from butadiene, which undergo a variety of insertion reactions with olefins or butadiene and ene reactions between ligands. The nickel acts as a template around which simple and puckered ring systems can build up, with phosphine (or other ligand) size and basicity as selectivity control parameters.

11.7.6 Base catalysis

The traditional alkaline catalysts are still used for aldol condensations; cross-Cannizzaro reactions occur when formaldehyde is one of the reactants, e.g. for 'pentaerythritol':

$$CH_3CHO + 4HCHO \rightarrow (HOCH_2)_3CCHO + HCHO$$
$$\rightarrow C(CH_2OH)_4 + HCO_2H$$
$$\text{pentaerythritol}$$

Alkali metal alkoxides catalyse the alcoholysis of esters, by a mechanism analogous to basic hydrolysis. Additionally, alkoxides catalyse the reaction of alcohols with carbon monoxide to give formate esters:

$$RO^- + CO \rightleftharpoons RO_2\bar{C} \underset{RO^-}{\overset{ROH}{\rightleftharpoons}} HCO_2R$$

There is now increasing commercial interest in the dimerization of olefins over supported alkali metals, via a carbanion mechanism. Propylene selectively produces 4-methylpent-1-ene and alkylaromatics are alkylated on the side-chain (α-carbon) with these materials[23].

$$C_3H_6 + Na \xrightarrow[-\frac{1}{2}H_2]{} Na^+\bar{C}H_2—CH=CH_2 \xrightarrow{C_3H_6} \bar{C}H_2 \cdot CHMeCH_2CH=CH_2$$
$$\text{(accepts proton from } C_3H_6\text{)}$$

11.7.7 Oxidations

11.7.7.1 *Catalysis in liquid-phase free-radical oxidations.* The conventional liquid-phase oxidation of hydrocarbons and their derivatives with air involves a free-radical chain mechanism[24]:

$$RH + O_2 \rightarrow R^{\cdot} + HO_2^{\cdot} \qquad \text{Slow initiation}$$

$$\left. \begin{array}{l} R^{\cdot} + O_2 \rightarrow RO_2^{\cdot} \\[4pt] RO_2^{\cdot} + RH \rightarrow RO_2H + R^{\cdot} \end{array} \right\} \quad \text{Propagation}$$

$$RO_2H \rightarrow RO^{\cdot} + HO^{\cdot} \qquad \text{Branching}$$

2 radicals $\rightarrow$ non-radical products   Termination

(R = alkyl, aralkyl, acyl, etc.)

Organic compounds which decompose to give free radicals (e.g. peroxides) may be added to accelerate initiation, and some metals in high oxidation states may also fulfil this role,

$$RH + M^{(n+1)+} \rightarrow R^{\cdot} + M^{n+} + H^+$$

With methyl aromatic substrates, the presence of bromide ions also aids hydrogen abstraction without significant formation of halogenated by-products.

$$M^{(n+1)+} + Br^- \rightarrow M^{n+} + Br^{\cdot}$$

$$ArCH_3 + Br^{\cdot} \rightarrow ArCH_2^{\cdot} + Br^- + H^+$$

However, the major role of most metal compounds, in solution or not, is to accelerate decomposition of the hydroperoxide intermediate in a branching mode:

$$RO_2H + M^{n+} \rightarrow RO^{\cdot} + M^{(n+1)+} + OH^-$$

$$RO_2H + M^{(n+1)+} \rightarrow RO_2^{\cdot} + M^{n+} + H^+$$

In general, metal concentrations of only 10–100 parts per million are required in the oxidation of alkanes and their oxygenated derivatives, somewhat higher levels for alkyl aromatics.

When the desired product is an alkyl or aralkyl hydroperoxide or peracid, careful control of metal concentration is required to minimize hydroperoxide cleavage. When the reactant is an alcohol, the hydroperoxide is the hydrogen peroxide adduct of an aldehyde or ketone, i.e.

$$\begin{array}{ccc} \overset{\displaystyle H}{\underset{\displaystyle OH}{R_1-C-R_2}} & \longrightarrow & \overset{\displaystyle OH}{\underset{\displaystyle O_2H}{R_1CR_2}} \rightleftharpoons R_1COR_2 + H_2O_2 \end{array}$$

Hence hydrogen peroxide may be a recoverable product from such oxidations under mild conditions.

Many of the above reactions of metal ions are reversible, and the kinetically favoured direction varies with the metal and type of radical. Hence, individual metals may cause acceleration or inhibition with different substrates, often accompanied by changes in product distribution.

Cobalt (III) appears to be unique, in that higher concentrations (0·5 to 2% w/w) permit high rates in aerial oxidation of hydrocarbons while maintaining specificity in the point of attack; free radicals, in the conventional sense, appear to be absent. Further evidence of a strong association between cobalt ions and reactive species is the exceptionally high reactivity of 'Co^{3+}' regenerated in situ compared with typical cobaltic compounds prepared by other means.

Finally, in the Bashkirov oxidation of normal paraffins to secondary alcohols, the boric oxide introduced is often referred to as a catalyst. This material does indeed catalyse the decomposition of hydroperoxides, but by a non-radical route to alcohol and oxygen, and therefore slows down the overall rate of reaction.

11.7.7.2 *Liquid-phase epoxidation.* Epoxides are produced in liquid-phase reactions of olefins with hydroperoxides in the presence of (soluble or insoluble) compounds of molybdenum, tungsten or vanadium.

$$\begin{array}{c} \diagup \\ C = C \diagdown \end{array} + RO_2H \longrightarrow \begin{array}{c} O \\ \diagup \diagdown \\ C - C \end{array} + ROH$$

The hydroperoxides (and hydrogen peroxide) form well-characterized peroxy-complexes with oxides of the above metals, but the mechanism whereby the olefin coordinates and reacts with these complexes is still unclear. (No catalyst is required if a peracid is used as the epoxidizing reagent.)

11.7.7.3 *Wacker-type oxidations.* Ethylene is oxidized to acetaldehyde in the presence of an aqueous solution of palladium (II) chloride and copper (II) chloride[1,25]. The initial reaction is believed to follow a sequence of the type

$$\| + \left[Pd^{II}\ Cl_n\ H_2O \right] \xrightarrow{-Cl^-} \left[\| \rightarrow Pd^{II}\ Cl_{n-1}\ H_2O \right] \xrightarrow{-H^+} \left[HOCH_2CH_2Pd^{II}\ Cl_{n-1} \right]$$

Although both chloride and hydrogen ions are essential for catalyst stability, the rate of reaction shows a negative order on both. β-Hydrogen abstraction then follows, nominally to give palladium metal by complete decomposition,

$$\left[HO-CH_2CH_2 - Pd^{II}\ Cl_{n-1} \right] \longrightarrow CH_3CHO + Pd^0 + H^+ + (n-1)Cl^-$$

However, in the presence of copper (II) chloride, the palladium is reoxidized without significant metal precipitation, possibily via a chloride bridged Cu–Pd species. The copper (I) chloride is reoxidized to copper (II) chloride in situ, or in a separate stage.

Substitution of acetic acid for water as solvent in the Wacker process leads to the formation of vinyl acetate (ethenyl ethanoate) from ethylene by an essentially identical mechanism (called 'acetoxylation'). This liquid phase system (chlorides in acetic acid) is exceedingly corrosive. However, the use of supported palladium catalysts in the liquid phase provides modest rates without chloride. Similarly, solid-catalyst systems (e.g. Pd/Te on carbon) are effective for the conversion of butadiene to 1,4-diacetoxybut-2-ene, which forms the basis of one commercialized route to 1,4-butanediol.

The commercial route to vinyl acetate is a gas-phase process, utilizing palladium or a palladium-gold alloy supported on alumina or a spinel[7]. The catalyst also carries potassium acetate, and sometimes other metal acetates, to provide a supported liquid phase (KOAc–HOAc) in which the reaction takes place. The role of palladium may be described by the reaction scheme

$$\| + Pd^0 + HOAc \longrightarrow \| \longrightarrow Pd \overset{(H)}{\underset{OAc}{\Big<}} \longrightarrow Pd \overset{(H)}{\underset{CH_2CH_2OAc}{\Big<}}$$

$$\longrightarrow Pd \overset{(H)}{\underset{(H)}{\Big<}} + CH_2{=}CHOAc \quad (\beta\text{-hydrogen abstraction})$$

Hydrides on the palladium are continually removed as water by oxygen present; no redox component is necessary. However, other oxidizing components were included in heterogeneous catalysts for the uncommercialized oxidation (Pd-phosphomolybdate) and 'oxycyanation' (olefin/HCN/O_2 over Pd-vanadia) of ethylene to acetaldehyde (ethanal)/acetic (ethanoic) acid and acrylonitrile (propenonitrile) respectively.

11.7.7.4 *General gas-phase oxidation over metals.* Only silver, gold and the noble Group VIII metals do not form bulk oxides in the presence of air at high temperatures, and of these only silver, as gauze or supported metal, finds application in a number of selective gas-phase oxidations of organic substances. (Pt and other noble metals catalyse total combustion for exhaust gas clean-up). The electrochemical measurement of oxygen ion activities on metal surfaces can throw light on the elementary steps in such processes[26].

Silver has a relatively strong affinity for oxygen, which is activated by anion formation.

$$Ag + O_2 \rightarrow Ag^+ + O_2^-$$

In most oxidation processes dioxygen ions are considered undesirable, but they appear to be essential in the oxidation of ethylene to ethylene oxide (oxirane).

$$C_2H_4 + O_2^- + Ag \rightarrow C_2H_4O + O^{2-} + Ag^+$$

Loss of selectivity by partial combustion, and its reduction by partial blocking of the metal surface with chloride ions, was for many years attributed to the formation of O^{2-} ions in the above reaction and on silver 'ensembles' (geometric theory), but this idea is no longer considered tenable[27].

11.7.7.5 *Gas-phase oxidation over metal oxides*. There is one particularly important finding common to essentially all metal oxide catalysts which provide rapid, moderately selective oxidations of organic substrates. If the supply of oxygen is interrupted, oxidation of the organic substrate continues, initially with little change in rate and selectivity. The oxide system becomes reduced, and tens, or even hundreds, of atomic layers below the surface become oxygen deficient. What are the implications?

The Redox Model of Mars and Van Krevelen describes the overall process as

$$MO_x + \text{substrate} \xrightarrow{r_1} MO_{x-n} + \text{products}$$

$$MO_{x-n} + \frac{n}{2}O_2 \xrightarrow{r_2} MO_x$$

Under steady-state conditions, r_1 equals r_2, and the oxide must be in a reduced state. Furthermore, the degree of reduction and catalytic performance will change with operating conditions.

Oxide anions (O^{2-}) must be mobile within the oxide structure, requiring both an appropriate spacing between the metal atoms in the crystal structure and stability of the metal atom lattice. Such structural stability occurs in vanadia provided that no more than two-thirds of the V(V) atoms are reduced to V(IV); further reduction causes changes in both structure and catalytic behaviour. The partially reduced oxides (and sometimes the fully oxidized states) therefore have vacancy (coordinatively deficient) structures.

Oxide ions also provide nearly all the oxygen present in the products and by-products, with little involvement of other chemisorbed oxygen species such as O_2^-. Most selective oxide catalysts show little or no oxygen isotopic exchange, again pointing to minimal involvement of such species. Hence substrate oxidation and oxygen reduction do not necessarily occur at the same site. In bismuth molybdate, the molybdenum and bismuth may respectively perform these separate functions. This feature requires electron transfers between metal atoms in the lattice; there must be partial covalency in the M—O—M bonds and near-degeneracy between some of the electronic states in the two metal atoms. The 4*d* states of molybdenum and the 6*p* band of bismuth are nearly degenerate, for example. It is also evident that other components can significantly affect both structural and electronic characteristics, and most commercial catalysts are indeed very complex. (One example is $K_{0.1}Ni_{2.5}Co_{4.5}Fe_3BiP_{0.5}Mo_{12}O_{55} + 17.5\%$ w/w SiO_2). Finally to achieve both reduction of atmospheric oxygen, yet selective oxidation of the organic substrate, requires a restricted range of metal–oxygen bond strengths.

Up to this point, we have only looked at general requirements of the oxide material, without considering the nature of the chemisorption and oxidation processes themselves. The situation proves complicated as a particular organic substrate can give different products over related families of oxides.

In the oxidation or ammoxidation of propylene on bismuth molybdate (and U–Sb, Sn–Sb oxides), all the evidence indicates that the first step is hydrogen abstraction to give a symmetrical intermediate. The first stages can possibly be written:

$$H-CH_2-CH=CH_2 \qquad\qquad \left[CH_2\!=\!\!=\!CH\!=\!\!=\!CH_2\right]^-$$
$$\downarrow \qquad\qquad\qquad\qquad \downarrow$$
$$\rightarrow O^{2-} - Mo - O^{2-}\; O^{2-} - \quad\longrightarrow\quad HO^- - Mo - O^{2-}\; O^{2-} -$$

Electrons must now drain away to permit formation of a C—O(N) bond before further hydrogen removal can occur. (The manner of nitrogen incorporation to form acrylonitrile is still under review[28]. The formation of acrolein (propenal), though until recently of lesser commercial importance, will be pursued).

$$\left[CH_2\!=\!\!=\!CH\!=\!\!=\!CH_2\right]^- \qquad CH_2=CH-CH_2 \qquad CH_2=CHCHO$$
$$\downarrow \qquad\qquad\qquad \overset{\displaystyle O}{\diagup} $$
$$HO^- - Mo - O^{2-}\; O^{2-} - \;\longrightarrow\; HO^- - Mo - O^{2-} - \;\longrightarrow\; HO^- - Mo - HO^- -$$
$$\Downarrow_{3e} \qquad\qquad\qquad \Downarrow_{1e}$$

Elimination of water and diffusion of two O^{2-} anions from the bulk complete the cycle. When the feedstock in an *n*-butene, the electron shift prompts the transfer of a second proton to surface oxygen, rather than C—O bond formation, leading to butadiene.

Little or no acrylic (propenoic) acid results from overoxidation of acrolein on bismuth molybdate. The addition of cobalt, nickel or vanadium oxides is necessary to induce a further transfer of oxygen (or OH) to carbon. However, if we oxidize propylene over tin or cobalt molybdates alone, the major product is acetone. The surfaces of these materials have high hydroxyl concentrations, with Brönsted acidity, leading to isopropoxy groups from which the α-hydrogen is readily abstracted[29]. *n*-Butenes give 2-butanone and acetic acid. In contrast, the incorporation of phosphorus into bismuth molybdate does not modify the initial stage, but butadiene formed from *n*-butenes is chemisorbed more strongly, and further stepwise oxidation to maleic anhydride occurs. (Modified vanadia catalysts are preferred for maleic anhydride production from C_4's).

Surprisingly perhaps, oxidation of C_3 and C_4 olefins over bismuth oxide alone leads mainly to oxidative dimerization to C_6 or C_8 dienes, and small amounts of cyclic hydrocarbons. The surface is possibly highly reduced, providing many Lewis acid centres for olefin coordination but few oxide ions for hydrogen abstraction and transfer. (The supported gallium oxide catalyst, developed by BP, leads to further hydrogen abstractions and significant yields of aromatics from lower olefins).

Finally vanadia catalysts are used extensively for oxidations of aromatic hydrocarbons. With benzene, the mechanism for ring breakage is not well defined, and the desorption of maleic anhydride itself appears to be rate-controlling. For the oxidation of *o*-xylene, the use of supported vanadia-titania catalysts limits ring cleavage. A well-defined major product sequence, *o*-xylene → *o*-tolualdehyde → *o*-toluic acid → phthalide → phthalic anhydride defines the main series of hydrogen abstraction/oxygen transfer processes which follow the initial coordination and hydrogen abstraction step.

11.7.7.6 *Halogen-mediated oxidations and oxychlorination.* A number of liquid-phase oxidation processes appear to be mediated by *in situ* oxidation of halide, halogen addition to an olefin, and solvolysis of the resulting intermediate. This comment applies particularly to Halcon's tellurium bromide system (in aqueous acetic acid) for glycol acetates from olefins. Halogenated organic products are co-produced, but are separated for recycle. (The process was commercialized as part of a manufacturing route to ethylene oxide, but apparently failed.) The mechanism is essentially:

$$TeBr_4 \rightleftarrows TeBr_2 + Br_2$$

$$TeBr_2 + 2H^+ + 2Br^- + \tfrac{1}{2}O_2 \rightarrow TeBr_4 + H_2O$$

Similarly, in the high-temperature gas-phase oxychlorination of alkanes over supported copper chloride melts, generation of the free chlorine appears to be necessary.

$$2CuCl_2 \rightarrow 2CuCl + Cl_2$$
$$Cl_2 + CH_4 \rightarrow CH_3Cl + HCl$$
$$\underline{2CuCl + 2HCl + \tfrac{1}{2}O_2 \rightarrow 2CuCl_2 + H_2O}$$
$$\text{overall } CH_4 + HCl + \tfrac{1}{2}O_2 \rightarrow CH_3Cl + H_2O$$

In contrast, the oxychlorination of ethylene occurs at much lower temperatures than those required to generate chlorine. Hence, direct transfer of halogen to coordinated olefin, with simultaneous reduction of Cu^{II} to Cu^{I}, has been proposed.

A direct halogen transfer and solvolysis mechanism has been put forward for the liquid phase 'oxycyanation' of butadiene to 1, 4-dicyano-but-2-ene by the system copper-iodine—HCN—air.

11.7.8 *Carbon monoxide chemistry*

11.7.8.1 *Heterogeneous catalysis.* In 1923, Fischer and Tropsch showed that carbon monoxide and hydrogen gave complex mixtures of hydrocarbons and oxygenated derivatives over iron catalysts at about 400°C and high pressures. Later cobalt and promoted iron catalysts gave predominantly linear hydrocarbons at much lower temperatures (200–300°C) and pressures (1–20 atm.). Under similar conditions, supported nickel gave mainly methane, but high-molecular-weight paraffins were formed on ruthenium.

Hydrocarbon formation is now believed to proceed via atomic carbon (carbide) to bridging methylene groups, and a chain polymerization process[30]:

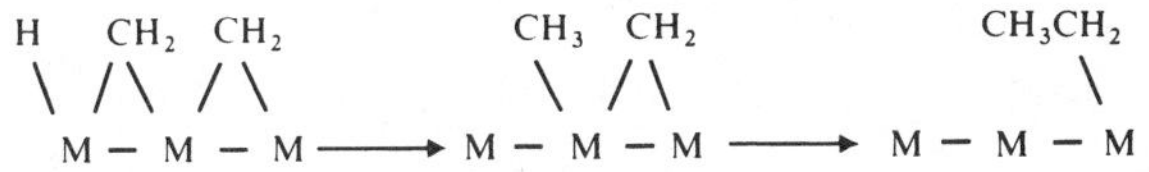

β-hydrogen abstraction and α-hydrogen addition give alkene and alkane respectively, while incorporation of unreduced carbon monoxide leads to oxygenates. The reverse, depolymerization, process is implicated in hydrogenolytic cleavage of hydrocarbons on these metals, with appreciable co-production of methane. The overall mechanism is retraced in the steam reforming of hydrocarbons over nickel to produce synthesis gas (CO/H_2 mixtures).

$$C_nH_{2n+2} + nH_2O \rightarrow nCO + (2n + 1)H_2$$

Developments in the 1930s led to zinc chromite for methanol synthesis, with pressures of about 300 atm-required for thermodynamic reasons at about 400°C. Reduction on a single metal centre seems a likely mechanism. The higher *initial* activity resulting from copper addition was known for many years, but the commercial success achieved by ICI and subsequent developers stems mainly from stabilization of the metal dispersion[31].

11.7.8.2 Hydroformylation.
Experiments with supported cobalt catalysts led to the hydroformylation (or OXO) process for the conversion of olefins and synthesis gas to aldehydes. Homogeneous catalysis followed, and is now used exclusively. The generally accepted mechanism involves the following reactions:

$$Co_2(CO)_8 + H_2 \rightleftharpoons 2HCo(CO)_4 \rightleftharpoons 2HCo(CO)_3 + 2CO$$

$$(CO)_3CoH + CH_2 = CHR \rightleftharpoons (CO)_3CoCH_2CH_2R \text{ (and } (CO)_3CoCHRCH_3)$$

$$CO + (CO)_2\overset{\diagup CO}{Co}-CH_2CH_2R \rightleftharpoons (CO)_3CoCOCH_2CH_2R$$

$$\overset{H_2}{\rightarrow} (CO)_3CoH + HCOCH_2CH_2R$$

The order on carbon monoxide is *negative*, but a high partial pressure is required to maintain catalyst solubility and limit olefin hydrogenation.

CO-insertion (or alkyl transfer to CO) occurs widely in organometallic chemistry; with cobalt, the addition of extra ligands drives the equilibrium to the right. (Substitution of water or alcohol for hydrogen in the process leads to carboxylic acid or ester, but the mechanism may then differ).

The selectivity to linear product is improved by using a cobalt trialkyl phosphine complex, but the major product is the corresponding *alcohol*. The use of rhodium triphenylphosphine hydrocarbonyl gives a still higher selectivity to linear aldehyde under milder conditions. These advantages are retained by using a water-soluble rhodium complex with sulphonated phosphane ligands, which provides a novel two-phase mode of operation (recently commercialized)[32].

11.7.8.3 *Methanol carbonylation.* BASF employ a cobalt catalyst, with iodide promoter, for the carbonylation of methanol to acetic acid under severe conditions (250°C, over 300 atm.). The mechanism is ill defined. Monsanto showed that rhodium again allows much milder conditions. In the following scheme the oxidative addition of methyl iodide to the rhodium(I) dicarbonyl diiodide anion appears to be rate-controlling[33]:

$$MeOH + HI \rightleftarrows MeI + H_2O$$

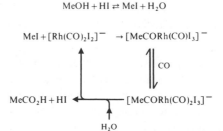

Similar intermediates are possibly involved in the Halcon/Eastman route to acetic anhydride from methyl acetate[34], and BP Chemicals' acetic acid/acetic anhydride co-production process, both now commercialized. However, all these cyclic mechanisms may be incomplete, as individual steps may themselves be complex or catalysed by other species present, and the author has omitted many (minor) side-reactions.

11.7.8.4 *New developments.* If hydrogen is introduced during the carbonylation of methanol with a soluble cobalt/iodide catalyst, acetaldehyde is produced. Similarly, ethylidene diacetate, formally an adduct of acetaldehyde and acetic anhydride, is formed by the reaction of methyl acetate with carbon monoxide and hydrogen in the presence of rhodium catalysts. In both these processes, hydrogenolysis of the metal-acyl is probably involved, linking together parts of the methanol carbonylation and hydroformylation cycles. At very high pressures (> 100 atm.), ethylene glycol is formed from synthesis gas with soluble rhodium and ruthenium catalysts, presumably via a hydroxymethyl ligand and CO-insertion.

The reaction of methanol, carbon monoxide and oxygen ('oxycarbonylation') with soluble palladium and copper salts gives dimethyl carbonate and/or dimethyl oxalate (dimethyl ethanedioate), for hydrogenation to ethylene glycol. Postulated mechanisms include the sequence

$$(CO)_2Pd(OMe)_2 \rightarrow Pd(CO_2Me)_2 \rightarrow Pd^0 + (CO_2Me)_2$$

Heterogeneous Fischer–Tropsch chemistry is coming under closer scrutiny, and there is increasing interest in the use of traditional catalysts and supported carbonyl clusters of Group VIII metals for the possible production of lower olefins or alcohols from synthesis gas. The heterogenization of homogeneous catalysts for hydroformylation, carbonylation, etc. is also attracting much attention. Conversely, several homogeneous systems have now been found to effect the water-gas shift reaction, though heterogeneous catalysts have yet to be challenged for industrial use.

11.8
The future

The ultimate objective of the industrial chemist is the ability to design catalyst systems, homogeneous or heterogeneous, which will

(a) effect known reactions with high selectivities;
(b) extend the range of possible chemical reactions; and
(c) provide commercially viable rates under the mildest conditions consistent with thermodynamic requirements.

Enzymes provide selectivity, and effect certain reactions which have proved difficult to duplicate in more conventional chemical systems, but with very specific reactants. While the active centre is often organometallic, the protein structure is important for steric and conformational control (co-operative distortion of ligands to control redox potential, etc.), but is subject to 'denaturation' at temperatures too low for acceptable rates in other than speciality uses (some large tonnage in the food industry). Furthermore, the systems remain too complex for any thought of catalyst design along these lines, though model systems are being studied.

Homogeneous catalysis provides the maximum opportunity for mechanistic interpretation. The blossoming field of organometallic chemistry and catalysis also seems to provide greater scope for achieving novel reactions, and possibly a better understanding of heterogeneous metal catalysis[35]. However, this very feature reflects the fact that organometallic chemistry is still young, and we are probably still on an early part of the learning curve. A considerable part of the effort directed at catalytic processes therefore retains the purely speculative or empirical approach.

Heterogeneous catalysis entails reactions between organic molecules and the surface of inorganic materials. Although a great deal of work has been carried out, and continues, to characterize the surfaces of materials, we remain far from being able to predict surface properties for all but the simplest. Even then, the interaction of an organic molecule with the surface may be characterizable, but is rarely predictable for any reaction, other than coordination. Nonetheless, the amount of information reported is exceedingly large, and by appropriate classification we can pick out short lists of possible candidates as starting points for a particular reaction. Theoretical considerations may point to those additives most likely to affect performance, but in the final analysis we follow a largely empirical approach yet again.

Only a few years ago, it was suggested that chemical technology was approaching maturity, but it now appears that a considerable number of possible new processes, many based on novel catalyst systems, are vying for consideration. However, the objective of catalyst design still seems remote.

References

1. J. Smidt, *C&I*, 1962, 54–61.
2. S. Fujiyama and T. Kasahara, *Hydrocarbon Processing*, 1978, (Nov.), 147–149.
3. E. S. Lewis (ed.), *Investigation of Rates and Mechanisms of Reactions (Techniques of Chemistry, Vol. VI)*, 3rd edn., Part I, Wiley Interscience, 1974.
4. D. C. Bailey and S. H. Langer, *Chem. Reviews*, 1981, **81** (2), 109–148.
5. M. J. Sundall and J. H. Näsman, *Chemtech*, 1993, (December), 16–23.
6. C. R. Nelson and M. L. Couter, *Chem. Eng. Progress*, 1954, **50** (10), 526–531.
7. *C&I*, 1968, 1559–1563.
8. O. Wiedermann and W. Gierer, *Chem. Eng.*, 1979, (Jan. 29), 62–63.
9. (a) J. J. Graham, *Chem. Eng. Progress*, 1970, (Sept.), 54–58.
   (b) *Kirk-Othmer Encyclopedia of Chemical Technology*, 3rd edn., Vol. 17 (Petroleum, Refinery Process Survey), Wiley, 1982.
   (c) M. L. Riekena *et al.*, *Chem. Eng. Progress*, 1982, (April), 86–90.
10. *Kirk-Othmer Encyclopedia of Chemical Technology*, 4th edn., Vol. 2 (Ammonia, pp. 638–691), Wiley, 1992.
11. (a) J. M. Thomas *et al.*, *New Scientist*, 1982, (18 Nov.), 435–438.
    (b) C. Naccache and Y. B. Taarit, *Pure & Appl. Chem.*, 1980, **52**, 2175–2189.
    (c) Ref. 15.
    (d) W. Hölderich, M. Hesse and F. Naumann, *Angew. Chemie Int. Ed. in English*, 1988, **27**, 247–260.
12. S. Brunauer, P. H. Emmet and T. Teller, *J. Amer. Chem. Soc.*, 1938, **60**, 309 (but see modern texts for current techniques and interpretation).
13. N. C. Saha and D. S. Mathur, *J. Chromatography*, 1973, **81**, 207–232.
14. J. T. Yates Jr., *C&EN*, 1974, (Aug. 26), 19–29.
15. S. L. Meisel, *Chemtech*, 1988, (January), 32–37.
16. M. Misono and T. Okuhara, *Chemtech*, 1993, (Nov.), 23–29.
17. K. Tanabe *et al.*, *Bull. Chem. Soc. Japan*, 1974, **47** (5), 1064–1066.
18. (a) Y. One and T. Mori, *J. Chem. Soc. Faraday Trans. I*, 1981, **77**, 2209–2221.
    (b) C. D. Chang and C. T.-W. Chu, *J. Catalysis*, 1982, **74**, 203–206.
    (c) C. D. Chang, C. T.-W. Chu and R. F. Socha, *J. Catalysis*, 1984, **86**, 289.
19. R. H. Crabtree *et al.*, *J. Amer. Chem. Soc.*, 1982, **104**, 107–113.
20. (a) A. A. Montagna and J. C. Floyd, *Hydrocarbon Processing*, 1994, (March), 57–62.
    (b) J. Chowdhury and S. Moore, *Chemical Engineering*, 1993, (April), 34–39.
21. A. K. Rappa and W. A. Goddard, *J. Amer. Chem. Soc.*, 1982, **104**, 446–456.
22. G. Wilke, *Angew. Chemie Int. Ed. in English*, 1988, **27**, 185–206.
23. (a) A. W. Shaw *et al.*, *J. Org. Chem.*, 1965, **30**, 3286–3289.
    (b) British Patent 932,342 to British Petroleum Company, 1963.
24. *Oxidation of Organic Compounds*, Vols. I and II, Advances in Chemistry Series, Nos. 75 and 76, American Chemical Society, 1968.
25. (a) P. M. Maitlis, *The Organic Chemistry of Palladium* (2 vols.), Academic Press, 1971.
    (b) J. E. Bäckvall *et al.*, *J. Amer. Chem. Soc.*, 1979, **101**, 2411–2416.
26. H.-G. Lintz and C. G. Vayenas, *Angew. Chemie Int. Ed. in English*, 1989, **28**, 708–715.

27. (a) *Kirk-Othmer Encyclopedia of Chemical Technology*, 4th edn., Vol. 9 (Ethylene Oxide, pp. 915–959), Wiley, 1994.
    (b) P. B. Grant and R. M. Lambert, *J. Catalysis*, 1985, **92**, 364–375.
28. J. D. Burrington *et al.*, *J. Catalysis*, 1983, **81**, 489–498.
29. Y. Takita *et al.*, *J. Catalysis*, 1972, **27**, 185–192.
30. W. A. Hermann, *Angew. Chemie Int. Ed.*, 1982, **21**, 117–130.
31. J. Ladebeck, *Hydrocarbon Processing*, 1993, (March), 89–91.
32. W. A. Hermann and C. W. Kohlpaintner, *Angew. Chemie Int. Ed. in English*, 1993, **32**, 1524–1544.
33. (a) J. F. Roth *et al.*, *Chemtech*, 1971, **1**, 600–605.
    (b) D. Forster, *Advances in Organometallic Chemistry*, 1979, **17**, 255–267.
34. (a) J. L. Ehrler and B. Juran, *Hydrocarbon Processing*, 1982, (Feb.), 109–113.
    (b) M. Schrod and G. Luft, *Ind. Eng. Chem. Product Research Dev.*, 1981, **20**, 649–653.
35. (a) D. L. King *et al.*, *Hydrocarbon Processing*, 1982, (Nov.), 109–113.
    (b) A. Aquillo *et al.*, *Hydrocarbon Processing*, 1983, (March), 57–65.
    (c) E. L. Muetterties and M. J. Krause, *Angew. Chemie Int. Ed.*, 1983, **22**, 135–148.
    (d) D. F. Shriver, Chem. Brit., 1983, (June), 482–487.
    (e) G. Süss-Fink, *Angew. Chemie Int. Ed. in English*, 1991, **30**, 72–73.
    (f) W. F. Maier, *Angew. Chemie Int. Ed. in English*, 1989, **28**, 135–145.

Bibliography

Thermochemical Kinetics, 2nd edn., S. W. Benson, John Wiley and Sons, 1976.
Thermochemistry of Organic and Organometallic Compounds, J. D. Cox and G. Pilcher, Academic Press, 1970.
The Chemical Thermodynamics of Organic Compounds, D. R. Stull, E.F. Westrum and G. C. Stinke, John Wiley and Sons, 1969.
Catalysis and Inhibition of Chemical Reactions, P. G. Ashmore, Butterworths, 1963.
Principles and Applications of Homogeneous Catalysis, A. Nakamura and M. Tsutsui, John Wiley and Sons, 1980.
Fundamental Research in Homogeneous Catalysis, M. Tsutsui (ed.), Plenum Press, 1979 (Incorporating the Proceedings of the 1st International Conference on Homogeneous Catalysis, Corpus Christi 1978, with some papers on polymer-supported complexes).
Catalysis by Supported Complexes, Yu. I. Yermankov, B. N. Kuznetsov and V. A. Zakharov, Elsevier Scientific Publishing, 1981.
Introduction to the Principles of Heterogeneous Catalysis, J. M. and W. J. Thomas, Academic Press, 1967.
Design of Industrial Catalysts, D. L. Trimm, Elsevier, 1980.
Heterogeneous Catalysis in Practice, C. N. Satterfield, McGraw-Hill, 1980.
Theoretical Aspects of Heterogeneous Catalysis, J. B. Moffat, Van Nostrand Reinhold, 1990.
Zeolite Technology and Applications, Recent Advances, Jeanette Scott (ed.), Noyes Data Corporation, 1980.
Zeolites: Science and Technology, F. R. Ribero, A.E. Rodrigues, L. D. Rollmann and C. Naccache (eds.), Martinus Nijhoff Publishers, 1984 (Proceedings of the NATO Advanced Institute on Zeolites; Science and Technology, Alcabideche, 1983).
New Developments in Selective Oxidation by Heterogeneous Catalysis, P. Ruiz and B. Delmon (eds.), Elsevier, 1992 (Proceedings of the Third European Workshop, Louvain-la-Neuve, 1991).
Physical and Chemical Aspects of Adsorbants and Catalysts, B. G. Linsen (ed.), Academic Press, 1970.
Characterization of Heterogeneous Catalysts, Francis Delannay (ed.), Marcel Dekker Inc., 1984.
1989/1990 Worldwide Catalyst Product, Process and Service Directory, Catalyst Consultants Publishing, 1989.

12 Petrochemicals

John Pennington

**12.1
Introduction**

12.1.1 *Layout*

With over 90% of synthetic organic materials presently produced from petroleum or natural gas, this chapter aims to show how we derive intermediate feedstocks or building blocks suitable for chemical processes from such raw materials and to illustrate the range of chemistry which can follow.

After the introduction and a brief section on refinery processes, sections 12.3 and 12.4 describe the major petrochemical cracking and reforming operations, in which energy requirements play an increasingly important role. Some immediate downstream products have also been included in these sections.

Thereafter, the operations of the petrochemical industry, or more accurately bulk organic chemical industry, centre on chemistry, with extensive use of catalysis to permit the use of the cheapest feedstock available and eliminate expensive reagents. However, potential changes in raw material availabilities and environmental pressures need a flexible response, possibly illustrated by the variety of production routes to acetic (ethanoic) acid and anhydride described in section 12.5. The remaining aliphatic chemical products are dealt with in carbon number sequence, followed by aromatics and nylon intermediates.

Petrochemicals are predominantly intermediates for polymers, and reference to these materials is essential to put the present topic into perspective, although plastics and polymers are not considered in detail here.

12.1.2 *The beginnings*

A major factor leading to initial developments in the petrochemical field was the enormous increase in popularity of the motor-car in the United States during the 1920s and 1930s. Henry Ford paved the way to mass ownership, and the first large finds of petroleum provided cheap gasoline. However, the separation of a gasoline cut from petroleum left refiners with increasing amounts of high-boiling fractions; thermal cracking gave additional gasoline, together with hydrocarbon gases. These were initially burned as refinery fuel, but an excess soon prompted the development of chemical processes for their use, mainly within the U.S. oil industry. Subsequently, the development of wet natural gas supplies, i.e. gas containing ethane, propane etc., which were available in the U.S.A., also attracted chemical interest, particularly in the use of ethane as a precursor for ethylene. As a result, the U.S. production of organic chemicals from these sources grew from about one hundred tonnes in 1925 to a million tonnes in 1940.

12.1.3 *Into the 1970s*

The wartime and postwar increases in demand for fuel, and the introduction of new synthetic materials, led to a very rapid growth in the U.S. petrochemical industry into the early 1970s, still largely (*c.* 80%) based on cheap refinery and natural gases.

In Europe, fuel oil requirements have dominated refinery operations. Gasoline (petrol) always represented less than 20% of petroleum consumption, and refinery cracking operations never figured prominently until the 1970s. However, naphtha (paraffinic, or low octane, motor spirit) became surplus to requirements, and the advent of progressively larger naphtha crackers into Europe after World War II (and later in Japan) allowed these countries to enter petrochemicals. From the 1950s, growth in Europe was higher than in the U.S.A., and the West European consumption of lower olefins caught up in 1973 at about 10 million tonnes (Mt) of ethylene and 5 Mt of propylene.

Until 1950 the coal-tar industries were the major source of aromatics in both Europe and the U.S. Thereafter, they contracted slowly, while demand for aromatics increased. The chemical industry again turned to petroleum, which eventually supplied some 90% of benzene (4–5 Mt) and other aromatic hydrocarbons worldwide.

The third major doorway into petrochemicals is via synthesis gas. Methanol synthesis is the major organic downstream operation, but is often integrated with still larger ammonia production facilities. U.S. methanol production in 1973 was about $2\frac{1}{2}$ Mt (cf. 14 Mt of ammonia), with a similar figure for Western Europe.

12.1.4 *The present*

Since 1973, progress in the petrochemical industry has been rather like a roller-coaster ride. Major rises in the price of crude oil in 1973–1974 and again in the late 1970s fed through most quickly into petrochemical products, affecting demand worldwide. This period saw many proposals to replace petroleum-based chemical feedstocks, and even hydrocarbon fuels, by methanol (from coal or remote natural gas supplies) or fermentation ethanol.

Despite some closures, postponements and cancellations of new projects during the 'downs', the 'ups' prompted new investment, much of it in the less well-developed countries of the Far East, Middle East, Africa and Central/South America to reduce their dependence on imports. New construction is almost inevitably in very large blocks; the total output from a petrochemical complex may exceed 1 million tonnes (Mt) per annum. The fall in export potential contributed to something like a 60% excess of nominal capacity over actual requirements in both Europe and the U.S.A. during 1982, and 1982 U.S. production figures for major chemicals were only slightly higher than for 1973.

The price of oil fell in 1986, providing a boost during a further period of growth, but the late 1980s saw the onset of another worldwide recession. New investment continued, especially in the Far East outside Japan; Japan's share of Asian ethylene capacity has fallen from 75% to 45% in the last 10 years. (South Korean ethylene production tripled between 1990 and 1993 to 1·5 Mt per annum, for example.) In contrast, the chemical industries of Eastern European countries and the former Soviet republics are heavily depressed, though we may soon see new investment targeted initially at export markets.

In 'the West' the situation remains rather fluid; while many downstream products remain depressed, polyolefins and other thermoplastic resins are showing continued growth, leading to increased demand for the major intermediates. 1993 saw U.S. production reach 18·7 Mt ethylene, 10·2 Mt propylene, 5·6 Mt benzene, 15·6 Mt ammonia and 4·8 Mt methanol[1]. In western Europe, now including all Germany, 1993 production figures for ethylene and propylene were over 15 Mt and 9 Mt, respectively.

12.1.5 *Individual feedstocks and routes*

Despite the high tonnages of petrochemicals, the chemical industry as a whole consumes rather less than 10% of available petroleum and natural gas hydrocarbons as feedstocks, with possibly a further 4–5% as fuel. For comparison, the current consumption of gasoline alone in Western Europe exceeds 120 Mt per annum, while the U.S. figure is over 300 Mt per annum. Hence, prices of individual hydrocarbon feedstocks are largely determined by other forces; the most economic feedstock/route combination has frequently changed with time, and may differ in different parts of the world. Furthermore, while a specific route may be preferred for new plants, older plants for which the capital is largely written off may well remain economically viable. Finally, special situations may prompt individual solutions. For example, Rhône–Poulenc in France derive the carbon monoxide for a very modern acetic acid plant, based on Monsanto's methanol carbonylation process, from the partial oxidation of methane to acetylene and synthesis gas. They will, therefore, continue to manufacture vinyl acetate by the otherwise obsolescent route from acetylene rather than ethylene (see section 12.3.3.1).

12.2.1 *Crude oil and natural gas*

Petroleum, or crude oil, is an extremely complex mixture derived like coal from prehistoric vegetation. The components range from gaseous to semi-solid or solid hydrocarbons, with compounds of sulphur, nitrogen, oxygen and various metals as impurities. Distillation gives roughly the fractions shown in Table 12.1. The hydrocarbons are almost entirely *saturated* paraffins, cyclo-paraffins (naphthenes) and aromatics/polyaromatics; the proportions vary enormously from one source to another.

Methane dissolved in, or trapped above, the oil is referred to as associated gas; ethane, propane, etc., are usually also present. Some of this gas separates at the well-head, and must be disengaged. Gas may also be found unassociated with oil. The gas may be 'wet', containing appreciable amounts of ethane and other hydrocarbons, but gas from the (southern) North Sea is 'dry', i.e. almost entirely methane.

12.2.2 *Refinery operations*[2]

The terms 'cracking' and 'reforming' are often used without qualification to describe a variety of refinery and primary petrochemical operations. It is hoped

Table 12.1 Petroleum fractions

Methane ⎱ Ethane ⎰	Associated gas	(Natural gas)
Propane ⎱ Butanes ⎰	Liquefied petroleum gases (LPG)	(Natural gas liquids—NGL)
C5(> 0°C)–70°C	Light gasoline or light naphtha	Feedstocks for motor spirit, also known as straight-run gasoline (US).
70°C–170°C	Naphtha (mid-range)	
170°C–250°C	Kerosene (UK)*	Vaporizing oil, jet fuel
250°C–340°C	Gas oil	Diesel fuel, light fuel oil
340°C–500°C	Heavy distillates (heavy gas oil in U.S.A.)	Feedstocks for lubricants and waxes, or heavy fuel oils.
Fluid residues (from light crudes)		
Semi-solid residues	Bitumen or asphalt	

* Depending upon projected use, fractions within this or a similar boiling range may be included within (full-range) naphtha or (light) gas oil classifications.

Table 12.2 Refinery and petrochemical cracking and reforming operations

Refinery	
Thermal cracking	Obsolete, thermal decomposition of middle/higher fractions to increase gasoline range hydrocarbons.
Thermal cracking/coking (delayed coking etc.)	Thermal decomposition of very heavy fractions to give mainly gases and a high coke yield.
Catalytic cracking	Accelerated decomposition, with some aromatization, of middle/higher fractions over solid acidic catalysts.
Hydrocracking	Accelerated hydrogenolysis/decomposition of heavy fractions to paraffinic hydrocarbons over metal/acid catalysts.
Thermal reforming	Obsolete, thermal rearrangement and aromatization of naphthas under high pressure.
Catalytic reforming (platforming etc.)	Metal/acid-catalysed rearrangement and aromatization of naphthas.
Petrochemical	
Steam cracking	Thermal cracking of C_2 + hydrocarbons to olefins in the presence of steam.
Thermal/autothermal cracking	More general term, including methane to acetylene— autothermal, with partial combustion.
Steam reforming	Nickel-catalysed formation of synthesis gases from hydrocarbons.

that Table 12.2 will provide a general guide to what is meant in a particular context.

12.2.2.1 *Catalytic cracking.* The old thermal cracking processes gave way to catalytic cracking—the 'cat-cracker'—during the 1940s in the U.S.A. The original clay catalysts have now been replaced by zeolites, operating at about 500°C and 2 atm. Carbon is laid down on the catalyst surface, and must be burned off at about 700°C; in the modern fluid (fluidized-bed) catalytic cracker (FCC unit), the particulate catalyst is transported continuously through the hydrocarbon reaction and regeneration zones. In the process, higher alkanes are cleaved, aromatics partially dealkylated and some naphthenes converted to aromatics by hydrogen transfer to olefins. The product mixture has a broad carbon number spread with some 60% by weight in the C_5 to 220°C cut and 15% as gases, mainly C_3 and C_4 alkenes and alkanes. The latter may be used in other refinery or chemical processes.

12.2.2.2 *Hydrocracking.* Hydrocracking is a more limited (and expensive) operation, aimed at converting still higher boiling fractions to naphtha. The process operates at about 450°C and 150–200 atm of hydrogen, with a metal (often palladium) on zeolite catalyst. Heavy aromatic and polyaromatic components are saturated and cleaved, to give a mainly paraffinic product (i.e. naphtha) suitable for reforming or petrochemical use.

12.2.2.3 *Catalytic reforming*. Older thermal reforming processes similarly gave way to catalytic reforming. A typical catalyst comprises platinum or a platinum-rhenium alloy (0·5–1%) on an acidic alumina, and is used at about 500°C and 7–30 atm pressure with a 80°–230°C naphtha feedstock. The most important reactions occurring are cyclization and dehydrogenation (known as 'dehydrocyclization') of alkanes, and dehydrogenation of naphthenes to aromatics. Various isomerization, alkylation, dealkylation, cleavage and hydrogen transfer reactions occur simultaneously. A typical feed and product composition might be:

	Feed	*Product* (c. 75% yield by volume)
Paraffins	60	32
Naphthenes	25	2
Aromatics	15	66
Individual aromatics (%)		
Benzene		4
Toluene		18
Xylenes and ethyl benzene		23
Higher		21

The rather low benzene content is fairly common and, with benzene in greatest demand, other sources of this material are required. Hydrogen and saturated C_{1-4} hydrocarbons (10–14% w/w) are also co-produced in reforming operations.

12.2.2.4 *Other refinery processes*. Some of the more intractable residues of refinery distillations are subjected to thermal coking, to provide small amounts of gas/liquid fuels while producing a more readily handled 'coke' fuel.

Olefins from cat-cracker operations (or even steam cracking of natural gas liquids in the U.S.A.), may be used to alkylate branched alkanes, particularly isobutane (c. H_2SO_4 or HF catalyst at near ambient temperature). Alternatively, propylene and butenes may be oligomerized, and the resulting C_6 + olefins hydrogenated to give 'polygasoline'.

The term 'hydrotreating' tends to be used to cover all refinery hydrogenation processes. These may range from hydrogenation of olefinic mixtures (liquid cracker products for example) under mild conditions, to the hydrodesulphurization of heavy fractions and the hydrogenation of (poly-) aromatic components.

12.2.3 *Energy consumption*

According to the number of operations carried out on a refinery site, some 7 to 10% of the fuel value of crude oil introduced is consumed in the separation and interconversion of product streams. If we momentarily take naphtha as a reference point, energy is required for the catalytic reforming process to provide higher octane gasoline. The subsequent separations of individual aromatic hydrocarbons consume still more energy; hence the 'chemical values' of these materials exceed their fuel values.

In later sections, the author has assigned a fuel value to naphtha. However, even this material requires the production, transportation and distillation of crude oil at the very least. Thus, any strict assessment of energy utilization for a particular product should ultimately include inputs to bring the raw material out of the ground and for all subsequent operations.

12.3
Lower olefins (alkenes) and acetylene (ethyne)

The paraffinic hydrocarbons in natural resources show very low reactivity under moderate conditions. Early petrochemical processes drew upon the greater reactivity of olefins, and when refinery supplies failed to meet demand means of converting alkanes to alkenes were developed. During the later period, low prices of naphtha (in Europe and Japan) and natural gas (in the U.S.A., and more recently Europe) also promoted petrochemical routes to acetylene.

The resulting processes all require high temperatures and are, therefore, both energy- and capital-intensive, so that very large scales of operation are desirable for efficiency and economic viability.

12.3.1 Cracking processes

12.3.1.1 Steam (thermal) cracking for lower olefins[3]. A typical naphtha for cracking has a carbon number range from 4 to 12 or more. The uncatalysed cracking reaction is carried out within tubes in a furnace enclosure at near atmospheric pressure (less than 3 atm). Steam makes up 30–45% w/w of the total feed to improve heat transfer, reduce the partial pressure of hydrocarbons (thermodynamically desirable) and remove carbon by the reaction

$$C + H_2O \rightarrow CO + H_2$$

Cracking temperatures were traditionally 750–850°C, but temperatures up to 900°C (high severity cracking) are becoming more common, in conjunction with shorter residence times (about 0·1 second).

Hydrocarbon dehydrogenation and cleavage reactions of the type

$$A \rightleftharpoons B + C + D \ldots$$

are always strongly endothermic, with ΔH values in the range 120–180 kJ for each additional molecule produced. Figure 12.1 shows the temperature dependence of $\log 10\, K_p$ (atm) values per g-atom of carbon for the formation of a number of relevant hydrocarbons from the elements. Either methane or benzene is the most stable hydrocarbon over the temperature range of interest, but *no* hydrocarbon is stable with respect to the elements between 800 and 4000 K. Hence the successful formation of olefins implies that kinetic and mechanistic factors are very important, and extensive use is now made of

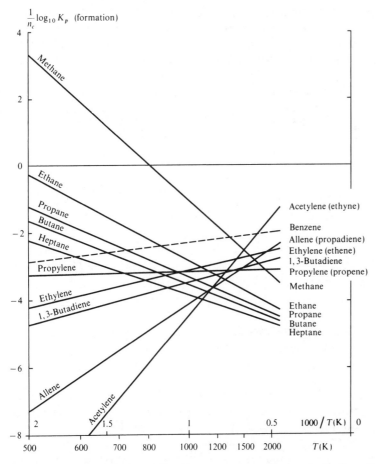

Figure 12.1   Temperature dependence of log K_p per gram atom of carbon for the formation of a series of hydrocarbons from their elements.

reaction and heat transfer models in the design of cracking furnaces and the selection of appropriate operating conditions by most licensing and operating companies.

The table on p. 358 shows typical overall yield patterns (%w/w), with ethane/propane recycle, for a number of cracker feedstocks; n-butane gives high ethylene yields, isobutane more propylene and methane. In all cases, the cracker product stream is cooled rapidly to below 400°C to minimize further reactions. After further cooling and separation of condensed hydrocarbons and water, the gases (H_2, C_1—C_4) are compressed, scrubbed with aqueous alkali to remove CO_2 and other acidic contaminants, and dried over solid beds. Thereafter, the C_2 and C_3 alkanes and alkenes are separated by distillations at pressures up to 35 atm., with refrigerated condensers for the early columns in the train. Selective hydrogenation to remove acetylenes and dienes (most frequently over a supported palladium catalyst) and further distillative purification may also be required. With ethane and propane feedstocks, the unconverted alkane is always recycled; the overall yield of ethylene on ethane may exceed 80%. In the cracking of higher feedstocks, ethane and propane may again be recycled, or recombined with the hydrogen and methane as fuel gas.

The C_4 fraction from the cracking of naphtha contains appreciable quantities of 1, 3-butadiene, and represents the major source of this material in western Europe and Japan (section 12.9.1).

Feed Product	Ethane	Propane	n/iso-Butane mixtures	Naphthas	Gas oil
Hydrogen	5–6	1–2	1	1	0.5
Methane	10–12	20–25	15–25	13–18	10–12
Ethylene	75–80	40–45	20–30	25–37	22–26
Propylene	2–3	15–20	15–25	12–16	14–16
Butadiene ⎱		1–2	0–2	3–5	3–5
Other C_4's ⎰	1	3–4	15–22	3–5	3–6
20(C_5's)–220°C	1	5–10	5–14	18–28	17–22
Fuel Oil	—	—	—	4–8	18–22

Finally, the gasoline fraction, frequently referred to as 'pyrolysis gasoline', is usually rich in aromatics, particularly benzene.

12.3.1.2 *Cracking processes for acetylene (ethyne)*[4]. Minor amounts of acetylene are formed in steam cracking, but rarely justify separation. However, if cracking temperatures are increased to over 1100°C acetylene becomes a significant component, and a different approach to reactor design is necessary. The Wulff process[5] furnaces contain stacks of tiles, and operate on 1-minute heating-cracking cycles. Fuel gas or oil is burned in air to heat the tiles to about 1200°C, and the feed is then switched to naphtha and steam, the temperature falling to about 1000°C to complete the cycle. The operation of 36 furnaces, with periodic flow reversals and coke burn-offs, is controlled by an IBM computer at Marathon Oil's plant in Germany.

At still higher temperatures (over 1500°C), methane can be partially converted into acetylene. An electric arc process has been operated, but the most successful and widespread approach is the BASF or Sachsse partial oxidation process. Typically methane (in excess) and oxygen are fed to a special burner, in which the temperature falls from a peak of about 2500°C to 1300°C in a very short time, at which point the gases enter a quench. A typical product stream (water-free basis) would contain some 8% (molar) acetylene, 25% carbon monoxide, 55% hydrogen and small amounts of methane, carbon dioxide and other gases. After acetylene extraction, the remaining gases may be integrated into a synthesis gas system. Société Belge d'Azote and Montecatini have commercialized their own, similar processes.

Finally, BASF have also developed a submerged flame autothermal cracking process for crude oil[6]. Partial combustion raises the temperature locally, to the point at which cracking to both ethylene and acetylene occur. A typical off-gas composition comprises approximately 1 mol acetylene, 1·15 mol ethylene, 7·5 mol carbon monoxide and 4·5 mol hydrogen, with minor amounts of C_3 and higher products.

12.3.2 *Energy balances and economics*

A standard economic analysis, as described in section 6.11, is obviously necessary when a company considers alternative primary raw materials and cracking process technologies. However, raw materials show significant variations in local availability and relative pricing, and primary petrochemical cracking operations give a variety of co-products which vary in value according to local demand. Hence, the relative economics for such alternatives vary quite markedly from one location to another. A rather more fundamental basis for comparison is the total net primary energy, including the energy value of the hydrocarbon (or other) feedstock(s), required per unit quantity of the major product. This figure determines the minimum non-renewable energy consumed in making all subsequent derivatives, and hence the status of materials and routes in attempts to conserve energy sources (see section 12.3.1)[7].

The following simplified examples are illustrative, rather than definitive. A full analysis requires detailed data for each individual operation. The author has used lower, or net, heats of combustion at 25°C for the energy (fuel) values of raw materials i.e. where all combustion products are in the gaseous state. Steam, whether generated or consumed, has been assigned an energy value of 42 kJ/mol, unless at high superheat. Electricity has been included at the primary fuel requirement based on a generation efficiency of 33%, which could be improved upon by future combined heat and power generation schemes. (Energy use in cooling water systems has been ignored).

The metals and other materials used for a plant, and fabrication of plant items, also entail energy consumption, which should strictly be added to the energy consumed by the process over the life of that equipment. The quantities involved are difficult to assess and have been ignored, but engineers may take such figures into account for new designs or the 'retrofitting' of additional equipment, particularly if only a marginal improvement in efficiency is expected. To some extent, the energy required for plant construction will be reflected in its capital cost, and other energy requirements being equal, lower capital cost options will be preferred on both energy and conventional economic grounds.

12.3.2.1 *Ethylene* (*ethene*) *and other olefins* (*alkenes*). We will start with the simpler cracking of ethane according to the following equation and theoretical

energy balance (in MJ/mol ethylene), based on energy (fuel) values as defined above.

$$C_2H_6(+\text{ heat}) \rightarrow C_2H_4 + H_2 \qquad (\Delta H_r^{\ominus} = 0.137)$$
$$1.428 + 0.137 \qquad 1.323 + 0.242 \qquad (\text{total } 1.565)$$

Although absent from the equation, generation of the co-fed steam requires energy, and extra heat is necessary to raise the temperature of the reactants to over 800°C. Some of this extra energy is recovered (by steam generation) from the furnace flue gases and the reactor outlet gases, but, with additional steam required for product separation and purification, a net steam consumption results. The overall yield of ethylene on ethane in a modern U.S. plant is about 85% molar, and an approximate energy balance might be as follows:

Inputs:		
	Ethane (1·18 mol)	1·68
	Furnace fuel	0·53
	Steam consumed (net)	0·23
	Electricity	0·01
		2·45 MJ/mol

By-products: (credited at fuel values)		
	Gases (H_2, CH_4)	0·33
	C_3, C_4	0·07
	Pyrolysis gasoline	0·05
		0·45 MJ/mol

Net input for ethylene	2·00 MJ/mol (71GJ/tonne)

Thus the net energy* utilization is about 1·5 times the fuel value of ethylene, indicative of significant inefficiency in the overall process. (As an alternative to cracking, Union Carbide have patented improved catalysts for the selective oxidation of ethane to ethylene, which may ultimately prove more energetically and economically favourable.)

The analysis of propane or naphtha cracking is rather more complex, owing to the variety of products formed, and changes in the distribution with operating conditions. However, some reasonably typical figures with ethane recycle (MJ/mol ethylene) are as follows:

Propane (64·2g)	2·98	—
Naphtha (84g)	—	3·62
Other fuel	0·58	0·80
Steam consumed	0·26	0·32
Electricity	0·01	0·01
Inputs	3·94	4·75

---

*Note: 'Energy' is used in a more limited sense in this chapter and is essentially only the chemical energy referred to in chapter 8.

Gases (CH_4, H_2, trC_2)	1·00	0·80
Propane	(recycled)	0·03
Propylene	0·37	0·66
C_4s	0·08	0·38
Pyrolysis gasoline	0·35	0·55
Fuel oil	—	0·08
By- and co-products (credited at fuel values)	1·80	2·50
Net input for ethylene	2.14 MJ/mol (76GJ/tonne)	2.25 MJ/mol (80GJ/tonne)

The energy requirements appear to be higher than for ethane cracking. But all energy losses have been assigned to ethylene by crediting the by- and co-products at fuel values. The deliberate production of propylene and other olefins as chemical feedstocks, and gasoline fractions of higher octane value than naphtha, would have incurred additional energy utilization and losses. Thus the useful co-products should be credited at a higher energy value, referred to as the chemical value in later analyses.

From an economic viewpoint, the capital costs of naphtha-crackers are higher; but naphtha is available worldwide whereas ethane availability is more limited, with the U.S. retaining a dominant, if no longer exclusive, position.

12.3.2.2 *Acetylene* (*ethyne*). Before evaluating the petrochemical routes to acetylene, we should look first at the long-established process for producing acetylene via calcium carbide and water:

$$CaO + 3C \rightarrow CaC_2 + CO$$
$$CaC_2 + H_2O \rightarrow C_2H_2 + CaO \text{ (or } Ca(OH)_2 \text{ with excess water)}$$

Essentially all the energy is consumed in calcium carbide manufacture. In the electrothermal process, the reaction is carried out at temperatures in excess of 1850°C. If we ignore the lime (recovered), and convert other consumption/by-product figures to energy values, we obtain (per mol acetylene equivalent):

Coke (3·5g-atoms C/mol)		1·4
Electricity—actual input	1·0	
Primary fuel value	—	3·0
	2·4 MJ/mol	4·4 MJ/mol
CO produced (1·32 mol, fuel value)		0·4
Net input	2·0 MJ/mol (77GJ/tonne)	4·0 MJ/mol (154 GJ/tonne)

Hence, with electricity generated by conventional means, the net primary energy consumption is very high at over three times the fuel value of acetylene, whereas use of hydroelectric power could possibly make acetylene production competitive. (Any energy losses in the conversion of coal, as mined, into coke have been ignored.)

Returning to petrochemical raw materials, some figures for a Wulff naphtha cracking operation are as follows (in MJ/mol acetylene):

Inputs:	Naphtha feed (113 g/mol)	5·07
	Furnace fuel	1·69
	Steam (net, for separations)	1·03
	Electricity (primary fuel)	0·11
		7·90 MJ/mol
By-products:	Gases (5 mol, H_2, CH_4, CO, C_2H_6)	1·97
	C_3 and higher	0·65
	Carbon	0·11
		2·73 MJ/mol
	Net	5·17 MJ/mol
Co-products:	C_2H_4(0·94 mol, @ 2 MJ/mol) chemical value	1·88
	Input for acetylene, net	3·29 MJ/mol (127GJ/tonne)

The 'chemical value' assigned to ethylene is based on the net energy requirement for ethylene production by steam cracking.

For the partial oxidation of methane, a generalized 'design case' gives the following set of approximate figures (again in MJ/mol acetylene).

Inputs:	Natural gas	5·45
	Additional fuels/steam	0·83
	Oxygen (4 mol/mol)	0·40
	Electricity (primary fuel value)	0·35
		7·03 MJ/mol
By-products:	Methane (in synthesis gas)	0·49
	C_2 and higher	0·27
		0·76 MJ/mol
	Net	6·27 MJ/mol
Co-products:	Synthesis gas (10 mol, c. $2H_2$:1CO); fuel value	2·55
	chemical value (say 1·2x)	3·06
	Input for acetylene, net	3·72 MJ/mol 3·21 MJ/mol
		(143 GJ/tonne) (123 GJ/tonne)

Somewhat similar net figures are obtained for the autothermal cracking of crude oil if the ethylene coproduct is credited at 2 MJ/mol. Even with chemical values attached to co-products, the petrochemical processes consume over 3 MJ/mol for the production of acetylene. Furthermore, the capital costs for these processes are appreciably higher than for ethylene crackers. Thus, while the production costs of acetylene show considerable variations from one location to another, it is invariably dearer, by 30% to over 100%, than ethylene.

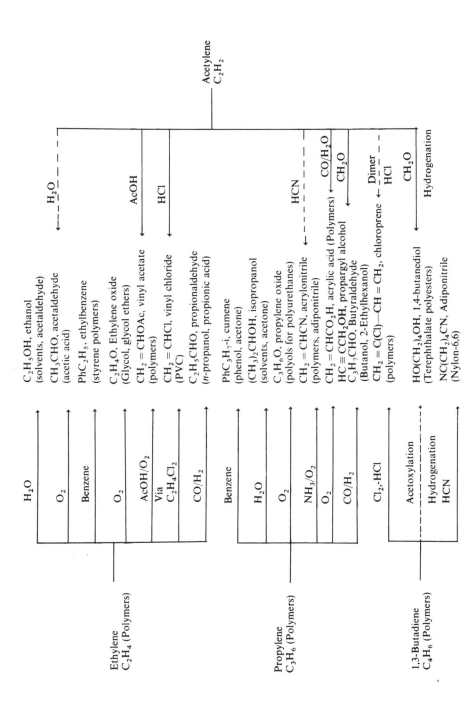

Figure 12.2 Major outlets for ethylene, propylene butadiene and acetylene. Parentheses indicate further uses of the intermediate product.

12.3.3 *Lower olefins (alkenes) versus acetylene (ethyne)*

The major outlets for ethylene, propylene, butadiene (section 12.9.1) and acetylene are shown in Figure 12.2.

12.3.3.1 *General considerations.* A number of organic products are potentially producible from either acetylene or a lower olefin, for example:

$$C_2H_2 + H_2O \qquad \rightarrow CH_3CHO \qquad\qquad \leftarrow C_2H_4 + \tfrac{1}{2}O_2$$
$$C_2H_2 + CH_3CO_2H \rightarrow CH_2 = CHO_2CCH_3 \leftarrow C_2H_4 + \tfrac{1}{2}O_2$$
$$+ CH_3CO_2H$$
$$C_2H_2 + HCN \qquad \rightarrow CH_2 = CH \cdot CN \qquad \leftarrow C_3H_6 + NH_3 + 1\tfrac{1}{2}O_2$$

The reactions of acetylene are simpler, and were important early commercial routes to the products indicated, acetaldehyde, vinyl acetate and acrylonitrile.

We have just seen that the net energy requirements for acetylene are higher than for ethylene, and the difference is reflected in the energy requirements for all downstream products. To give an indication of how energy balances can continue to be applied the following table has been compiled from fairly modern sources. Energy inputs of 2·0 and 3·0 MJ/mol have been used for ethylene and acetylene respectively.

	Acetaldehyde		Vinyl acetate	
	C_2H_2	C_2H_4	C_2H_2	C_2H_4
C_2H_2 (1·05 mol)	3·15	—	3·15	—
C_2H_4	— (1·05 mol)	2·10	— (1·08 mol)	2·16
Oxygen	—	0·06	—	0·07
Steam	0·21	0·12	0·46	0·48
Electricity	0·08	0·02	0·05	0·08
Acetic acid			say 2·55	2·55
Total input	3·44 MJ/mol	2·30 MJ/mol	6·21 MJ/mol	5·34 MJ/mol
	(78 GJ/tonne)	(52 GJ/tonne)	(72 GJ/tonne)	(62·1 GJ/tonne)

However, the main driving force to discover and develop olefin based routes was the dramatic increase in scale of steam cracking operations in the post-war years. Early increases in demand for synthetic ethanol and ethylene oxide/glycol were followed by a rapid escalation in polyethylene production, now the major outlet for ethylene. In contrast, the group of products for which acetylene was the preferred feedstock grew more modestly, and a disparity in required production scales and prices developed. The steam cracking of naphtha in Europe and Japan also gave propylene in large quantities, for which new outlets had to be found. Thus acetylene was slowly squeezed out, including later displacement by butadiene for chloroprene manufacture, and the ratio of ethylene production to acetylene production is now 100-fold in the U.S.A., if somewhat less elsewhere.

12.3.3.2 *Vinyl chloride (chloroethene)*. The production of vinyl chloride naturally attracted the attention of those companies already involved in chlorination processes. The quantities of by-product hydrogen chloride available were often an embarrassment and the relatively facile production of vinyl chloride from acetylene and anhydrous HCl, over a supported mercuric chloride catalyst, provided a valuable outlet.

As the market for vinyl chloride grew, major petrochemical companies without existing chlorination capabilities looked to other possible routes. With falling ethylene prices, the route pursued was the addition of chlorine to ethylene, to give ethylene dichloride (EDC, 1, 2-dichloroethane), and the thermal cracking of EDC at about 550°C to vinyl chloride and HCl.

$$CH_2 = CH_2 + Cl_2 \rightarrow ClCH_2CH_2Cl \rightarrow CH_2 = CHCl + HCl$$

A snag was that these companies now had to dispose of hydrogen chloride. The most elegant solution was to react this with acetylene in a separate stage, to give what became known as the 'Balanced Process'. This scheme became a driving force for naphtha cracking to mixtures of ethylene and acetylene, as exemplified by the Wulff process.

Nevertheless these special crackers proved expensive, and many gave serious operational problems. The 'final solution' came with the development of the ethylene oxychlorination process. The reaction

$$CH_2 = CH_2 + 2HCl + \tfrac{1}{2}O_2 \rightarrow ClCH_2CH_2Cl + H_2O$$

is carried out at 250–300°C over supported copper chloride (melt) catalysts. This process may be combined with chlorine addition and EDC cracking to provide a balanced operation, if required; the overall stoichiometry then becomes

$$2CH_2 = CH_2 + Cl_2 + \tfrac{1}{2}O_2 \rightarrow 2CH_2 = CHCl + H_2O$$

As a guide, a rough energy balance for a modern vinyl chloride production/polymerization operation is as follows (MJ/mol or GJ/tonne as indicated)

Ethylene (1·05 mol)	2·1	MJ/mol
Chlorine	0·8	
Processing inputs	0·9	
Vinyl chloride	3·8	MJ/mol

which gives

Vinyl chloride monomer	62	GJ/tonne
Processing inputs	10	
Polyvinyl chloride	72	GJ/tonne

The polymer referred to here is rigid PVC. The flexible material contains appreciable amounts of 'plasticizers' (mainly phthalate esters), which obviously affects the overall energy utilization.

12.3.4 *Polyethylene (polyethene) and polypropylene (polypropene)*

Low-density polyethylene (LDPE) is produced by a free radical process, initiated with traces of oxygen or peroxides, at temperatures of 200°C to over 300°C and pressures up to 3000 atm. Over half of this low-melting point, flexible product is used for packaging film manufacture.

High density polyethylene (HDPE) is now produced mainly by polymer growth on microscopic particles of catalytic materials (chromium or Ziegler systems) suspended in a non-solvent or carried along in the gas-phase, at only 70–125°C and 15–40 atm. The product has a higher melting point and is more rigid, and finds major uses in containers, moulded items and pipes. Polypropylene, with both molding and fibre uses, ethylene-propylene copolymers and the newest product line, linear low-density polyethylene (LLDPE), are produced by similar methods. A C_4/C_6 terminal olefin co-monomer is used in the latter to provide a tough, flexible film forming product.

Yields in all these processes are in the range 97–99% of theory overall, giving feedstock energy inputs of 72–73 GJ/tonne with ethylene at 2 MJ/mol. Processing requirements add 10–12 GJ/tonne for the high pressure (LDPE) process and 6–7 GJ/tonne for the low pressure (HDPE, LLDPE) processes.

12.3.5 *Production and use statistics*

The U.S. production of ethylene is expected to exceed 19 Mt in 1994. Although well over half the capacity can use naphtha or heavier feedstocks, almost half the ethylene produced is still derived from ethane and a quarter from propane.

In western Europe, ethylene production is running at over 16 Mt per annum. Naphtha and a small quantity of heavier fractions have been the traditional cracker feedstocks. However, since the commissioning of the 500 kt per annum Esso–Mossmoran ethane cracker in the mid-1980s, considerable additions have been made to ethane/LPG cracking capacity in the northern U.K. (including BP)[8], which, with similar operations elsewhere in Europe, have significantly changed the feedstock pattern.

Japanese ethylene production slipped below 6 Mt in 1993, almost entirely based on naphtha

The following table gives an indication of the major end uses of ethylene in the U.S.A.

LDPE (high pressure)	16%
HDPE	24%
LLDPE	13%
Ethylene oxide	14%
Dichloroethane	13% (for vinyl chloride)
Ethylbenzene	7% (for styrene)
Linear olefins and alcohols	6%
Vinyl acetate	3%
Acetaldehyde	1%
Ethanol	1%

Outside the U.S.A. the pattern varies, with LLDPE capacity continuing to lag behind, despite rapid expansions. In western Europe, LDPE still represents over half and LLDPE only 12% of all polyethylene produced (8.8 Mt per annum in total); vinyl chloride production is about 5 Mt per annum (cf. some 6 Mt per annum in the U.S.A., despite environmental concern regarding chlorine-containing chemicals).

The ratio of propylene to ethylene production is over 0.7 in Japan, but slightly less than 0·6 in Europe, reflecting differences in cracker feedstocks. In the U.S.A., additional propylene is recovered from refinery streams to meet the 10 Mt per annum demand. Some recent figures for propylene usages in the U.S.A. are:

Polypropylene	39%
Acrylonitrile	14%
Propylene oxide	11%
Cumene	10%
Oxo alcohols	8%
Isopropanol	7%
Oligomers	5%
Acrylic acid	3%

The proportion converted to polypropylene is about 45% in Japan and approaching 50% in western Europe, where isopropanol is rather less important.

The U.S. production of acetylene fell from a peak of 450 kt per annum in the late 1960s, but has recently risen slowly from 130 to 160 kt per annum. The largest producers use methane oxidation, with carbide in second place. Some 45% is used for vinyl chloride (to give only 3% of all vinyl chloride produced); the more important use is for 1,4-butanediol (section 12.9.8).

The displacement of acetylene was somewhat slower in Europe. With recent ups and downs, the author has lost track of the status of the various production facilities, but there are continuing outlets into vinyl acetate and BASF's growing C_4 diols production. Japan still produced some 100 kt of acetylene from carbide in 1993, possibly supplemented by cracker co-product.

12.4
Synthesis gas, ammonia
and methanol

(See also section 3.5.1)

The production of mixtures of carbon monoxide and hydrogen from coal or coke was the basis for 'town gas' manufacture from well before the turn of the century, and the processes were adapted to provide appropriate feeds for ammonia and methanol synthesis in Germany and elsewhere.

$$C + H_2O \rightarrow CO + H_2 \quad \text{(endothermic)}$$
$$C + \tfrac{1}{2}O_2 \rightarrow CO \quad \text{(exothermic)}$$
$$CO + H_2O \rightleftharpoons CO_2 + H_2 \quad \text{(water gas shift)}$$

Low prices for natural gas in the U.S. and naphtha in Europe prompted a change in feedstock. In the U.K., 'town gas' was also produced by the 'steam reforming' of naphtha, which provided about 90% of supplies in 1968, but was totally displaced by the newly found natural gas during the 1970s. European production of synthesis gas for chemical processing is now almost solely based on steam reforming of natural gas.

12.4.1 *Process descriptions*

12.4.1.1 *Synthesis gas processes.* The methanation of carbon monoxide

$$CO + 3H_2 \rightarrow CH_4 + H_2O$$

over supported nickel catalysts has long been known to be very thermodynamically favourable at temperatures up to 400°C. In fact, the reaction is reversible, the free energy change passing through zero at about 620°C. The endothermic reverse reaction, the 'steam reforming' of methane, is carried out with a H_2O/CH_4 molar ratio of about 3:1 at 800–850°C, within catalyst filled tubes in a fuel-fired furnace[9]. The feedstock must first be freed from sulphur compounds. Although thermodynamically undesirable, the process may operate at 10–40 atm. to take advantage of gas supply pressures and minimize subsequent compression. An important secondary reaction is the water-gas shift reaction, which forms CO_2 and increases the H_2/CO ratio to *c.* 5:1 molar (the equilibrium constant is roughly 1 at reformer temperature). The CO_2 may be scrubbed out, but if higher concentrations of CO are required, this CO_2 and possibly additional CO_2 recovered from the furnace flue gas may be recycled to the reformer inlet. Alternatively, if very high H_2/CO ratios are required, the exit gas is subjected to further water-gas shift reaction over iron or copper catalysts, and the CO_2 is scrubbed out and rejected.

The steam reforming of naphtha is, in essence, the reverse of the Fischer–Tropsch reaction, and is carried out under very similar conditions to methane reforming.

$$(CH_2)_n + nH_2O \rightarrow nCO + 2nH_2$$

The feedstock must again be desulphurized, most frequently by 'hydrodesulphurization' over a cobalt-molybdenum oxide catalyst. The nickel reforming catalyst is usually doped with alkali or supported on magnesia to reduce carbon deposition. The stoichiometry leads to somewhat lower H_2/CO ratios, but CO_2 recycle or addition and further shift reactions can again be applied if required.

If the gas mixture is destined for ammonia synthesis, natural gas is the preferred feedstock, and a high steam/hydrocarbon ratio may be used to achieve the highest possible ratio of H_2 to $CO + CO_2$. As an option to conventional reforming, the initial steam reforming stage may be effected at lower temperatures, with incomplete conversion of the hydrocarbon, and the process stream then subjected to 'secondary reforming', wherein air is introduced and the combined stream passed through a compact catalyst-filled reaction vessel. Here, partial combustion raises the temperature to complete the reforming process, while consuming oxygen to leave the requisite quantity of nitrogen. (Similar schemes, but with oxygen injection into the secondary- or post-reformer, or even the principal reformer, are claimed to save energy in methanol synthesis gas production[10]). Multiple shift reaction and CO_2 removal stages would follow, with a final methanation step, or contact with a solid adsorbent in the Linde process, to eliminate traces of carbon oxides.

Finally, any hydrocarbon can be converted to CO/H_2 mixtures by partial oxidation, which may be the preferred commercial option, if a higher CO/H_2 ratio is required from the most readily available feedstock, e.g.

$$CH_4 + \tfrac{1}{2}O_2 \rightarrow CO + 2H_2$$

$$(CH_2)_n + \frac{n}{2}O_2 \rightarrow nCO + nH_2$$

When naphtha prices were high, there was considerable activity on developing and piloting improved partial combustion processes for heavier oil fractions and coal (especially in Germany and the U.S.A.). But with lower efficiencies, the need for expensive gas purification and the handling problems of coal and residual ash, the work has lost impetus for the time being.

If required, relatively pure carbon monoxide can be recovered by cryogenic separation (i.e. condensation at very low temperatures), use of selective scrubbing solutions (for example $CuAlCl_4$ in toluene, the COSORB process) or solid adsorbents, and by other methods. Some CO recovery techniques are also applicable to the CO-rich (c. 70% molar) off-gases from basic oxygen furnaces and the leaner off-gases from air blast furnaces produced in steel manufacture, though operations of this type are still limited.

12.4.1.2 *Ammonia synthesis*[11]. Although inorganic, ammonia is currently dependent on hydrocarbon feedstocks, with production figures of a similar order to ethylene from plants making 1000 tonnes/day or more. The reaction

$$3H_2 + N_2 \rightleftharpoons 2NH_3$$

is exothermic to the extent of over 45 kJ/mol NH_3. Ammonia formation becomes quite unfavourable at the temperatures of about 450°C required to achieve suitable reaction rates over iron catalysts, and pressures of 250–300 atm. are necessary to achieve conversions of about 20–25% per pass. Thus, a hydrogen-rich, CO-free synthesis gas will be combined with extra nitrogen, if necessary, pressurized by means of a very large centrifugal compressor, and passed through a series of adiabatic catalyst beds, with inter-cooling by heat exchange or cold gas injection. The ammonia may be condensed by cooling under pressure, and the gases recycled, with a purge to limit the build-up of 'inerts'.

12.4.1.3 *Methanol synthesis*[12]. The thermodynamics of methanol synthesis

$$CO + 2H_2 \rightleftharpoons CH_3OH \qquad \Delta H^\circ = -100 \, kJ/mol.$$

show some similarities to ammonia synthesis. The following table shows the effect of temperature on the equilibrium constant and the calculated pressure (uncorrected for departures from ideality) for an equilibrium concentration of 10 mol % methanol in a 1 $CO:2H_2$ mixture.

Temperature (°C)	$K_p (atm^{-2})$	Pressure for 0·1 mole fraction (atm.)
250	$2·09 \times 10^{-3}$	21·6
300	$2·85 \times 10^{-4}$	58·3
350	$5·29 \times 10^{-5}$	136
400	$1·24 \times 10^{-5}$	281

Any carbon dioxide present can react directly with hydrogen to give methanol

$$CO_2 + 3H_2 \rightarrow CH_3OH + H_2O$$

or undergo the water-gas shift reaction. However, larger amounts detract from the overall rate and equilibrium conversion, as do any inert components. Other minor side-reactions give higher alcohol impurities.

The earlier zinc-chromium oxide catalysts require reaction temperatures of about 400°C, and hence pressures of 300 atm and above are commonly used. For very large-scale operation (500 tonnes/day or more), the compression/reaction system is very similar to that described for ammonia synthesis. However, large methanol plants are not required in less well-developed areas. The characteristics of compressors are such that smaller high-pressure plants must use reciprocating machinery, which is costlier to instal and maintain.

The introduction of copper into methanol catalysts by ICI in the late 1960s, with other modifications to provide acceptable catalyst life, revolutionized this situation by permitting reaction temperatures of about 250°C. Thus pressures down to 50 atm. became usable, allowing the retention of centrifugal compressors, with smaller economic penalties, for modest scales of operation. Methanol yield also improved by 2–3%. Other process licensing companies have now developed copper-containing catalysts, and these form the basis for essentially all new methanol plants, but with somewhat higher pressures (up to 100 atm.) in the larger plants. This approach has allowed Lurgi to design multitubular reactors, cooled by steam generation in the shell, to provide improved temperature control and energy recovery.

12.4.2 *Energy balances and economics*

For synthesis gas mixtures, or syngas for short, energy inputs have been expressed per mole of reducing gas, that is the total quantity of carbon monoxide and hydrogen. This quantity is unaffected by CO_2 addition, recycle or removal and any water-gas shift operations, though the fuel value of the mixture changes with composition.

The table below, based on data from the late 1960s to the mid 1970s, shows energy requirements (in kJ/mol) for the steam reforming of natural gas (NG) and naphtha feedstocks to give either a hydrogen-rich stream, suitable for ammonia synthesis, or a stream destined for methanol synthesis. The latter would typically comprise H_2, CO and CO_2 in molar proportions of approximately 3:1:0·3. Although still slightly hydrogen-rich compared with the stoichiometric requirements for the parallel routes from $CO + 2H_2$ and $CO_2 + 3H_2$ to methanol, CO_2 injection would be required with natural gas as feedstock.

| Feedstock | H_2-rich syngas | | Syngas for methanol production | |
	Natural Gas	Naphtha	Natural Gas (at 10 atm)	Naphtha
Reformer feed	217	220	221	217
Reformer fuel	146	150	125	152
CH_4 in syngas	(16)	(15)	(20)	(16)
	347	355	326	353
Electricity	2	2	3	3
Steam generated	(36)	(27)	(34)	(36)
Net energy input	313 kJ/mol	330 kJ/mol	295	320 kJ/mol
CO_2 recovery: steam			12	
compression			4	
(electricity)				
			311 kJ/mol	

Later data for ammonia manufacture (Haldor Topsøe and Uhde)[13] suggest that the energy input for hydrogen production by the steam reforming of natural gas can be reduced to approx. 300 kJ/mol; the total inputs are then about 0·5 MJ/mol ammonia, i.e. less than 30 GJ/tonne. (For comparison, the inputs for a modern coal-based ammonia plant would be approx. 50 GJ/tonne, with much higher capital costs.)

To show the impact of the ICI low-pressure catalyst on methanol synthesis, the figures for syngas generation from naphtha and some additional data from early comparisons of the high-pressure (HP) and low-pressure (LP) processes have been taken, with no other changes. The energy required for syngas compression was reduced. The higher selectivity of the ICI catalyst reduced not only the quantity of syngas required to form 1 mole of methanol (from 3·21 to 3·13 moles), but also the quantity that had to be purged from the methanol synthesis gas-recycle loop to avoid the build-up of impurities. Finally, purification of the crude methanol was easier. This notional exercise gives the figures below (in MJ/mol methanol).

| | Notional synthesis | | Recent Lurgi data | |
| | (HP) | (LP) | (LP) | (LP) |
Feedstock	Naphtha	Naphtha	Naphtha	Natural Gas
Reformer feed	0·781	0·729	0·680	(Not
Reformer fuel	0·547	0·510 ⎫		reported)
Purge gases	(0·144)	(0·134) ⎬	0·277 net	
Net feed/fuel	1·184	1·105	0·957	0·938
Steam				
reformer	(0·130)	(0·121) ⎫	Steam generation in the	
syngas compr.	0·166	0·119 ⎬	reformer and methanol	
purification	0·107	0·096 ⎭	synthesis balances usages	
Net use	0·143	0·094	0	0
Electricity	0·010	0·009	v. small	0·017
Total energy input	1,337	1,208	0·957	0·955 MJ/mol
	(42)	(38)	(30)	(30) (GJ/tonne)

Thus, the ICI catalyst could reduce energy requirements by about 10%. Furthermore, the lower compressor duty reduced capital costs for the methanol synthesis plant by about 20%, and a slightly smaller reformer was required. For small scales of operation, the elimination of reciprocating machinery meant even larger notional savings.

In the meantime, each company has tackled the question of energy efficiency within the integrated synthesis gas generation/methanol production system. Lurgi's somewhat more expensive multi-tubular synthesis reactor permits one of the most efficient operations. Some comparative figures have been included in the above table, while the higher energy utilizations entailed in using heavy residue or coal as feedstocks are demonstrated by the following Lurgi data (in a more conventional economic form, albeit with feedstock and fuel requirements expressed in terms of their lower heats of combustion rather than weight or volume)[14].

Feedstock	Natural gas	Heavy residue (oxidation)	Coal
Feed and fuel, GJ/tonne	29·7	38·3	40·8
Electric power, kWh	—	—	—
Raw water, tonne/tonne	3·1	2·5	3·8
Catalyst and chemicals, $ per tonne	1·0	0·5	0·6

12.4.3 Urea (carbamide), formaldehyde (methanol), amino resins and polyacetal

Ammonia and methanol are starting points for amino-resins, which are therefore syngas derivatives. Urea-formaldehyde resins are used in particle board (chipboard) and plywood manufacture; 'UF-foam' is also widely used for cavity wall insulation in the U.K. Melamine-formaldehyde resins are possibly more familar from the dinnerware and heat-resistant wood finishes produced.

Urea is manufactured from ammonia and carbon dioxide at about 200°C and pressures from 150 atm. upwards. Melamine (2,4,6-triaminotriazine) is formed by heating urea to a high temperature (over 350°C) at modest pressures (1–7 atm.) over an alumina or silica-alumina catalyst. The predominant overall reaction is:

$$6\,NH_2CONH_2 \longrightarrow \text{Melamine} + 3\,CO_2 + 6\,NH_3$$

Urea   Melamine

Two process variants are used for the manufacture of formaldehyde from methanol, partial oxidative dehydrogenation or oxidation. In the first, methanol vapour is mixed with a stoichiometrically deficient quantity of air and passed over a silver catalyst at temperatures of 400–600°C. The two reactions

$$CH_3OH + \tfrac{1}{2}O_2 \rightarrow HCHO + H_2O \text{ (exothermic)}$$

and

$$CH_3OH \rightarrow HCHO + H_2 \text{ (endothermic)}$$

occur in parallel. In the oxidation process, methanol and excess air are passed over an iron molybdate catalyst at 350–450°C. Overall selectivities have improved from about 89% to 93% over the years.

In resin formation, the major reaction is of the type

$$-NH_2 + \underset{O}{\overset{H}{\underset{\|}{C}}}{}^{H} + H_2N\!-\!-\!\rightarrow -NH-CH_2-NH- + H_2O$$

A ratio of formaldehyde molecules to amino groups of about 0·8 is typical. Excluding the final processing inputs, energy requirements for urea-formaldehyde resins are now quite low at about 40 GJ/tonne.

Formaldehyde (anhydrous) is also the major reactant for polyacetal resins, important for engineering components. The chemical structure is essentially $RO(CH_2O)_nR'$, with co-monomers or 'end-capping' reagents incorporated to prevent slow reversion to formaldehyde.

12.4.4 *Production and use statistics*

Production figures for ammonia in the U.S.A. and western Europe are about 16 Mt and 11 Mt per annum respectively.

Urea and nitric acid production consume about 20% and 15% respectively. However, like most ammonia derivatives, these are predominantly used in fertilizers which, together with the uses of hydrocarbon fuels and electricity in farm machinery, contribute to a considerable consumption of fossil fuel resources in modern agriculture. Plastics and synthetic fibre outlets account for less than 15%, including the urea-formaldehyde (0·8 Mt in the U.S.A., higher in Europe) and melamine resins.

U.S. methanol production in 1993 amounted to 4·8 Mt, but demand possibly exceeded 8 Mt, the difference being met by imports. The increased production of MTBE (methyl *t*-butyl ether, see section 12.9.3) appears to have consumed some 4 Mt of methanol, compared with 1·45 Mt for formaldehyde (1·3 Mt, 100% bases) and possibly 0·7–0·8 Mt for the carbonylation routes to acetic acid anhydride. (Amino-, phenolic- and polyacetal-resins accounted for 27%, 22% and 12% of formaldehyde use respectively, and C_4 diols about 12%.)

Almost half the 5+ Mt per annum demand for methanol in western Europe is now met by imports, but formaldehyde production requires some 2·4 Mt of this; MTBE production appears to require a little over 1 Mt methanol at present.

12.5
Acetic (ethanoic) acid and anhydride

12.5.1 *Acetic acid production*

The various routes to acetic acid warrant a brief review, to illustrate the chemical industry's response to change and provide an introductory guide to future options.

The earliest route to acetic acid was the bacterial souring of wine, which eventually became the basis of vinegar manufacture. Wood distillation yielded stronger solutions of acetic acid (15 to over 90%). Surprisingly, some 10 000 tonnes per annum of acetic acid were still produced by this means in the U.S.A. in the mid-1960s.

However, the major routes to acetic acid, developed during and after the

First World War, were based on the oxidation of acetaldehyde derived from either acetylene or fermentation ethanol. The latter could well return to favour in countries such as Brazil (section 12.7.1.). After the Second World War, fermentation ethanol gave way to synthetic ethanol, via the direct hydration of ethylene. (Synthetic ethanol made by the sulphuric acid process had already made some inroads in the U.S.A.). From 1960 onwards, the Wacker oxidation of ethylene added a further option for acetaldehyde manufacture.

When we turn to energetic aspects, acetylene hydration and ethylene oxidation have already been discussed in section 12.3.3.1. The appropriate energy input for fermentation ethanol is somewhat difficult to establish (see section 12.7.1.), but for the hydration of ethylene and subsequent oxidation of synthetic ethanol (in modern plants) we have:

Ethylene (1·03 mol)	2·06
Processing energy	0·51
Input for ethanol	2·57 MJ/mol (56 GJ/tonne)
Ethanol (1·03 mol)	2·65
Processing energy	0·29
Input for acetaldehyde	2·94 MJ/mol (67 GJ/tonne)

In practice, crude ethanol can be fed to oxidation, with some overall energy savings (possibly 0·2 MJ/mol), but there is little doubt that the Wacker process is energetically advantageous.

The oxidation to acetic acid is carried out by feeding acetaldehyde and air or oxygen into acetic acid containing a soluble manganese compound at 40–60°C. (With an ester solvent, and no catalyst, peracetic (perethanoic) acid is formed. Although hazardous, this material finds use in a number of epoxidation processes). An approximate energy balance, based on the most favourable cracking and Wacker oxidation data, gives:

Acetaldehyde (1·04 mol)	2·39
Oxygen (0·55 mol)	0·06
Steam and power	0·10
Input for acetic acid	2·55 MJ/mol (42·5 GJ/tonne)

(A number of companies have sought to oxidize ethylene directly to acetic acid, with some success, but no process has been commercialized.)

Processes for the liquid-phase oxidation of hydrocarbons were conceived to by-pass the cracker. Until the early 1970s, fixed costs dominated primary petrochemical operations, and represented a significant part of the price increment in each successive chemical conversion; acetic acid was well down the line. With very low feedstock prices, selectivity was of lesser concern. *n*-Butane was selected as a preferred feedstock by Celanese and Union Carbide in the U.S.A., providing acetic acid in rather more than 50% selectivity (carbon basis), with only minor amounts of formic acid (c. 10% w/w) and propionic acids as co-products; some 2-butanone could be withdrawn if desired. (The Union Carbide plant was shut down some years ago, but the Celanese plant was reconstructed, and expanded, following a serious fire in the late 1980s). BP Chemicals' 'DF Process' oxidizes light naphtha to give acetic acid as the major product, but with quite significant quantities of formic and propionic acids, acetone and other products if required. All these oxidation processes operate at temperatures of 170–200°C and with air at pressures of 50–60 atm. Over the years, optimization of operating parameters, the use of catalysts and improved hydrocarbon recoveries have provided progressive improvements in selectivities, with scope for modest changes in product ratios in the DF Process. Of course, a multiplicity of distillations are required to separate and purify the oxidation products.

Turning to energy considerations, part of the loss of selectivity is to carbon dioxide, and the recovery of energy from this oxidation/partial combustion process is important. Fortunately, the designers of the two 90 + kilotonnes per annum DF plants, commissioned in 1967 and 1972 before the oil crisis, also had the foresight to incorporate features which permit efficient energy recovery from the system as a whole. For example, air compressors are driven by gas turbines, with steam generation by exhaust gases, while off-gas turbines generate electricity.

The final route considered is the carbonylation of methanol. British Celanese reported experiments in the late 1920s, but German workers were more persistent. BASF finally commissioned a small plant in 1960, operating at about 250°C and well over 300 atm. initially, with a cobalt/iodide catalyst system. Nevertheless production was expanded several times, and the process licensed to Borden in the U.S.A. In 1970, Monsanto commissioned their own 150 000 tonnes per annum methanol carbonylation plant at Texas City, with a rhodium/iodide catalyst system at 150–200°C and only 30–50 atm[15]. In the last 15 years, plants based on Monsanto technology (now owned and licensed by BP Chemicals) have come onstream in a number of countries, including BP's own unit in the U.K. operating alongside the DF Process since the early 1980s.

The reaction is described by the simple equation:

$$CH_3OH + CO \rightarrow CH_3CO_2H$$

and the selectivity is very high. Therefore, with the most efficient reforming, methanol synthesis and gas separation processes, it should be possible to produce acetic acid with an energy utilization of under 2 MJ/mol (33 GJ/tonne). Furthermore, as a further coal-based option, the process is seen as providing the flexibility needed for the future, and has prompted wider attention to the feasibility of producing other C_2, and higher, products from syngas.

12.5.2 *Acetic anhydride production*

The mechanism of acetaldehyde oxidation is relatively complex. Considering only molecular species, the main steps appear to be

$$CH_3CHO \overset{O_2}{\to} CH_3COO_2H \underset{}{\overset{CH_3CHO}{\rightleftharpoons}} CH_3COO_2CHOH$$
$$\text{peracetic acid} \qquad\qquad \downarrow \quad \overset{|}{CH_3}$$
$$2\ CH_3CO_2H \qquad\qquad \leftarrow \qquad (CH_3CO)_2O + H_2O$$

If the oxidation is carried out with a metal catalyst and a hydrocarbon diluent, and the mixture is separated rapidly, an appreciable proportion of the acetic anhydride can be recovered. Thus mixtures of acetic anhydride (50–70% w/w) and acid are obtained in a total selectivity of over 90%.

An alternative commercial route is the cracking of acetic acid to ketene, in the presence of a phosphate ester at over 700°C. After separating the gaseous ketene from condensed water, it is absorbed into acetic acid.

$$CH_3CO_2H \to CH_2 = C = O + H_2O$$
$$CH_2 = C = O + CH_3CO_2H \to (CH_3CO)_2O$$

The selectivity in the cracking step is some 90%, giving an overall selectivity to anhydride based on acetic acid of about 95%.

However, Halcon have now developed a process, catalysed by rhodium (or nickel) with iodine and other promoters, for the carbonylation of methyl acetate (or dimethyl ether) to acetic anhydride. Like the ketene route, this technology fits in well with acetylation processes.

$$CH_3O_2CCH_3 + CO \to (CH_3CO)_2O$$
$$\text{substrate} - OH + (CH_3CO)_2O \to \text{substrate} - O_2CCH_3 + CH_3CO_2H$$
$$CH_3OH + CH_3CO_2H \to CH_3O_2CCH_3 + H_2O$$

In 1983, Eastman Kodak commissioned a 250 kt per annum acetic anhydride plant based on this technology in Kingsport, Tennessee; a second plant was commissioned in late 1991, and this route now provides about 60% of the U.S. acetic anhydride capacity. The sting in the tail is that the whole complex is based on the gasification of coal[16]. In contrast, BP Chemicals' newest 200 kt per annum carbonylation plant in the U.K., commissioned in 1989 for the simultaneous production of acetic acid and anhydride, is based on natural gas.

12.5.3 *Production and use statistics*

The reported U.S. production figures for 1993, 1·66 Mt of acetic acid and approx. 0·9 Mt anhydride, give some idea of the importance of these products. Over 0·9 Mt of the acetic acid was used for vinyl acetate manufacture. The production of intermediate acetic anhydride (via ketene) for acetylations involved a net use of some 250 kilotonnes of acetic acid, while acetate esters and the net consumption of acetic acid solvent in terephthalic acid manufacture each accounted for about 150 kt. Some 80% of the anhydride referred to above or produced by carbonylation was used for cellulose acetate manufacture. European figures are roughly half those for the U.S.A.

12.6
C₁ products

For methanol and formaldehyde, see section 12.4.

12.6.1 *Formic (methanoic) acid and derivatives*

The modest U.S. requirements for formic acid (*c.* 30 kilotonnes per annum) appear to be met mainly by co-product material from butane oxidation (when available) and imports. Outlets are into a variety of speciality areas.

In Europe, demand is swollen to over 100 kilotonnes per annum by use for the preservation of damp silage (grass) and other animal feedstuffs—a means of saving the energy otherwise required for drying. Co-product material from naphtha oxidation is now exceeded by BASF's recently expanded (200 kilotonnes per annum capacity) methyl formate (methyl methanoate) production from syngas. The selective reaction

$$CH_3OH + CO \rightarrow HCO_2CH_3$$

is catalysed by sodium methoxide at 80–100°C and 30–60 atm. BASF can hydrolyse about 2/3 of the ester directly to formic acid (100,000 tonnes per annum capacity) by a recently improved process. A Finnish plant is based on by-product CO.

N, N-dimethylformamide (DMF) is an important solvent for acrylic fibres and polyurethane leather production (and for separating acetylene from ethylene). Methyl formate reacts readily with dimethylamine.

$$\underset{\text{DMF}}{HCO_2CH_3 + (CH_3)_2NH \rightarrow HCON(CH_3)_2} + CH_3OH$$

Alternatively the amine can be introduced directly into methanol/carbon monoxide reaction systems. European production (possibly 60–70 kilotonnes per annum) is much higher than in the U.S.A. (about 15 kilotonnes per annum).

Mitsubishi have developed a process for methyl formate by the dehydrogenation of methanol. However, the major objective is the production of separate hydrogen and carbon monoxide streams for other uses from imported methanol.

$$2CH_3OH \rightarrow HCO_2CH_3 + H_2$$
$$HCO_2CH_3 \rightarrow CO + CH_3OH$$

12.6.2 *Hydrogen cyanide*

The reported demand for hydrogen cyanide in the U.S.A. is now over 550 kt per annum. Possibly one quarter is by-product from acrylonitrile manufacture; the remainder is produced by the oxidation of methane/ammonia mixtures over platinum at about 1100°C. The major use (40–45%) is in DuPont's adiponitrile production, with some 30–35% used for methyl methacrylate (MMA) (methyl 2-methylpropenoate) manufacture and 10% for sodium cyanide.

Acrylonitrile manufacture probably contributes a major proportion of the (much smaller) demand in western Europe for MMA and NaCN production, though first-intent production is practised; for example Degussa dehydrogenate methane/ammonia mixtures over platinum at 1200–1300°C.

12.6.3 *Chloromethanes*

Passage of methanol and hydrogen chloride over a Lewis acid catalyst at about 350°C provides methyl chloride.

$$CH_3OH + HCl \rightarrow CH_3Cl + H_2O$$

This product can be chlorinated to give the more highly substituted derivatives.

However, more operations are now based on the oxychlorination of methane, over copper chloride catalysts, at temperatures of over 500°C.

$$2CH_4 + Cl_2 + \tfrac{1}{2}O_2 \rightarrow 2CH_3Cl + H_2O$$
$$2CH_3Cl + Cl_2 + \tfrac{1}{2}O_2 \rightarrow 2CH_2Cl_2 + H_2O, \text{etc.}$$

Mixtures are obtained with proportions dependent on the initial ratio of chlorine to methane. Some carbon tetrachloride is also produced by the chlorinolysis (i.e. C—C cleavage) of higher carbon number materials, and chlorination of carbon disulphide.

The total U.S. production of chloromethanes is over 0·9 Mt per annum. Chloroform (trichloromethane) and carbon tetrachloride (tetrachloromethane) are precursors for fluorocarbons; their elimination from the aerosol can market has been partially offset by increased use as refrigerants and intermediates for fluoroplastics. Methylene dichloride (dichloromethane) has picked up some of the aerosol business, and is used in paint removal and degreasing. Methyl chloride (chloromethane) is used for silicones and the waning production of tetramethyl lead.

12.7.1 *Ethanol*

**12.7
C₂ products**

Synthetic ethanol is produced almost entirely by the direct hydration of ethylene over supported phosphoric acid catalysts at 270–280°C and approx. 70 atm. After falling during the 1980s, the U.S. production of synthetic alcohol had risen to 330 kt in 1993, but was supplemented by imports and use of fermentation alcohol to meet the 730 kt demand for solvents and chemical outlets (e.g. ethyl esters).

Meanwhile, U.S. fermentation ethanol production has risen to 3·8 Mt per annum, 90% of which finds (subsidized) use in gasoline in connection with lead phase-out. A 1994 proposal by the Environmental Protection Agency, that 30% of oxygenates used as fuel additives should be based on renewable resources by 1996, may lead to a marked increase in ethanol usage, either directly or as the *t*-butyl ether (section 12.9.3).

The future potential for ethanol as an *alternative* to gasoline is a rather different story. In Europe, the fermentation of molasses requires a processing input of about 1·4 MJ/mol ethanol (30 GJ/tonne) from conventional fossil sources, which is slightly higher than the fuel value of the ethanol produced (27 GJ/tonne). In the U.S.A., the pretreatment and fermentation of grain reportedly requires about 40 GJ/tonne for processing, while the energy inputs for grain production are equivalent to about 16 GJ/tonne ethanol, giving a total similar to that for synthetic ethanol (56 GJ/tonne, section 12.5.1). While there is little doubt that these energy requirements are being reduced, the main difference is that, by appropriate location of a fermentation plant, an associated power unit could generate a significant proportion of the energy input from agricultural wastes; but field machinery would still need a liquid fuel and net gains (from solar energy) are expected to be modest[17]. Nevertheless, Brazil started a programme in the 1980s to increase fermentation ethanol production capacity to 9 million tonnes per annum, still only 4% of fuel requirements, but a faster growth in demand has made Brazil the world's largest ethanol importer.

Synthetic ethanol production in Europe is believed to be on a similar scale to the U.S.A., but the total market including fermentation ethanol is probably still short of 1 Mt per annum.

12.7.2 *Acetaldehyde (ethanol)*

The conversion of ethanol to acetaldehyde can be effected by dehydrogenation over copper at 250–300°C or by (partially) oxidative dehydrogenation over silver at 450–500°C. However, this route was largely superseded by the Wacker process for the direct oxidation of ethylene in aqueous solutions of Pd/Cu chlorides. Rhône–Poulenc and BP have also patented potential processes for the homologation of methanol to acetaldehyde:

$$CH_3OH + CO + H_2 \rightarrow CH_3CHO + H_2O$$

U.S. production fell rapidly as a result of incursions by the carbonylation routes to acetic acid and anhydride, but appears to have levelled off at about

300 kt per annum (for use mainly as an intermediate for speciality products).

In Europe, acetaldehyde has also been largely displaced as a feedstock for acetic acid, though some producers have sought to retain a share indirectly by producing ethyl acetate (ethyl ethanoate) via the low-temperature Tishchenko reaction catalysed by aluminium ethoxide:

$$2CH_3CHO \rightarrow CH_3CO_2CH_2CH_3$$

12.7.3 Ethylene oxide (oxirane) and glycol (ethane-1,2-diol)

Ethylene oxide is manufactured by the silver-catalysed oxidation of ethylene at 200–250°C, with yields of over 75% in modern plants. In the presence of alkalis, ethylene oxide reacts with water to give glycol and/or polyglycols, with alcohols and (alkyl) phenols to give the corresponding polyglycol ethers (nonionic detergents) and with ammonia to give ethanolamines.

$$C_2H_4 + [O] \longrightarrow \overset{O}{\overset{\diagup \diagdown}{CH_2-CH_2}}$$

$$ROH \xrightarrow{C_2H_4O} ROCH_2CH_2OH \xrightarrow{C_2H_4O} RO(CH_2CH_2O)_2H \text{ etc.}$$

$$NH_3 \xrightarrow{C_2H_4O} H_2NCH_2CH_2OH \xrightarrow{C_2H_4O} HN(CH_2CH_2OH)_2 \text{ etc.}$$

The U.S. production of ethylene oxide is over 2·5 Mt, of which about 65% is hydrolysed to ethylene glycol (ethane-1,2-diol). Nearly 60% of the glycol is used in the production of polyethylene terephthalate and the rest in antifreeze. Di- and triethylene glycols are used in resins and for gas drying; ethanolamines are used in detergents and for gas separations.

European production figures for ethylene oxide are about 1·5 Mt per annum, with less than half now hydrolysed to glycol.

Several companies have sought direct routes from ethylene to ethylene glycol, to achieve higher selectivity, while Ube (Japan) and Union Carbide (U.S.A.) are jointly developing a syngas based route:

$$2CO + 2CH_3OH + \tfrac{1}{2}O_2 \xrightarrow[CuCl_2]{PdCl_2} \underset{\text{dimethyl oxalate}}{CH_3O_2CCO_2CH_3} + H_2O$$

$$\xrightarrow{H_2} HOCH_2CH_2OH(+ 2CH_3OH)$$

The carbonylation or hydroformylation of formaldehyde may also provide a future synthetic route[18].

12.7.4 Vinyl acetate (ethenyl ethanoate)

The last U.S. plant for the production of vinyl acetate from acetylene shut down in the late 1980s, though this simple reaction, over zinc acetate at about 200°C, is still in use in Europe and Japan. The production of vinyl acetate by the 'acetoxylation' of ethylene entails passage of ethylene, acetic acid vapour and oxygen over a supported palladium catalyst at about 160–200°C and 10 atm. Energetic aspects have been dealt with in section 12.3.3.1.

U.S. production of vinyl acetate is over 1·3 Mt, while western Europe and Japan each produce over half a million tonnes. In the U.S. and Europe, polymers are used mainly for adhesives (e.g. plywood) and paints. In Japan, however, nearly 70% of vinyl acetate polymer is hydrolysed to polyvinyl alcohol (with recovery of acetic acid), for use in fibres.

Halcon have reported the reaction of methyl acetate with carbon monoxide and hydrogen to ethylidene diacetate $CH_3CH{\displaystyle\bigg\langle}{\scriptstyle O_2CCH_3 \atop O_2CCH_3}$, an adduct of acetaldehyde and acetic anhydride. This material can be thermally decomposed to vinyl acetate, providing a route based on syngas.

$$2CH_3CO_2CH_3 + 2CO + H_2 \rightarrow CH_3CH(O_2CCH_3)_2 + CH_3CO_2H$$

12.7.5 Chloroethylenes (chloroethenes) and chloroethanes

The production of ethyl chloride from ethylene and hydrogen chloride, mainly for tetraethyl lead manufacture, is expected to fall rapidly. Ethylene dichloride (1, 2-dichloroethane) and vinyl chloride (chloroethene) have already been discussed in section 12.3.3.2. These two materials are also starting points for the production of more highly chlorinated C_2 products. The following scheme illustrates routes used for vinylidene dichloride (1, 1-dichloroethene) and 1, 1, 1-trichloroethane (also known as methylchloroform in the U.S.A.);

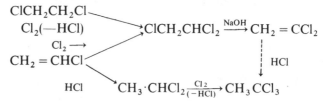

Tri- and per- chloroethylene (tri- and tetra- chloroethene) are produced together by the oxychlorination of 1,2-dichloroethane. An alternative route to the latter material is the high temperature (550–700°C) chlorinolysis of propane (or other hydrocarbons), e.g.

$$C_3H_8 + 8Cl_2 \rightarrow Cl_2C = CCl_2 + CCl_4 + 8HCl$$
$$2CCl_4 \rightarrow Cl_2C = CCl_2 + Cl_2$$

In the U.S.A., environmental pressures are causing a continuing fall in demand for the two major products, perchloroethylene (120 kt per annum), used for dry cleaning and chlorofluorocarbon manufacture, and 1, 1, 1-trichloroethane (200 kt per annum), for cold cleaning and vapour degreasing of metals. European production of 1, 1, 1-trichloroethane is believed to be on a similar scale.

12.8.1 *Isopropanol (2-propanol) and acetone (propanone)*

Isopropanol (2-propanol) is now manufactured largely by direct hydration of propylene over supported phosphoric acid catalysts at about 180°C and 50–60 atm., though Deutsche Texaco has commercialized the use of an ion-exchange resin as catalyst.

Some 95% of the U.S. market for acetone (2-propanone) is now supplied by the cumene–phenol process (section 12.11.2), with the remainder produced mainly by dehydrogenation of isopropanol (over a copper or zinc oxide catalyst). In Europe, co-product acetone from BP's naphtha oxidation supplements that from cumene–phenol and isopropanol.

In 1993 U.S. production of isopropanol had slipped to 0·55 Mt, with only 8% converted to acetone. The major use is in solvents, with some converted to derivatives (e.g. isopropyl esters). European production is believed to be about 0·5 Mt per annum.

The U.S. and European production figures for acetone are approximately 1·1 and 0·5 Mt per annum respectively. The largest single chemical use (approx. 30%) is in methyl methacrylate manufacture. Of the aldol derivatives, methyl isobutyl ketone (MIBK) (4-methylpentan-2-one) takes some 10–15%, but several others, such as diacetone alcohol and isophorone, have individual sales of 10 to 25 kt per annum (for brake fluids, speciality solvents, etc.).

12.8.2 *Propylene oxide (1-methyloxirane) and glycol (propane-1, 2-diol)*

Until about 1970, the chlorohydrin route to propylene oxide predominated worldwide.

$$Cl_2 + H_2O \rightleftharpoons HCl + HOCl$$
$$C_3H_6 + HOCl \rightarrow CH_3CHOHCH_2Cl$$
$$2CH_3CHOHCH_2Cl + Ca(OH)_2 \rightarrow 2CH_3CH\overset{O}{\overset{\diagup\diagdown}{-\!}}CH_2 + CaCl_2 \cdot 2H_2O$$

During the 1970s, Oxirane (now Arco) built four indirect oxidation plants in the U.S.A. (400 kt per annum). Holland (200 kt per annum), Spain (30 kt per annum) and Japan (90 kt per annum), to which they have since added. The stoichiometry of the original U.S. and Dutch operations is approximated by the equations:

$$2(CH_3)_3CH + 1\tfrac{1}{2}O_2 \rightarrow (CH_3)_3COOH + (CH_3)_3COH$$
$$(CH_3)_3COOH + C_3H_6 \rightarrow (CH_3)_3COH + C_3H_6O$$

Liquid isobutane is oxidized to a mixture of *t*-butylhydroperoxide and *t*-butanol at about 120°C and 25–35 atm., with no catalyst present. A toluene solution of the hydroperoxide is then contacted with excess propylene gas at about 160°C, in the presence of a molybdenum compound. Although some hydroperoxide decomposes, the selectivity on propylene is very high (98%). The primary outlet for *t*-butanol is into gasoline additives (section 12.9.3).

The operations in Spain and Japan use ethylbenzene as the hydrocarbon feedstock for oxidation. The co-produced alcohol, 1-phenylethanol, is dehydrated to styrene (phenylethene, section 12.11.1). This version is now also used in Arco's latest U.S. plant, commissioned in late 1991.

The U.S. production of propylene oxide now exceeds 1·3 Mt per annum, with a somewhat smaller figure for Europe. The chemistry is similar to that of ethylene oxide; the glycol, polyethers and polyether derivatives of other polyols, such as glycerol and trimethylol propane, are used extensively in polyurethanes and polyester resins.

12.8.3 *Acrylonitrile (propenonitrile)*

The ammoxidation of propylene is carried out by feeding the olefin, air, ammonia and steam over a fixed or fluidized bed of catalyst at between 420 and 500°C. Several binary oxide systems, Bi-Mo, U-Sb (Sohio) and Sn-Sb (BP Chemicals), form the basis of commercial catalysts. Selectivities are now about 70% on propylene, with acetonitrile and HCN as byproducts.

$$C_3H_6 + NH_3 + 1·5O_2 \rightarrow CH_2 = CHCN + 3H_2O$$

U.S. and western European production figures are 1·1 and 0·9 Mt per annum respectively, with about 0·6 Mt per annum produced in Japan. Use in fibres (about 60%) and as a co-monomer with styrene and butadiene in resins (ABS and SAN) accounts for most of the production.

12.8.4 *Acrylic (propenoic) acid and acrolein (propenal)*

A variety of processes have been used for the production of esters of acrylic (propenoic) acid, including the solvolysis of acrylonitrile with c. H_2SO_4 and an alcohol. The 'Reppe' process, commercialized by BASF and associated companies, is based on the reaction of acetylene with carbon monoxide, nickel carbonyl and an alcohol. However, the last U.S. operator of this process commissioned a propylene oxidation plant in October, 1982.

Several companies have oxidized propylene to acrolein (propenal) on a modest scale, mainly for the production of methionine. All but Shell (copper catalyst) used catalysts similar to those for acrylonitrile production. Two-reactor or dual-catalyst systems have now been introduced to oxidize acrolein further to acrylic acid, most frequently over cobalt and molybdenum oxides.

Growth in the traditional coating and textile applications of acrylic acid and acrylates is being extended by the use of polyacrylic acid as an absorbent (in diapers!); demand now exceeds 500 kilotonnes per annum in the U.S. and 250 kilotonnes per annum in western Europe.

12.8.5 *Allylic (propenyl) derivatives*

Allyl alcohol has been produced by the selective reduction of acrolein and the isomerization of propylene oxide. However, most allylic derivatives are produced via allyl chloride, obtained by the chlorination or oxychlorination of propylene.

A more significant use of allyl chloride is for the production of epi-chlorohydrin (3-chloro-1, 2-epoxypropane), an important precursor for epoxy-resins.

12.8.6 *n-Propanol, propionaldehyde (propanal) and propionic (propanoic) acid*

Historically, these materials had modest outlets, e.g. in speciality solvents or as intermediates for polyols, agricultural chemicals, etc. From the late 1980s, BP established the use of propionic acid as a grain preservative, now of major importance in Europe, to utilize the significant quantities co-produced in naphtha oxidation.

In the U.S.A., all these materials are produced via the hydroformylation of ethylene. Union Carbide chose the rhodium/phosphine catalyst system for their early 1980s expansion (and another planned); older plants use cobalt catalysts.

$$C_2H_4 + CO + H_2$$
$$\downarrow$$
$$C_2H_5CHO \quad \text{Propionaldehyde (Propanal)}$$

$$H_2 \quad\quad \text{Oxidation}$$

$$C_2H_5CH_2OH \quad\quad C_2H_5CO_2H$$

n-(1-) Propanol   Propionic (Propanoic) Acid

Growing use of propionic acid as a preservative in the U.S. required some 40% of the 90 kt produced in 1993.

In Europe, BP's co-production of propionic acid is supplemented by the 'hydrocarboxylation' of ethylene:

$$C_2H_4 + CO + H_2O \xrightarrow{\text{Ni catalyst}} C_2H_5CO_2H$$

The propionaldehyde route is operated on only a small scale.

Texaco (U.S.A.) are seeking to homologate acetic acid:

$$CH_3CO_2H + CO + 2H_2 \rightarrow CH_3CH_2CO_2H + H_2O$$

12.9

C$_4$ products

12.9.1 *Butenes and butadiene*

The use of naphtha for ethylene production provides about 30–35% w/w C$_4$ hydrocarbons relative to ethylene, with the following approximate composition

> 40% butadiene (1,3-)
> 30% isobutene (2-methylpropene)
> 20% *n*-butenes
> 10% *n*- and isobutanes

Butadiene is first recovered by extractive distillation, employing furfural or an amide, such as dimethylformamide. Some 1·5 Mt and 0·8 Mt per annum of butadiene are obtained in western Europe and Japan respectively. With less naphtha and gas oil cracking in the U.S.A., over 20% of the 1·7 Mt demand for butadiene in 1993 was met by imports. Butadiene can also be made by catalytic or oxidative dehydrogenation of *n*-butenes, or even *n*-butane. The major outlets (approx. 65%) are for tyre rubbers, as polybutadiene or co-polymers with styrene or acrylonitrile. Some 15% is used for adiponitrile in the U.S.A. Chloroprene and co-polymer resins account for the remainder.

The remaining components are common to many refinery streams, and thus isobutene and *n*-butenes may be derived from either source. Isobutene is generally recovered from the butadiene-free 'raffinate' by chemical means. For example, by contacting the mixture with 65% sulphuric acid, *t*-butanol is formed in the acid phase, from which fairly pure isobutene (for 'butyl rubber') may be regenerated. Alternatively, the isobutene within the mixed C$_4$ stream may be converted directly to the dimer and trimer by contact with solid acidic catalysts, to viscous 'polybutenes' by treatment with Lewis acid catalysts, or to a *t*-butyl ether (section 12.9.3). The reported 1993 U.S. production of isobutene was about 500 kt, but this refers only to isolated material (including dehydration of Arco's co-product *t*-butanol); derivatives possibly consumed 7–8 Mt in the U.S. compared with roughly 2 Mt in Europe.

Mixed *n*-butenes are used for sec-(2-)butanol manufacture, but the separation of 1-butene, primarily for co-monomer use in LLDPE (section 12.3.4), is of growing importance (approaching 300 kt per annum in the U.S.A.).

12.9.2 *Sec-butanol (2-butanol) and methyl ethyl ketone (2-butanone)*

After isobutene removal, *n*-butenes in the mixture with butanes may be converted to 2-butanol by the old two-stage sulphuric acid process or by using an acidic ion-exchange resin catalyst. Essentially all 2-butanol so produced is converted to 2-butanone by dehydrogenation (cf. isopropanol to acetone). 2-Butanone is a major solvent in the preparation of coatings. Production figures are about 250 kilotonnes and 200 kilotonnes in the U.S.A. and western Europe respectively.

12.9.3 *Tert-butanol and t-butyl ethers*

Most *t*-butanol is now obtained as a co-product of Arco's propylene oxide manufacture (section 12.8.2). In addition to minor speciality uses, some is used directly in gasoline, but most is dehydrated to isobutene for conversion into MTBE (methyl *t*-butyl ether), a preferred octane improver in 'reformulated' (lead-free) gasoline.

The demand for MTBE has now far outstripped this source of isobutene, and most is obtained by contacting mixed C_4 streams with methanol and an acidic ion-exchange resin[19]:

$$(CH_3)_2C{=}CH_2 + CH_3OH \rightarrow (CH_3)_3COCH_3$$

C&EN[1] estimated that U.S. production of MTBE had more than doubled between 1992 and 1993 to nearly 11 Mt per annum (cf. nearly 3 Mt per annum in western Europe). Although much lower U.S. figures for 1993–1994 (4–5 Mt) are reported elsewhere, this may reflect the problem of accounting for refinery operations in chemical production statistics. Nonetheless, all sources predict that worldwide demand will increase dramatically over the next few years, with some projections of over 20 Mt per annum for the U.S.A. alone. This will put pressure on the availability of both methanol and isobutene-containing streams; additional isobutene could be produced by commercializing known technology for the isomerization of mixed butanes to isobutane and dehydrogenation, albeit at considerable expense. An alternative, TAME (*t*-amyl methyl ether), is currently produced in much smaller quantities from 'isoamylenes', produced by the isomerization and dehydrogenation/demethylation of C_5/C_6 streams.

The EPA proposal on use of renewable resources, referred to in section 12.7.1, could mean some displacement of MTBE by ethanol addition or substitution of ethanol for methanol in the above reaction, to give ETBE (ethyl *t*-butyl ether) in future.

12.9.4 *Maleic anhydride (cis-butenedioic anhydride)*

Maleic anhydride (MA) is important in the production of unsaturated polyesters, used mainly for glass-fibre resin products. The long-established oxidation of benzene over vanadia catalysts at 400–450°C is very wasteful, with only 40% of the carbon incorporated into the product at a typical selectivity of 60% of theory. Mitsubishi in Japan (and a small European plant) oxidize a stream rich in *n*-butenes, but with a selectivity of only 45%.

However, a number of companies in the U.S.A. developed catalysts to give better yields from *n*-butane (including BP Sohio in fluid-bed reactors), and this feedstock is now used for all U.S. production (190 kt per annum) and new plants worldwide.

12.9.5 *Chloroprene (2-chlorobuta-1, 3-diene)*

Butadiene is chlorinated in the gas phase to give mixtures of 1, 2-dichlorobut-3-ene and 1, 4-dichlorobut-2-ene, which are interconvertible in the presence of copper catalysts. The former is dehydrochlorinated by contact with aqueous sodium hydroxide, to give chloroprene (2-chlorobuta-1, 3-diene) for polymer production (neoprene). The earlier acetylene-dimer route has been phased out. U.S. production of neoprene has fallen to about 120 kt per annum, of which over 40% is exported; European production is somewhat lower.

12.9.6 *Methacrylates (2-methylpropenoates)*

Methyl methacrylate (methyl 2-methylpropenoate) is still manufactured by the old acetone cyanhydrin route:

$$(CH_3)_2CO + HCN \rightarrow (CH_3)_2C - CN \xrightarrow[c.H_2SO_4]{CH_3OH} CH_2 = C - CO_2CH_3$$

$$\underset{OH}{|} \qquad\qquad\qquad \underset{CH_3 \ (+NH_4HSO_4)}{|}$$

The oxidation of isobutene to methacrylic acid has been commercialized in Japan, and several alternative routes have attracted serious attention of late.

The homopolymer provides clear sheet products (Perspex, Lucite), and copolymers are used in coatings and paints. U.S. production is approaching 500 kilotonnes per annum.

12.9.7 *Butyraldehydes (butanals) and primary butanols*

During the 1960s, older routes based on acetaldehyde (section 12.7.2) largely gave way to the hydroformylation of propylene, and displacement is now virtually complete. The use of cobalt catalysts at 145–180°C and 100–400 atm. still predominates, giving *n*- and iso-butyraldehydes in ratios of 3 or 4 : 1.

$$C_3H_6 + CO + H_2 \rightarrow CH_3CH_2CH_2CHO \text{ or } (CH_3)_2CHCHO$$

In addition to the C_4 alcohols, most companies produce 2-ethylhexanol by the aldolization of *n*-butyraldehyde and subsequent hydrogenation; in the

Aldox process, the aldolization occurs within the propylene hydroformylation reactor.

The co-production of isobutyraldehyde has always proved embarrassing, but the introduction of rhodium liganded with a phosphine (Union Carbide) or phosphane (Ruhrchemie/Rhône–Poulenc) as catalyst permits milder reaction conditions and gives n-/iso-ratios of 8:1 or more (see section 11.7.8.2).

U.S. production of n-butanol has increased to 600 kt per annum, largely for conversion to unsaturated esters and saturated ester and ether solvents, while 2-ethylhexanol (about 300 kt per annum in the U.S., but much higher in Europe with exports at 250 kt per annum) is used mainly for the phthalate ester as a plasticizer for PVC. The n- and iso-butyraldehydes are also subjected to aldol condensation/crossed Cannizzaro reactions with formaldehyde, to give the polyols trimethylol-propane (2, 2-bishydroxymethyl-1-butanol) and neopentyl glycol (2,2-dimethyl-1,3-propanediol).

12.9.8 C_4 diols and related products

1,4-Butanediol is the main intermediate to tetrahydrofuran and γ-butyrolactone, but is finding increasing use in the production of polybutylene terephthalate, an engineering plastic. Production in the U.S.A. (over 200 kt per annum) and Europe (approx. 100 kt per annum) is mainly via 'Reppe' chemistry, based on acetylene:

$$C_2H_2 \xrightarrow[\text{CuCat.}]{CH_2O} HC \equiv CCH_2OH \xrightarrow[\text{CuCat.}]{CH_2O} HOCH_2C \equiv CCH_2OH$$

propargyl alcohol   1, 4-butynediol

$$\Big\downarrow H_2$$

$$HO(CH_2)_4OH \xleftarrow{H_2} HOCH_2CH = CHCH_2OH$$

1, 4-butanediol   1, 4-butenediol

The intermediates have a number of specialist applications, such as corrosion inhibitors.

Davy have recently commercialized the hydrogenation of dimethyl maleate to 1,4-butanediol (via γ-butyrolactone, with some polymerization-grade tetrahydrofuran as co-product)[20]. Routes based on the hydrolysis of 1,4-dichlorobut-2-ene (see chloroprene), hydrogenation of maleic anhydride (smallish scale) and diacetoxylation of butadiene were all commercialized in Japan, while ARCO announced their intention to hydroformylate allyl alcohol, obtainable by isomerization of propylene oxide. Some tetrahydrofuran is also produced by hydrogenating furan, derived from natural pentoses.

12.10.1 *Isoprene*

Isoprene (2-methylbuta-1, 3-diene) is produced mainly by the dehydrogenation of 'isoamylenes' (2-methylbutenes) or isopentane (2-methylbutane), in processes similar to those used in the U.S.A. for butadiene manufacture.

In France and Japan, isoprene is also produced from isobutene (or *t*-butanol) and formaldehyde in 1 or 2 stage processes (via a 1, 3-dioxan intermediate). The overall reaction is:

$$(CH_3)_2C = CH_2 + CH_2O \rightarrow CH_2 = \overset{\overset{\displaystyle CH_3}{\displaystyle |}}{C} - CH = CH_2 + H_2O$$

U.S. and Western European production figures for polyisoprene, the closest synthetic to natural rubber, appear to be quite small at less than 100 kilotonnes per annum in each area. However, there is a much higher capacity, and presumably greater use, for polyisoprene in Eastern Europe and the former Soviet republics.

12.10.2 *Plasticizer alcohols*

Plasticizers for polyvinyl chloride are predominantly the C_7–C_{13} primary alkyl diesters of phthalic acid; diesters of aliphatic diacids are also used as plasticizers and in synthetic lubricants. 2-Ethylhexanol, via *n*-butyraldehyde, was the first alcohol used; now, many others, produced by the hydroformylation of C_6–C_{12} olefins, have entered the scene.

Over acid catalysts, propylene and butenes give homo- and co-oligomers within the required molecular weight range. The derived primary alcohols are highly branched. (The prefix iso- is used for commercial names e.g. iso-octanol). However, linear alpha-olefins (terminal) have found greater demand. Their production by the thermal cracking of waxes, at 500°C upwards, has largely given way to the oligomerization of ethylene (approx. 1 Mt per annum in the U.S.A.), with either aluminium alkyls or Ni(0) complexes as catalysts. The latter are the basis for SHOP, the Shell Higher Olefin Process. These processes produce olefins with a broad carbon number range, the lower molecular weight fractions are used for plasticizer alcohol production, and the higher fractions for detergents (next section). Shell can tailor the range of olefins produced, by converting unwanted higher fractions to internal olefins and dismutation with ethylene.

In addition to cobalt and Union Carbide's rhodium-phosphine catalysts for the hydroformylation process, Shell's cobalt/phosphine system has potential advantages, providing higher *n*-/iso-ratios than cobalt alone while producing mainly the *alcohol* in one step.

$$RCH = CH_2 + CO + 2H_2 \xrightarrow{\text{Co/phosphine}} RCH_2CH_2CH_2OH$$

12.10.3 *Detergent intermediates*

The scheme below provides some guidance on the type of chemistry involved in the production of the major petrochemical components of most detergents (excluding the more specialized amine-based cationics).

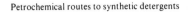

Petrochemical routes to synthetic detergents

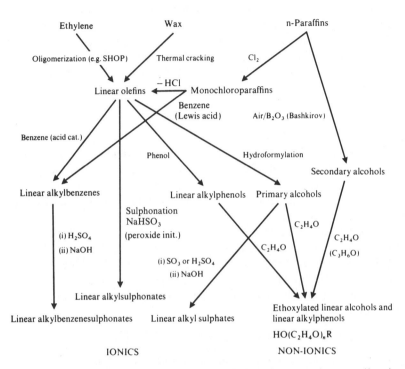

The use of highly branched olefins to alkylate benzene has totally given way to the production of (secondary) linear alkylbenzene (LAB), by alkylation with a linear olefin or a chlorinated paraffin, to achieve biodegradability. The LAB is subsequently sulphonated to produce a linear alkylbenzenesulphonate (LAS, as the sodium salt), the 'workhorse' of the detergent market with a worldwide production of about 2 Mt per annum.

Synthetic C_{12+} primary alcohols, produced mainly by the hydroformylation of ethylene oligomers (previous section), compete with 'fatty alcohols' derived from natural oils. They may be sulphated (to give an anionic product) or reacted with ethylene oxide to produce alcohol ethoxylates, the major non-ionic detergent components (possibly approaching 1 Mt per annum worldwide):

$$RCH_2OH + SO_3 \rightarrow RCH_2OSO_3H \xrightarrow{NaOH} RCH_2OSO_3Na$$

$$RCH_2OH + nC_2H_4O \rightarrow RCH_2O(C_2H_4O)_nH$$

In Japan, secondary alcohols for ethoxylation are also produced by Bashkirov oxidation of *n*-paraffins (liquid phase, in the presence of boron oxides at around 160°C). Some ethoxylates are also sulphated, while olefins can be sulphonated with SO_3 or converted to sodium alkanesulphonates directly by free-radical addition of sodium bisulphite:

$$RCH = CH_2 + NaHSO_3 \rightarrow RCH_2CH_2SO_3Na$$

Alkylphenol ethoxylates, still mainly derived from branched nonylphenol, now have only modest uses.

Changes in the domestic markets, including the introduction of 'compact' powders and liquid concentrates, are expected to result in progressive changes in the demand for the above and alternative components[21].

12.11 Aromatics

12.11.1 *Hydrocarbons*

The separation of aromatic hydrocarbons (as a group) from mixtures with close-boiling paraffins and olefins, such as 'catalytic reformate' or 'pyrolysis gasoline', is generally effected by extraction with furfural (2-formylfuran), liquid SO_2 or other solvents. Benzene, toluene and mixed xylenes are then separated by distillation.

The requirement for benzene is over 6 Mt per annum in the U.S.A., somewhat lower in western Europe. Coal-tar operations make only modest contributions. In western Europe the main source is pyrolysis gasoline, whereas catalytic reforming (section 12.2.2.3) is the main source in the U.S.A., where the catalytic (clay) or thermal hydrodealkylation of toluene is also more important.

$$PhCH_3 + H_2 \xrightarrow{550-650°C} PhH + CH_4$$

By far the largest outlet for benzene (approx. 60%) is styrene (phenylethene), produced by the reaction of benzene with ethylene; a variety of liquid and gas phase processes, with mineral or Lewis acid catalysts, are used. The ethylbenzene is then dehydrogenated to styrene at 600–650°C over iron or other metal oxide catalysts in over 90% selectivity. Co-production with propylene oxide (section 12.8.2) also requires ethylbenzene, but a route involving the cyclodimerization of 1,3-butadiene to 4-vinyl-(ethenyl-)cyclohexene, for (oxidative) dehydrogenation to styrene, is being developed by both DSM (in Holland) and Dow[22]. 60–70% of all styrene is used for homopolymers, the remainder for co-polymer resins. Other major uses of benzene are cumene (20%, see phenol), cyclohexane (13%) and nitrobenzene (5%). Major outlets for toluene (over 2·5 Mt per annum) are for solvent use and conversion to dinitrotoluene.

Of the remaining aromatics, only *p*-xylene and *o*-xylene have major chemical uses (requiring 2·6 and 0·4 Mt per annum respectively in the U.S.A.). *o*-Xylene is generally separated first by distillation; the use of low-temperature ($-70°C$) crystallization for recovery of p-xylene has largely given way to selective adsorption onto molecular sieves (Zeolites; Parex process). With periodic shortages, the remaining material rich in m-xylene may be isomerized, or toluene disproportionated into benzene and xylenes; in both cases, newer Zeolite catalysts can give p-xylene-rich C_8 fractions.

Partial displacement of aromatics by oxygenates in reformulated gasoline may lead to a downturn in refinery reforming operations. A number of alternative processes for producing mixed aromatics, including the BP 'Cyclar' process for propane/butane feedstocks, have been developed and offered for licence[23].

12.11.2 *Phenol*

Alkylation of benzene with propylene produces cumene (isopropylbenzene). Oxidation of cumene at about 110°C to modest conversions (*c.* 20%) gives the hydroperoxide, which on treatment with sulphuric acid at 70–80°C produces phenol and acetone in equimolar quantities (90–92% of theory).

$$PhH + C_3H_6 \xrightarrow{Acid} Ph-\underset{\underset{CH_3}{|}}{\overset{\overset{CH_3}{|}}{CH}} \xrightarrow{O_2} Ph-\underset{\underset{CH_3}{|}}{\overset{\overset{CH_3}{|}}{COOH}} \xrightarrow{Acid} PhOH + \underset{\underset{CH_3}{|}}{\overset{\overset{CH_3}{|}}{CO}}$$

Dow has commercialized a 2-stage process from toluene. Benzoic acid is produced by liquid-phase oxidation in the presence of cobalt, and is then subjected to 'oxidative decarboxylation' with a copper catalyst, again in the liquid phase:

$$PhCH_3 \xrightarrow{O_2} PhCO_2H \xrightarrow{O_2} PhOH + CO_2$$

Of the $1.7\,Mt$ of phenol produced in the U.S.A. in 1983, 34% went into phenol–formaldehyde resins (for plywood adhesives), 15% to caprolactam and 5% to aniline. However, an increasing proportion (presently 35%) is being converted into 'bisphenol A', the basis for polycarbonate glazing and high-performance engineering resins[24]:

bisphenol A   polycarbonate

In western Europe, phenol production is $1\,Mt$ per annum, with less into phenol-formaldehyde resins (more urea-formaldehyde is used) and rather more into cyclohexanone for nylons.

12.11.3 *Benzyls*

Several other companies oxidize toluene to benzoic acid, as above. Except for Dow and Snia Viscosa (nylon-6), outlets are mainly into speciality uses. Benzaldehyde is often recovered as a by-product, while Rhône–Poulenc and others oxidize toluene to give mainly benzaldehyde and benzyl alcohol (phenylmethanol). However, many derivatives, including the major benzyl esters, are produced via the side-chain chlorination of toluene.

$$PhCH_3 \overset{Cl_2}{\rightarrow} PhCH_2Cl \overset{Cl_2}{\rightarrow} PhCHCl_2 \overset{Cl_2}{\rightarrow} PhCCl_3$$

(Butylbenzyl phthalate – a plasticizer)

12.11.4 *Nitro-compounds and amines*

Traditional processes are still used for the nitration of benzene and toluene, and the hydrogenation of the nitro-derivatives to amines. Some aniline is also produced by amination of phenol. The production of polyurethanes, for foams or coatings, is a major outlet for aniline (*c.* 60%) and virtually the sole use for 2, 4/2, 6-toluenediamines.

Aniline is coupled by reaction with formaldehyde to produce 4,4'-diaminodiphenylmethane:

Reaction of a diamine with phosgene then produces a di-isocyanate:

$$R(NH_2)_2 + 2COCl_2 \rightarrow R(NCO)_2 + 4HCl$$

The di-isocyanates finally react with 'polyols', either simple (e.g. glycerol, trimethylolpropane) or, more generally, adducts with propylene oxide, to give the cross-linked addition polymers; for each group:

$$RNCO + HOR' \rightarrow RNH \cdot CO_2R'$$

U.S. production figures for aniline and toluenediamines exceed 450 kilotonnes and 200 kilotonnes per annum respectively.

12.11.5 *Phthalic* (*benzene*-1,2-*dicarboxylic*) *anhydride*

Phthalic anhydride is manufactured largely by the gas-phase oxidation of *o*-xylene at 350–400°C over vanadia catalysts. Selectivities are now about 80% of theory.

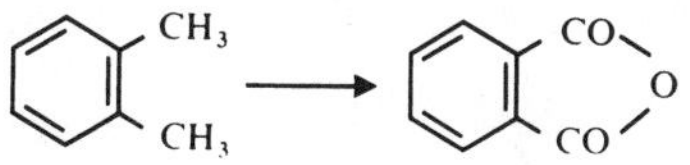

In the U.S.A., some plants for the fluid-bed oxidation of naphthalene, of coal tar or petroleum origin, are presently idle.

Over half the phthalic anhydride is used to produce diesters as plasticizers for flexible PVC. The remainder is used mainly to produce unsaturated polyester and alkyd resins. U.S. production has drifted down to 380 kt per annum; western European production is about 50% higher.

12.11.6 *Terephthalic* (*benzene*-1, 4-*dicarboxylic*) *acid*

In the U.S.A., the Far East and the U.K. (ICI), most terephthalic acid (1,4-benzenedicarboxylic acid) is produced by the liquid-phase oxidation of *p*-xylene in acetic acid with a cobalt/manganese/bromide catalyst system, originated by Amoco. Several non-bromide variants, requiring the addition of readily oxidizable precursors of acetic acid, were commercialized, but most, if not all, have been shut down. Increasingly, the final product is the high-purity diacid, whereas most older plants converted the crude acid to the dimethyl ester for purification and sale. (Small quantities of isophthalic acid are manufactured in a similar manner.)

The major alternative, still very important in Europe, is the Hercules–Witten process; *p*-xylene is oxidized without a solvent to *p*-toluic (4-methylbenzoic) acid, which is esterified before further oxidation:

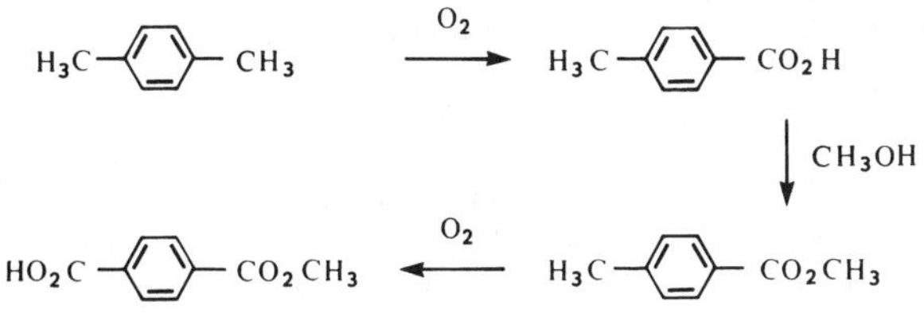

Final esterification again gives dimethyl terephthalate for purification. Finally, two Henkel processes were commercialized in Japan, but the current status is again unclear:

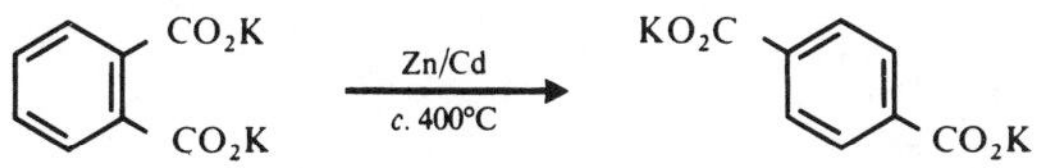

Combined 1993 production figures for pure terephthalic acid and dimethyl terephthalate were about 3·5 Mt in the U.S.A., 2 Mt in western Europe and 1·5 Mt in Japan. Over 95% is converted to polyethylene terephthalate by reaction with ethylene glycol. While fibre production remains the largest use, the major area of activity is the production of 'PET' bottle resins, with world capacity scheduled to grow rapidly (presently over 2 Mt per annum, of which 0·9 Mt per annum is in the U.S.A.).

12.12
Nylon intermediates

Although nylon intermediates are aliphatic, the main starting materials are aromatic hydrocarbons. In the earliest processes, phenol was hydrogenated (over nickel or palladium) to cyclohexanone/ol mixtures, known as 'KA oil', which gave adipic (hexanedioic) acid on oxidation with nitric acid. Half of this adipic acid was converted to the dinitrile, by vapour phase reaction with ammonia. Finally the dinitrile was hydrogenated to hexamethylenediamine (1, 6-diaminohexane). Nylon-6, 6 is produced by the condensation of essentially equimolar quantities of adipic acid and the diamine.

In subsequent processes, now predominating, benzene is hydrogenated to cyclohexane for aerial oxidation (cobalt catalysed) to 'KA oil' (or Bashkirov oxidation to cyclohexanol). One or two companies appear to have substituted aerial oxidation of KA oil for the nitric acid oxidation process, while Celanese produce their 1, 6-diaminohexane via 1, 6-hexanediol, rather than adiponitrile.

Routes to nylon 6, 6 and nylon 6 intermediates

However, the most significant changes in nylon-6, 6 production lie in the newer routes to adiponitrile introduced by Monsanto and Du Pont into both the U.S.A. and Western Europe. The former entails the electroreductive dimerization of acrylonitrile, while Du Pont use the addition of HCN to 1,3-butadiene catalysed by Ni(O) complexes:

$$CH_2 = CH \cdot CH = CH_2 \xrightarrow{HCN} CH_2 = CHCH_2CH_2CN \xrightarrow{HCN} NC(CH_2)_4CN$$

This process is by no means as simple as this equation implies, with the formation of several isomers, and hence the need for interconversions at each stage. Both processes take the pressure off benzene, from which the multistage processes gave poor yields of dinitrile (c. 50% of theory). In a somewhat similar manner, BASF have operated a semi-commercial plant to produce adipic acid by the bis-hydrocarboxylation of 1,3-butadiene:

$$CH_2 = CH \cdot CH = CH_2 + 2CO + 2H_2O \xrightarrow{Gp\ VIII\ metal} HO_2C(CH_2)_4CO_2H$$

Cyclohexanone alone is the main starting point for the newer nylon-6; the hydrogenation of phenol can be directed to give this material, but KA oil (or cyclohexanol) must be dehydrogenated. The sulphate or, more recently, phosphate of the oxime, formed by reaction with hydroxylamine (as its salt), undergoes the Beckmann rearrangement in the presence of oleum. (Excess acid is neutralized with ammonia for sale into fertilizers.)

Caprolactam

Snia Viscosa (Italy) operate an entirely different process, wherein benzoic acid is hydrogenated to cyclohexanecarboxylic acid, which gives caprolactam by reaction with nitrosyl hydrogen sulphate:

Finally, Toray have reported the pilot-scale photoinitiated nitrosation of cyclohexane to cyclohexanone oxime:

U.S. production figures for nylon-6,6 and -6 are over 0·7 Mt and 0·5 Mt per annum respectively. Non-cyclic feedstocks now provide practically all the diamine for nylon-6,6 and phenol well under half the cyclohexanone for caprolactam. In Western Europe, the figures are about 0·7 Mt per annum for each, with again only a modest dependence on phenol, while Japanese production appears to be solely nylon-6 (0·5 Mt). Fibre use predominates, but use as an engineering resin may become more important. Small quantities of higher nylon analogues are produced.

12.13
The future

12.13.1 *The products*

The future production level of all material goods is a case for the crystal ball, but reduced material and energy usages, and longer life, will certainly be sought in all applications. In such circumstances, the major products of the petrochemical industry, the plastics, will only remain in use if they are producible with lower energy utilizations, or conserve energy in other ways, when compared with the competitive traditional materials for the same applications.

The following table contains energy input data for a number of fabrication and construction materials (in cast form, unless otherwise indicated). They are derived from a variety of sources[25], and the bases for individual figures may vary. The renewable solar energy, represented by the fuel value of wood, is not included for cellulosic products.

	Density tonne/m^3	Energy Utilization, GJ per tonne	per m^3
Titanium	4·5	c. 550	2 500
Copper	8·9	c. 130	1 150
Aluminium	2·7	250–300	700–800
Magnesium	1.74	270–400	470–700
Zinc	7·1	c. 55	400
Steel (carbon)	7·8	28(45)	220(350)
Polyacetal	1·43	240	340
Polyphenylene oxide			290
Nylons-6, 6 and 6	1.14	235	270
Polyester			240
Polycarbonate	1.2	180	220
Acrylic	1·2	180	220
Polypropylene (film and cast)	0.90	110–180	100–160
ABS resin	1.07	120	130
PVC (rigid)	1·38	65–80	90–110
Polystyrene	1·06	75–105	80–110
HDPE	0·95	85–120	80–115
LDPE (film and cast)	0·92	80–110	75–100
Cellulose film		180	
Paper (sack)		60	
Glass (sheet)	(2·5)	29	(70)
(containers)	(2·5)	19–20	(50)
Portland cement	(2·1)	5–9	(10–20)
Bricks	(2·5)	4–5	(10–12)
Concrete		c. 1	

Even the most energy intensive 'engineering' resins are competitive with light casting alloy components on a volumetric basis, the most relevant in many applications. The tough film polymers, such as LLDPE and polypropylene, are competitive with cellulosic films and even paper sacks, which must be over double the weight for comparable strength. A more striking example is the plastic bottle, which need weigh only 50 g for 1-litre contents compared with 500 g in glass; the plastic bottle is also subject to far fewer breakages and

saves considerable energy in transportation. Longer life and weather resistance (e.g. PVC versus wood for window frames), general resistance to
corrosion and the ability to produce light, rigid composites (such as glass-
fibres reinforced resin panels for lighter vehicles), all possibly support recent
predictions that plastics will continue to displace other materials in some uses.

In the long term, some of the predictions based on current energy inputs
may prove a little shaky; alternative technologies for electrical power
generation could affect some values, for example. The recycle of selected
scrap, well established for aluminium, glass and paper, has possibly reduced
overall energy use for these materials, but is now also being extended to
some plastics.

When we come to the second rung of petrochemical products, other
environmental factors are more likely to have a greater impact on their
continued use, such as concerns over emissions into the atmosphere of
solvents (in general) and chlorine- and fluorine-containing fluids (affecting
the ozone layer)[26]. The need to reduce pollutant emissions and land or water
contamination by chemical processes is a separate matter, but could again
affect the future selection of both products and process routes.

12.13.2 *Future raw materials and production routes*

As we are all too aware, supplies of petroleum and natural gas are expected
to run out in a matter of tens of years at current consumption rates, despite
continuing exploration activities in ever more remote and inhospitable
areas. Total consumption is the most important factor and, despite current
'low' oil and gas prices, there must be greater emphasis on conservation.

Unfortunately, concern about global warming by CO_2 emission (the
'greenhouse' effect) has resulted in backtracking on earlier moves towards
greater use of coal for power generation. Nuclear power is under a cloud.
'Cleaner' means of generating electricity by hydroelectric schemes, wind
power or tides are gaining ground, but there are geographic limitations, and
potential environmental problems, in seeking to greatly extend their use in
the more densely populated areas of the world.

Of course, organic products require a carbon source and, for maximum
efficiency, must presently be derived from light hydrocarbon liquids and
gases; it may be logical to reserve supplies for the more modest requirements
of petrochemicals over a longer time-scale. In this context, the automobile
may prove to be either friend or foe. Gasoline production presently takes the
naphtha fraction but, foreseeing a continuing need for suitable liquid fuels,
many companies are working on gasoline extenders (and substitutes), which
may free more naphtha or themselves be amenable to chemical conversions.

Hydrogenation and solvent extraction of coal can provide liquid hydrocarbons, but the presently favoured schemes start with the steam reforming
of natural gas (in more remote areas, as commercialized in New Zealand[27])

and the gasification of coal (Germany and the U.S.A.) to produce syngas for the following options:

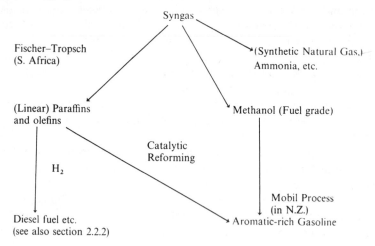

Methanol/syngas chemistry for acetic acid/anhydride could possibly be extended to some related intermediates, while Fischer–Tropsch products may fit in with detergent requirements (or provide a cracker feedstock). The Mobil methanol-to-gasoline process is potentially attractive for aromatic hydrocarbons. In view of the importance of polyethylene, the direct conversion of syngas or methanol to lower olefins has also received considerable attention[28], such that many current cracker-based routes could possibly continue. The possibility of fermentation ethanol re-emerging as an intermediate for chemicals is less clear.

However, it is unlikely that major changes will be seen in the short term. The unreliability of long-term market forecasts and availability of good plants for current routes point to a fairly modest rate of constructional activity, except possibly where the occasional new process shows an immediate and significant economic benefit. Nonetheless, there remains considerable scope for progressive improvement, both in the efficiency of the energy-intensive processes for converting primary feedstocks into the major intermediates and in the selectivity of their subsequent reactions[29].

The more dramatic changes in the petrochemical industry will need to be in attitudes and research. In view of the uncertainties of the long-term future, the watchwords must be adaptability and feedstock flexibility, be it by establishing new routes or by revitalizing the old.

References

1. *C&EN*, 1994, July 4, 30–74 (Facts & Figures issue).
2. (a) *Our Industry Petroleum*, 5th edn., British Petroleum Company, 1977.
   (b) *Kirk-Othmer Encyclopedia of Chemical Technology*, 3rd edn., Vol. 17 (Petroleum, Refinery Processes Survey), Wiley, 1982.
3. *Kirk-Othmer Encyclopedia of Chemical Technology*, 4th edn., Vol. 9 (Ethylene), Wiley, 1994.

4. (a) *Kirk-Othmer Encyclopedia of Chemical Technology*, 3rd edn., Vol. 1 (Acetylene), Wiley, 1978.
   (b) *Kirk-Othmer Encyclopedia of Chemical Technology*, 4th edn., Vol. 1 (Acetylene-derived Chemicals), Wiley, 1991 (Acetylene manufacture will be published under Hydrocarbons, Acetylene in this new edition).
5. *The Oil and Gas Journal*, 1972, (Sept. 4), 103–110.
6. *Hydrocarbon Processing*, 1975, (November), 104.
7. (a) *Energy in the 80's*, The Institution of Chemical Engineers, Symposium Series No. 48, 1977.
   (b) *Energy Conservation in the Chemical and Process Industries*, The Institution of Chemical Engineers, G. Godwin Ltd., 1979.
   (c) Process Energy Conservation, *Chem. Eng. Magazine*, McGraw-Hill, 1982.
8. *The Oil and Gas Journal*, 1993, (March 8), 37–44.
9. H. D. Marsch and H.J. Herbort, *Hydrocarbon Processing*, 1982, (June), 101–105.
10. (a) R. V. Schneider III and J. R. LeBlanc Jr., *Hydrocarbon Processing*, 1992, (March), 51–57.
    (b) G. L. Farina and E. Supp, *Hydrocarbon Processing*, 1992, (March), 77–79.
    (c) T. S. Christensen and I. I. Primdahl, *Hydrocarbon Processing*, 1994, (March), 39–46.
11. *Kirk-Othmer Encyclopedia of Chemical Technology*, 4th edn., Vol. 2 (Ammonia), Wiley, 1992.
12. *Kirk-Othmer Encyclopedia of Chemical Technology*, 3rd edn., Vol. 15 (Methanol), Wiley, 1981.
13. *Hydrocarbon Processing*, 1981, (November), 129–132.
14. (a) E. Supp, *Hydrocarbon Processing*, 1981, (March), 71–75.
    (b) *Hydrocarbon Processing*, 1987, (November), 79.
15. J. F. Roth *et al.*, *Chemtech*, 1971, **1**, 600–605.
16. (a) J. L. Ehrler and B. Juran, *Hydrocarbon Processing*, 1982, (Feb.), 109–113.
    (b) H. W. Coover Jr. and R. C. Hart, *Chem. Eng. Progress*, 1982, (April), 72–75.
17. (a) G. H. Emert and R. Katzen, *Chemtech*, 1980, (October), 610–614.
    (b) P. B. Weisz, *Chemtech*, 1980, (November), 653–654.
18. *C&EN*, 1983, (April 11), 41–42.
19. *Hydrocarbon Processing*, 1993, (March), 193–195.
20. *Hydrocarbon Processing*, 1993, (March), 170.
21. *Chemical Week*, 1994, (January 26), 35–48 (6 short papers).
22. (a) *The Oil and Gas Journal*, 1993, (March 29), 58–59.
    (b) *Chemical Week*, 1994, (May 4), 11.
23. (a) *ECN Process Review* (An *ECN* Supplement), 1994, (April), 7–9.
    (b) *Hydrocarbon Processing*, 1993, (March), 168.
24. D. Freitag, G. Fengler and L. Morbitzer, *Angewandte Chemie Int. Edn. in English*, **30**, 1991, 1598–1610.
25. (a) Ref 7(a)
    (b) R.M. Ringwald, *C&I*, 1982, (May 10), 281–286.
    (c) *Chemtech.*, 1983, (February), 128.
    (d) *Chemtech*, 1980, (September), 550–551 and 557.
    (e) *Engineering Materials and Design*, 1980, (April).
26. *Chemical Week*, 1994, October 5, 31–33.
27. (a) S. L. Meisel, *Chemtech*, 1988, (January), 32–37.
    (b) C. J. Maiden, *Chemtech*, 1988, (January), 38–41.
    (c) J. Z. Bem, *Chemtech*, 1988, (January), 42–46.
28. (a) T. Inui and Y. Takegami, *Hydrocarbon Processing*, 1982, (November), 117–120.
    (b) C. B. Murchison and D. A. Murdick, *Hydrocarbon Processing*, 1981, (January), 159–164.
29. Wolfgang Jentzsch, *Angewandte Chemie Int. Edn. in English*. **29**, 1990, 1228–1234.

Bibliography

Basic Organic Chemistry, Part 5, *Industrial Products* (ed. J. M. Teddar, A. Nechvatal and A. H. Jubb), John Wiley and Sons, 1976.
Chemicals from Petroleum, 4th edn., A. L. Waddams, John Murray, 1980.
Faith, Keyes and Clark's Industrial Chemicals, 4th edn., F. A. Lowenheim and M. K. Moran, John Wiley and Sons, 1975.
Industrial Organic Chemicals in Perspective, Part 1, *Raw Materials and Manufacture*, H. A. Witcoff and B. G. Reuben, John Wiley and Sons, 1980.

Kirk-Othmer Encyclopedia of Chemical Technology, Wiley Interscience. (Complete 3rd edn. from 1978 to 1984; publication of the 4th edn. commenced in 1991).

Directory of Chemical Producers, Western Europe, 8th edn., Vol 1 (Index to Companies) and Vol 2 (Products and Regions), SRI International, 1985. (Volume 2 lists companies' nominal production capacities/locations for many bulk chemicals. New editions are fairly frequent.)

Periodical Special Issues and Supplements

Hydrocarbon Processing publishes the *Petrochemical Handbook* in odd-numbered years (in the November issue up to 1989 and the March issue since 1991). This provides very brief descriptions, with simplified flow-sheets and some information on economics and commercial status, for a considerable number of processes being offered for licence.

C&EN (*Chemical & Engineering News*, a U.S. weekly) publishes 'Facts and Figures' as part of a June or July issue every year. This lists the previous year's production figures for over 50 chemicals, polymers and chemical sectors (and some company information) for the U.S., and for a somewhat smaller range of materials from many other countries.

ECN (*European Chemical News*, a U.K. weekly) periodically issues separately-bound supplements, under the title *Chemscope*, on specific topics, e.g. chemical sectors, the chemical industries in particular European countries and related matters (plant construction, distribution etc.).

Every issue of the U.S. weekly *Chemical Marketing Reporter* contains a 'Chemical Profile' for a specific chemical, giving U.S. companies' capacities/locations and a brief rundown on production, price, outlets and growth potential (for the U.S. only). (There is a 3 year gap between updates on any particular chemical.)

Index